AF696209

THE NONLINEAR OPTICS OF SEMICONDUCTOR LASERS

The Proceedings of the Lebedev Physics Institute of the Academy of Sciences of the USSR

Series Editor: Academician N.G. Basov

Volume 165 **Solitons and instantons, operator quantization,** Edited by V.L. Ginzburg.

Volume 166 **The nonlinear optics of semiconductor lasers,** Edited by N.G. Basov.

Volume 167 **Group theory and the gravitation and physics of elementary particles,** Edited by A.A. Komar.

Volume 168 **The quantum electrodynamics of a high intensity field,** Edited by V.L. Ginzburg.

Volume 169 **The physical effects in the gravitational field of black holes,** Edited by M.A. Markov.

Volume 170 **The theory of target compression by longwave laser emission,** Edited by G.V. Sklizkov.

Volume 171 **Research on laser theory,** Edited by A.N. Orayevskiy.

Volume 172 **Phase conjugation of laser emission,** Edited by N.G. Basov.

Volume 173 **Quantum mechanics and statistical methods,** Edited by M.M. Sushchinsky.

Volume 174 **Nonequilibrium superconductivity,** Edited by V.L. Ginzburg.

Volume 175 **Luminescence centers of rare earth ions in crystal phosphors,** Edited by M.D. Galanin.

Volume 176 **Classical and quantum effects in electrodynamics,** Edited by A.A. Komar.

Proceedings of the Lebedev Physics Institute of the Academy of Sciences of the USSR
Series Editor N.G. Basov
Volume 166

THE NONLINEAR OPTICS OF SEMICONDUCTOR LASERS

Edited by N.G. Basov

Translated by Kevin S. Hendzel

NOVA SCIENCE PUBLISHERS, INC.
COMMACK

NOVA SCIENCE PUBLISHERS
283 Commack Road
Suite 300
Commack, New York 11725

This book is being published under exclusive English Language rights granted to Nova Science Publishers, Inc. by the All-Union Copyright Agency of the USSR. (VAAP)

Library of Congress Cataloging-in-Publication Data

Nelineĭnaiá optika poluprovodnikovykh lazerov. English.

(Proceedings of the Lebedev Physics Institute of the Academy of Sciences of the USSR ; v. 166)
Translation of: Nelineĭnaiá optika poluprovodnikovykh lazerov.
Bibliography: p.
Includes index.
1. Semiconductor lasers. 2. Nonlinear optics.
3. Quantum electronics. 4. Optoelectronics. I. Basov, N. G. (Nikolaĭ Gennadievich), 1922- . II. Title.
III. Series: Trudy Ordena Lenia Fizicheskogo instituta im. P.N. Lebedeva. English ; v. 166.
QC1.A4114 vol. 166 530 s 87-12262
[QC689] [621.36'61]
ISBN 0-941743-01-2

ISBN 0-941743-01-2

Copyright 1987 Nova Science Publishers, Inc.

All Rights Reserved. No part of this book may be reproduced, stored in a retrieval system or transmitted in any form or by any means: electronic, electrostatic, magnetic, tape, mechanical, photocopying, recording or otherwise without permission from the publishers.

Printed in the United States of America

Table of Contents

AN INVESTIGATION OF THE DYNAMICS OF STRIPE LASERS: SELF-MODULATION, NONLINEAR LOSSES, MODE "CHIRPING" 123

Yu. M. Popov, I.A. Skopin

MODES AND MODE CONVERSION IN ACTIVE SEMICONDUCTOR WAVEGUIDES 169

Yu. M. Popov, I.A. Skopin

INVESTIGATION OF THE WAVEFRONT OF PLANAR STRIPE GaAlAs/GaAs-HETEROLASERS 208

M.A. Man'ko, B.I. Makhsudov, Fam Van Khoy

INVESTIGATION OF THE SPECTRAL AND SPATIAL CHARACTERISTICS OF NARROW MESA-STRIPE CONTACT HETEROLASERS 224

Yu. L. Bessonov, N.B. Kornilova, V.D. Kurnosov, M.A. Man'ko, V.N. Morozov, Chan Min'Tkhay, Fam Van Khoy, V.R. Shidlovskiy

ACHIEVEMENTS AND ISSUES IN THE PHYSICS OF INJECTION LASERS

N.G. Basov, P.G. Eliseev, Yu. M. Popov

Abstract: A brief survey of the developmental stages of the physics of semiconductor lasers is given together with a discussion of recent achievements in the field of single-mode lasers, quaternary heterosystems for lasers, reliability, lasers with an ultrathin active layer and quantum dimensional structures. Results from the development of a nonlinear optical model of a semiconductor laser are discussed. The most important developmental trend is the increase in the emission power of injection lasers while maintaining or increasing coherence. Several aspects of research on lasers with compound resonators are also examined.

Almost a quarter of a century has passed since the first theoretical concepts and ideas relating to semiconductor lasers were advanced [1-4]. The study published in 1961 relating to the possibility of obtaining a population inversion at the *p-n*-junctions of degenerate semiconductors [3] produced the most important results; this served as the basis for the research and development of injection semiconductor lasers. Today semiconductor lasers are already finding broad practical applications. Their advantages in speed, compact size, economy and design simplicity are being exploited in such applications as laser communications by optical cable, disk recording and playback systems, the spectroscopic diagnostics of gaseous media, atmospheric pollution monitoring, printers, laser projection television, etc. [5]. Fundamental research and practical designs produced at the P. N. Lebedev Physics Institute of the Academy of Sciences of the USSR [6] and the A. F. Joffe Institute of Applied Physics of the Academy of Sciences of the USSR as well as a number of other leading physics laboratories around the world have played a role in this development.

The primary stages in the creation and development of semiconductor lasers have included:

- the theoretical formulation and substantiation of the principle of operation [1-4, 7];

- the development and improvement on *p-n*-junction injection lasers ("homolasers") [8-13];

- investigations of the speed and dynamics of lasing in the 10^{-10} to 10^{-9} second range [14-19];

- the development of semiconductor lasers employing various pumping methods (electron, optical, breakdown) [20-23];

- the development of tunable infrared lasers for high resolution spectroscopy [24-26];

- the development of heterostructure lasers and the achievement of continuous lasing at room temperature [27-31];

- the development of heterostructures for a broad range of wavelengths based on multicomponent systems and the development of lasers based on InGaAsP [32-34];

- the development of single-frequency lasers and the investigation of optical heterodyning using semiconductor lasers for coherent optical communications [35-39].

Important results in solving the practical problems for the broad utilization of semiconductor lasers in the national economy should be added to this list. Primarily this focuses on the fundamental solution to the problem of the reliability of injection lasers and the layout for fabrication of lasers of this type as well as modules and devices with a service life of more than 10,000 hours and then further extending the service life characteristics. Today a wide range of injection lasers are currently manufactured on a broad scale; such lasers are based primarily on two types of heterostructures, GaAlAs/GaAs and InGaAs/InP. These cover the important spectral ranges of 0.83, 1.3 and 1.55 μm used in fiber-optic systems.

The physics of semiconductor lasers is continuing to expand with the addition of new areas and ideas. Recent popular areas include such aspects as the nonlinear optics of semiconductor active media and structures, including optical bistability, quantum-dimensional effects, the problems of frequency stabilization and the stabilization of other radiative characteristics, new technological problems (molecular-beam epitaxy, metallo-organic synthesis and the epitaxy of semiconductors, crystallographic growing on relief substrates, manufacturing structures with anisotropic gain, multistripe structures, the achievement of efficient lasing in the shortwave ranges of the visible spectrum, etc.).

Important areas include semiconductor integrated optics, where practical results are still few in spite of the extensive research and constructive formulations. The early stages in this field involved attempts in monolithic integration of circuits from discrete elements which were most successfully implemented in the form of hybrid and discrete interconnected designs. A more fundamental stage has been the development of qualitatively new circuits performing data processing operations by means of optoelectronics. As early as the 1960's, semiconductor laser circuits were examined and constructed in prototype form; these made it possible to perform all primary and necessary logic operations with input optical or electrical signals by virtue of the nonlinear interaction of optical fluxes in the semiconductor medium [15-18, 40, 41]. This field, which developed in a number of laboratories, made it possible to develop the first integrated optical circuits in the world [41]. The possibility for creating serial links between lasers in a monolithic design using selective etching technology for fabricating resonator mirrors was demonstrated. The high speed of such circuits capable of executing 10^{-9} operations/second was also demonstrated. The small size of injection lasers with an active region of approximately 10^{-10} cm^3 makes it possible to integrate a large number of elements within a single monolithic circuit, although, clearly this involves overcoming significant technological difficulties. In the last decade semiconductor lasers have been supplemented by a wide variety of devices such as periodic feedback structures (distributed feedback, Bragg reflectors) based on heterostructures, single-mode stripe structures, multistripe structures employing diffraction feedback, etc. This expands the range of devices that may be used to develop new integrated optical circuits.

The nonlinear optics of semiconductors plays an important role in these discussions. The new field in nonlinear optics, unlike the traditional field (utilizing the nonlinearity of the medium for harmonic generation and frequency conversion) is based on the use of nonlinear refraction to control optical fluxes (optical bistability, the "transphaser" amplifier, beam deflection, etc.). Bistable devices based on semiconductors, i.e., the same materials used in lasers, are very small in size, have a high speed (approximately 1 ns) and operate in the same spectral ranges as semiconductor lasers. The same technology as described above for building resonators with end mirrors without destruction of the monolithic wafer (liquid phase or ion etching, etc.) is required for fabricating such devices in the integrated optical design. Evidently the combination of laser and nonlinear devices for beam control may solve a variety of problems in integrated optics.

In the physics of semiconductor lasers, it is critical to understand and utilize the phenomena of nonlinear refraction. It is impossible to formulate a reliable optical model of a laser ignoring such phenomena. As noted previously in study [3], in semiconductors the influence of free carriers on the dielectric constant and, consequently, on the refractive index is significant. The carrier concentration in turn depends on the emission intensity. With photoelectric absorption, the concentration increases while with stimulated emission the concentration drops. This causes two different types of nonlinear refraction. In the laser medium this refraction manifests a self-focusing nature and generates the corresponding optical inhomogeneities and instabilities. In certain conditions when external resonators are used, this nonlinear refraction makes it possible to obtain bistable modes as is the case in nonlinear resonators, although in this case external sources are not needed and continuous lasing may be achieved at room temperature. Bistable modes in an injection laser with an external spectral selector are investigated in detail and explained in study [43]. A necessary element here is a compound resonator, i.e., a combination of two or more flat resonators whose relative tuning depends on emission intensity due to nonlinear refraction.

Another important aspect of using nonlinear refraction involves suppressing undesired modes in the spectral vicinity of the laser line in single-frequency lasing by means of parametric mode interaction [44]. The physical image of the phenomenon accounts for the fact that the space-time (dynamic) lattice of the optical parameters of the medium arising in mode beating produces an energy exchange between the interfering modes. Two or three modes producing beating at the common frequency Ω are introduced to the process. With two-mode interaction a sharp spectral asymmetry is discovered with energy transfer towards the longwave mode. One exception is small values of $\Omega \leq 1/\tau$, where τ is the relaxation time of the induced dynamic lattice. Here, suppression of lateral modes on both sides of the "strong" mode predominates. If the Q of the resonator suppresses "remote" modes, an internal stabilization mechanism of single-frequency lasing is activated in a laser with such a resonator. Here, due to the optical nonlinear scattering in the dynamic lattice, antiphase oscillations (a phase inversion) arise at the side frequency, which suppresses oscillations and fluctuations at the side frequencies.

The majority of materials in this volume of the "Proceedings of the Physics Institute of the Academy of Sciences" are devoted to an examination of the phenomena resulting from the nonlinear refraction and influence of carriers on the refractive index of a semiconductor as well as other nonlinear phenomena in semiconductor lasers.

We will discuss a number of important areas in the physics of semiconductor lasers.

NEW HETEROSTRUCTURES

The advantages of using heterostructures in semiconductor lasers was proven convincingly using a sample GaAlAs/GaAs system covering the 0.75-0.90 μm wavelength range (300 K) [28-30]. New materials were required to cover a broader wavelength range, specifically the important transparency and minimum dispersion ranges in fiber optic light guides. The InGaAsP/InP laser heterosystem was introduced for research and development purposes in response to this requirement in 1974 [31-34] together with a number of other quaternary systems based on the A^3B^5 semiconductor compounds. The decisive factor in converting to quaternary systems has been the possibility for expanding the spectral range and at the same time satisfying the requirement to match the lattice constants in the heterojunctions, including the lattice constant of the substrate. The principle of isoperiodic substitution in multicomponent systems of solid solutions [31] used in the design of new heterostructures involves the proportional interchanging of atoms of different types in such systems without changing the lattice constant, where the influence of one additive is compensated for by another. As a result, a number of compositions are generated, each pair of which is suitable for forming a defect-free heterojunction with a matched lattice constant (a "coherent" heterojunction).

Isoperiodic (i.e., having a fixed lattice constant) heterostructures from quaternary materials may be grown by epitaxical methods, primarily liquid-phase epitaxy. This includes systems formed by elements appearing as neighbors on Mendeleev's Periodic Table of Elements. The further expansion in the selection of new materials has resulted in the conversion to systems with remote neighbors on the Periodic Table, such as Ga-Al-In-As, In-Sb-As-P etc. Their development has encountered more serious difficulties, since there is a strong segregation of the atoms in this case in quasiequilibrium conditions. Evidently in this field the future belongs to molecular beam epitaxy, which produces systems considered inaccessible to liquid-phase and other quasiequilibrium methods.

In conjunction with photolithography technology, ion etching and laser technology, epitaxical methods (liquid-phase, molecular-beam and metallo-organic methods) make it possible to develop a number of new heterostructure designs (multilayered and buried heterostructures, heterostructures with periodic feedback structures and with ultrathin layers, etc.) which have found a number of practical applications today. One example is a InGaAsP/InP mesa stripe structure grown on a *p*-InP substrate for injection lasers at 1.3 [42] and 1.55 μm. This structure has served as a prototype for a family of new lasers with various mode composition and emission power characteristics. A single-mode low-threshold variant makes it possible to significantly reduce working currents (with a lasing threshold of ~10 milliamperes) and to obtain stable single-frequency emission. The enhanced power design provides a level of up to 100 mw in the single-mode condition. One significant fact is that the new InGaAsP/InP system has a number of physiochemical features that provide an advantage compared to more traditional systems. These include:

- chemical, photochemical and electrochemical strength of the crystal surfaces;

- high optical strength;

- the absence of the degradation "dark line disease";

- good wettability of surfaces in the liquid-phase process and ease of mass transfer in the gas-phase process;

- dislocations play only a minor role in degradation mechanisms and internal degradation is slow.

As a result, several problems such as providing reliability and long service lives, eliminating mirror erosion and mirror "catastrophic" degradation are not as critical in InGaAsP/InP as in, for example, the GaAlAs/GaAs system.

InGaSbAs/GaSb and GaAlSbAs/GaSb systems may also be classified as new promising heterostructures [45]. These cover the longer wave portion of the spectrum, where the next generation of fiber-optic waveguides will operate with even lower losses than silicon dioxide waveguides. New stages have been achieved in this field. The first continuous lasers (300 K) operating at 1.75 μm have been developed together with pulsed lasers (300 K) in the 2.3 μm range.

LASER HETEROSTRUCTURES WITH AN ULTRATHIN ACTIVE LAYER

The most impressive achievements in recent times have been the incorporation of methods of fabricating heterostructures with an ultrathin layer using new epitaxical methods. Specifically, the metallo-organic variant of gas-phase epitaxy (the so-called MOC-hydride method) made it possible to obtain uniformly thick and sufficiently planar layers in a GaAlAs/GaAs system in which quantum dimensional effects are reproduced in spite of the fact that previously gas-phase epitaxy was virtually unused due to difficulties in monitoring thicknesses and incomplete morphology of the heteroboundaries. High-quality structures are also obtained by using molecular-beam epitaxy.

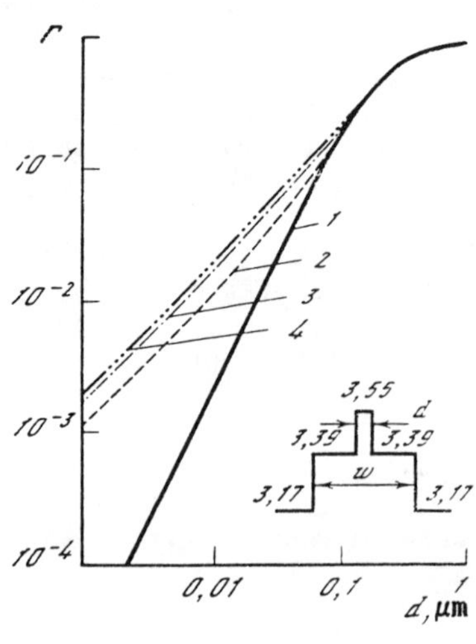

Fig. 1. The optical limiting parameter Γ plotted as a function of the thickness *d* of the active layer.
1 - in a regular double heterostructure, $w = d$; 2-4 - in a heterostructure with a "three-layer waveguide" for different total thickness values: w (μm) = 0.1 (2); 0.2 (3); 0.3 (4). The insert shows the refractive index profile in the three-layer waveguide at 1.55 μm.

The advisability of reducing the thickness of the active layer in laser heterostructures to 200-500 Å was discussed previously [46] from a design optimization to minimize the threshold current density. Within the scope of a model of an ideal planar bilateral heterostructure (with a single-layer waveguide) a reduction in the thickness to these "quantum-dimensional" levels is advisable only because the increase in gain with an increase in pumping current occurs more rapidly than the second power of the current density. Otherwise narrowing the active layer will either fail to reduce the threshold or will cause the threshold to increase due to a reduction in the optical limiting parameter Γ (see [46, 47]). In the general case the curve of the gain factor $g(j)$ plotted as a function of current has a "superquadratic" section. Here the optimum corresponds to the upper end of this section. It is possible to determine Γ and the anticipated optimum thickness $d_{\text{опт}}$ by the volumetrical gain g and the necessary mode amplification g_m at threshold. Such a calculation was performed in study [46] which also noted the low values $d_{\text{опт}}$ (~200 Å) for GaAlAs/-GaAs room temperature lasers (with a threshold of ~500 Å/cm^2 and lower).

Experience has demonstrated that bilateral heterostructures with a gradient or three-layer waveguide make it possible to reduce the threshold current to ~100 Å/cm^2 [48-50]. A calculation of the parameter Γ for a three-layer waveguide (i.e., for double heterostructures with an expanded waveguide or with a separate threshold) has demonstrated that the form of the relation $\Gamma(d)$ changes (Fig. 1) and with small values of d the value of Γ may be increased by near an order of magnitude with the same total heterobarrier value. This means that it is possible to increase the mode amplification with the same pumping level (assuming total electron confinement) or to reduce the threshold current. As a result, as indicated by the corresponding estimate, the threshold current density may be fundamentally increased to values of 50-70 Å/cm^2 at room temperature.

Fig. 1 indicates that as $d \to 0$ the dependence of Γ on the thickness in a three-layer waveguide is significantly weaker than in a single-layer waveguide, and approaches a linear relation. Thus, thickness optimization serves to further reduce $d_{\text{опт}}$ to values of 40-50 Å, and at the same time it is possible to also improve the second optimized parameter: the total thickness of the three-layer waveguide w.

QUANTUM-DIMENSIONAL EFFECTS

The spatial confinement of carriers perpendicular to the plane of the narrowband layer serves to redistribute the energy levels (quantization) and in the final analysis leads to a modification of the state density function and the matrix elements for various quantum transitions. The state density for the case of single-dimensional quantization takes a graded form, while in the quantum-dimensional case it appears as a series of peaks as shown in Fig. 2. The lower level is elevated over the nominal bottom of the band by the "kinetic localization energy" of $\sim \hbar^2/2md^2$, where m is the effective mass and d is the thickness of the potential well. The quantum dimensional effect in the energy spectrum becomes evident when the characteristic quantization energies (in the order of the kinetic localization energy noted above) exceeds the blurring of energy levels due to the attenua-

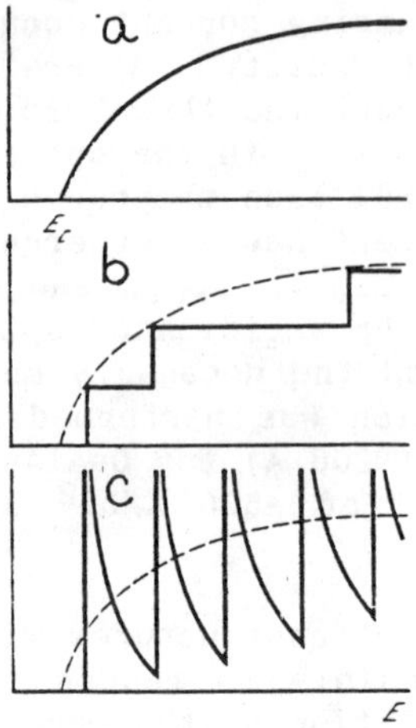

Fig. 2. A schematic representation of the modification of the state density plotted as a function of energy in quantum-dimensional structures. a - the three-dimensional case, $\rho(E) = \sqrt{E - E_C}$; b - single-dimensional quantization; c - two-dimensional quantization.

tion of the corresponding quantum states (in the order of h/τ, where τ is the intraband relaxation time). The spectral characteristics of the quantum dimensional effect in GaAlAs/GaAs, InGaAsP/InP heterostructures become manifest at thicknesses of less than 500 Å. It is also determined that in single-layer structures, carrier trapping at the lower quantum levels is retarded as the thickness decreases, since at narrow thicknesses (less than ~150 Å) the intensity of luminescence from the lower levels is significantly reduced. Consequently we may anticipate a weakening of the electron confinement effect, since the majority of trapped carriers are present at levels significantly higher than the bottom of the potential well.

The favorable change in the form of the state density $\rho(E)$ in quantum-dimensional layers led to certain hopes for improving the temperature dependence of the threshold current (i.e., for reducing the threshold at room temperature). In principle the use of a discrete spectrum, as shown in Fig. 2, may eliminate the influence of such an important factor of the temperature dependence of the threshold as the thermal spreading of the population of the band in the semiconductor, and at the same time will make it possible to eliminate the difficulties associated with the rapid (exponential) temperature increase in the threshold in injection lasers. For the case of a graded form of the spectrum of $\rho(E)$ we may also anticipate a reduction in the temperature influence. However, we may immediately identify the limits on the possible improvement. It is qualitatively clear that populating the "nonworking" levels above the working level region is undesirable; if the thermal population of the next stage (or the peak of $\rho(E)$) begins, this corresponds to an acceleration of threshold growth even compared to a laser with a thick active layer. Thus, the possibility for reducing the threshold current by modification of $\rho(E)$ is limited on the upper end by the temperature at which the characteristic quantization energy becomes insufficient with respect to the thermal energy.

Regarding the trapping of carriers by the ultrathin narrowband layer, multilayered quantum-dimensional "superlattice" structures are used to accelerate this process. It is demonstrated that the lower quantum levels operate efficiently in such structures and lasing is comparatively easily achieved. With respect to the optical structure, such lasers may be calculated by analogy to regular double heterostructures with a certain effective (total) thickness of the waveguide layer. If we return to the threshold current optimization procedure, in this case we must also deal with the total thickness of the active layers and, consequently, it may be expected that, compared to the single-layer structure, the threshold current will be greater by a factor equal to the extent of pumping of the active layers (with approximately equal conditions). Evidently another route is implemented in structures with a three layer (graded or gradient) waveguide, where the potential well has a graded or "funnel" shape. It is possible that two-stage carrier trapping is more efficient. In any case, in studies with low-threshold ultrathin laser structures, the problem of carrier trapping is not noted even at a thickness of a single active layer equal to 60-80 Å.

Other important areas in this field of research and development include the application of the technology of ultranarrow and quantum-dimensional structures to the important practical wavelength ranges of 1.3 and 1.55 μm, determining the influence of the quantum-dimensional effect on various radiative and nonradiative mechanisms as well as other processes playing a role in the physics of an injection laser and the further optimization of laser structure (accounting for the lateral barrier, integrated optical elements, insulation technology, etc.).

INCREASING THE EMISSION POWER OF INJECTION LASERS

It is necessary to increase emission power in order to expand the areas of application (printing and other methods of optical data recording, nonlinear spectroscopy, long-range communications, illumination, etc.). However, our discussion concerns increasing power while preserving the high quality (coherence) of the emission. For example, for "bit" recording on disks, we require focusing of 20-40 mw on a spot 10^{-4} cm in diameter. Single-mode emission with low wavefront distortion is needed for this application. Coherent communications engineering requires dynamic single-frequency (modulated) emission with a spectral width corresponding to the modulation band.

We will examine several methods for increasing the coherent emission power from injection lasers.

1. Cascade amplification. Optical amplifiers based on laser diodes have not yet found broad application. Interest has increased in recent years in connection with their possible application in receiver preamplifiers, in a cascade transmitter and in repeaters. The primary obstacle has been the low efficiency in introducing external emission to the active region of the amplifier. A gain of approximately 10-20 dB is lost in compensating for losses at the input and, consequently, only amplifiers with a very high gain are promising in this application. This high gain may be achieved in a traveling wave amplifier if the end reflection is reduced, or in a regenerative amplifier at the cost of a sharp decrease in the signal bandwidth.

Study [51] developed and investigated a laser active element with inclined mirrors with respect to the active stripe providing lasing in a ring resonator (i.e., deliberately producing "external" amplification). An analysis of the tuning curves has revealed that the residual end reflection has been reduced to 10^{-4} (at 10°). This makes it possible to obtain an "internal" gain in the traveling wave amplifier mode of up to 40 dB. Such a continuous traveling wave amplifier was coupled to a two-stage laser transmitter design (Fig. 3) producing narrowband emission (with a bandwidth of ~10 MHz) preserving coherence under modulation (although maintaining stable single-frequency conditions in the master oscillator). Under modulation at frequencies of 0.1-2 GHz, a clear image of lateral lines was achieved at the corresponding distance from the carrier. In this case with moderate gain (~5-10 dB) an increase in available emission of "radio" quality [51] was achieved.

2. Locking in a multistripe structure. The phenomenon of mutual oscillation trapping in injection lasers when optical coupling exists between such lasers was investigated theoretically [52] and experimentally [53] long ago. There was insufficient stimulus to expand studies in this field, since the optical coupling needed could be created only by cumbersome external optics. Regarding the diffraction coupling resulting from the close proximity of the elements, this could not be successfully utilized due to the mutual thermal influence of the diodes. The decisive change occurred recently after optimization (narrowing) of the active layer was used to reduce the threshold currents. In the multistripe structure (period of 10 μm, active stripe width of 3-4 μm), described in study [54] continuous lasing was achieved significantly over the threshold and with efficient locking of optical oscillations (with the neighboring stripes in phase or out of phase). As we know, the maximum power in such diodes was increased to 2.6 watts while the high degree of relative coherence of the oscillations was preserved up to 0.8 watts, i.e., near single-mode lasing was achieved. An objective evaluation of emission quality is the capability to concentrate the emission in a small spot. Such measurements in study [55] made it possible to focus up to 90 mw on a spot 2.5 μm in diameter (at 770 nm).

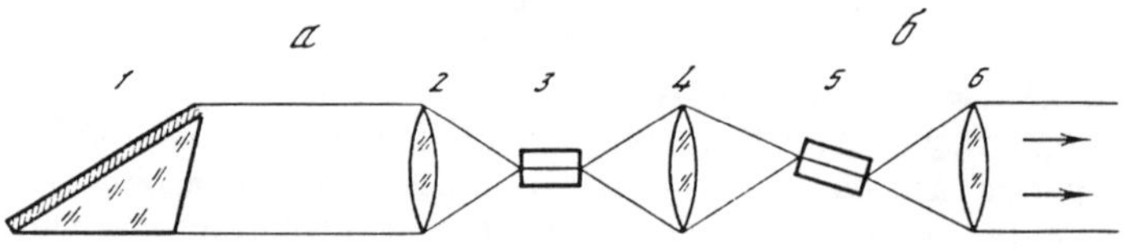

Fig. 3. A two-stage laser transmitter scheme with a bandwidth of ~2 GHz
a - single mode oscillator; b - modulator-traveling wave amplifier. 1 - spectral selector; 2, 4, 6 - lenses; 3 - laser diode; 5 - traveling wave diode in a "Brewster" configuration.

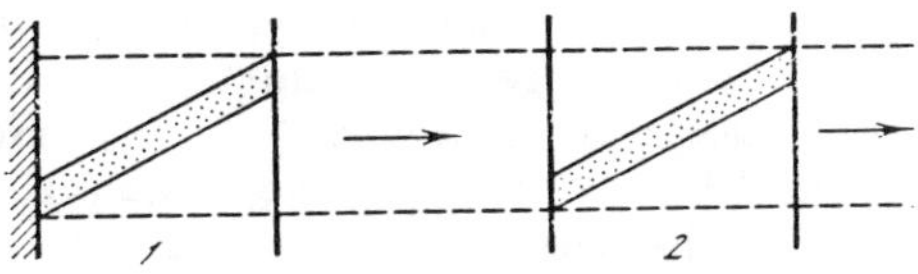

Fig. 4. The optical series interconnection of "nonwaveguide" laser diodes (1, 2)
The active regions are indicated by the dotted line.

3. Optical strength. Experience has shown that the limitations on peak power in semiconductor lasers are highly dependent on the material. In GaAs the critical fluxes, which depend on the pulse duration (300-30 ns) amounted to 1-10 Mw/cm^2 [56]. Nonetheless, for lasers at 1.3 μm these limits have not been established with certainty. It has been reported that catastrophic degradation has not been observed at flux densities 3-5 times greater than these values. Thus, it is possible to select a more optimum semiconductor to increase the pulse power. From the viewpoint of the optical breakdown mechanism (a thermal "microburst" model [56]) the reason for the discrepancy may be different surface states (surface recombination) predetermining the initial "primer" source of heat release.

4. Nonwaveguide lasers. From the viewpoint of the engineering of pulsed lasers with improved collimation of the output beam, "nonwaveguide" resonators [56] are of interest; the active plane in such resonators is inclined with respect to the mirrors at an angle that eliminates regular lasing along the waveguide layer. Such injection lasers (at an angle of 10°) are used in study [58] operating at 300° K in a pulsed mode with a threshold current density of 18 ka/cm^2 (at 1.3 μm). The divergence was improved by at least a factor of 5-8 (according to the directional pattern in the plane perpendicular to the active layer). Another variety of nonwaveguide laser includes lasers radiating perpendicular to the plane of the active layer (without Bragg structures). These also have higher threshold current values [59]. However, both types of nonwaveguide resonators are promising for further improvement. Due to the expanded beam aperture at the exit, such lasers are comparatively easily coupled optically to serial structures (Fig. 4) without intermediate optics or with minimum optics. In this case, it is possible to reduce the threshold by adding the active length and converting to a laser geometry better known for other types of lasers (such as disk glass lasers). Attempts are already being made to develop multilayer nonwaveguide lasers by heteroepitaxy methods.

COMPOUND RESONATORS

For applications in long-range broadband communications, semiconductor lasers are needed that do not lose emission coherence in the presence of significant RF modulation. Such a requirement is not satisfied by regular laser diodes. In these diodes microwave modulation of the pumping current easily disrupts single-mode lasing as, for example, was demonstrated in study [60]. The problem is that with narrow pulse emission, the natural mechanism of uniform mode competition cannot suppress undesirable modes, as in the steady-state case.

As a result, the coherence length is reduced to $c\Delta t$, where Δt is the length of a separate emission pulse. A factor is introduced for dividing the power among many modes with powerful fluctuations in each mode, which generates a new source of residual noise in the communications links. Analogous "mode splitting" phenomena arise as a result of self-modulated instabilities (peak modes).

In order to maintain dynamic single-frequency conditions (in other words in order to obtain spectrally-matched modulation whose bandwidth determines the width of the line in the optical band) a higher spectral selectivity is needed than is found in diode resonators. Such methods of improving selectivity as external dispersion resonators [36-38] and Bragg structures exist. In both cases either the laser device or the laser technology is complicated. One method that is simple and suitable for an integrated design involves using the selectivity of so-called compound (multiple-mirror) resonators. In the general case, the principle of operation of a multiple-mirror system is similar to the principle used in multilayer interferential filters and mirrors, however in this case three to four mirrors are sufficient. With an aperiodic configuration of the mirrors, the highest Q is achieved at modes with an identical frequency in partial resonators while the same modes with different frequencies have the lowest Q.

Compound resonators were proposed for semiconductor lasers and were investigated in a number of studies of the Physics Institute of the Academy of Sciences as early as the 1960's [35, 36, 61]. It was demonstrated that piezoelectric or electrooptic tuning and spectral tuning are possible in such resonators (using the "vernier" effect, i.e., the coincidence of frequencies in two mode series with a different period) [62]. Moreover, digital frequency modulation of the signal is possible, together with spectral multiplexing of channels, etc.

Today interest in such structures has increased sharply. Specifically, a number of injection laser designs employing compound resonators with a passive nontunable external resonator and a tunable resonator have been built. A compound resonator consisting of two diode sections obtained by cleaving a single original diode (a "C^3" resonator) was built at the Physics Institute of the Academy of Sciences as early as 1965 [63].

In this brief survey we have not discussed all important issues such as those associated with the use of semiconductor lasers in frequency standards, or in projection laser television, nor issues dealing with streamer lasers, the dynamics of picosecond pulses, bistable modes, reliability, etc. However, we may conclude on the basis of the aspects considered that the physics of semiconductor lasers is continuing to expand and grow and is on a new productive level. This field will continue to spawn new interesting and important practical fields as indicated by the publication of this volume which is devoted primarily to the nonlinear optics of semiconductor lasers.

BIBLIOGRAPHY

1. Basov, N. G., Vul, B. M., Popov, Yu. M. "Quantum mechanical semiconductor oscillators and electromagnetic amplifiers" ZHETF, 1959, V. 37, p. 587.

2. Basov, N. G., Krokhin, O. N., Popov, Yu. M. "The generation, amplification and display of infrared and optical emission by means of quantum systems" UFN, 1961, V. 72, p. 161-209.

3. Basov, N. G., Krokhin, O. N., Popov, Yu. M. "Obtaining negative temperature states at the *p-n*-junctions of degenerate semiconductors" ZHETF, 1961, V. 40, p. 1879-1880.

4. Basov, N. G. "Semiconductor lasers: Nobel Lecture" UFN, 1965, V. 85, p. 585-598.

5. Eliseev, P. G. "The application of semiconductor lasers" In: Itogi nauki i tekhniki [Achievements in science and technology] Moscow: VINITI, 1982, V. 28, p. 3-124.

6. Eliseev, P. G. "Investigations of injection lasers at the Physics Institute of the Academy of Sciences" TR. FIAN, 1983, V. 141, p. 3-17.

7. Bernard, M., Durrafourg, G. "Laser condition in semiconductors" PHYS. STATUS SOLIDI, 1961, V. 1, No. 2, p. 699-703.

8. Hall, R. N., Fenner, G. E., Kingsley, J. D. et al. "Coherent light emission from GaAs junctions" PHYS. REV. LETT., 1962, V. 9, No. 9, p. 366-378.

9. Nathan, M. I., Dumke, W. P., Burns, G. et al. "Stimulated emission of radiation from GaAs *p-n*-junctions" APPL. PHYS. LETT., 1962, V. 1, No. 3, p. 62-64.

10. Holonyak, N., Jr., Bevacqua, S. F. "Coherent (visible) light emission from GaAs ($As_{1-x}P_x$) junction" APPL. PHYS. LETT., 1962, V. 1, No. 4, p. 82-83.

11. Quist, T. M., Rediker, R. H., Keyes, R. J. et al. "Semiconductor maser of GaAs. APPL. PHYS. LETT., 1962, V. 1, No. 4, p. 91-92.

12. Bagaev, V. S., Basov, N. G., Vul, B. M., et al. "A semiconductor laser operating at the *p-n*-junction in GaAs" DAN SSSR, 1963, V. 150, No. 2, p. 275-278.

13. Basov, N. G., Eliseev, P. G., Istailov, I. et al. "An investigation of the properties of injection lasers in the 0.8-1.1 μm range" ZHTF, 1967, V. 37, No. 2, p. 349-359.

14. Kurnosov, V. D., Magalyas, V. I., Pleshkov, A. A. et al. "Emission self-modulation of an injection semiconductor laser" PIS'MA V ZHETF, 1966, V. 4, p. 449-450.

15. Basov, N. G., Morozov, V. N., Nikimin, V. V., Setenov, A. S. "An investigation of emission fluctuation conditions of a GaAs laser" FTPA, 1967, V. 1., p. 1570-1574.

16. Drozhbin, Yu. A., Zakharov, Yu. P., Nikimin, V. V., et al "The generation of ultranarrow light pulses in a gallium arsenide laser" PIS'MA V ZHETF, 1967, V. 5, No. 6, p. 180-182.

17. Basov, N. G., Morozov, V. N. "Theory of the dynamics of injection lasers" ZHETF, 1969, V. 567, p. 617-627.

18. Basov, N. G., Nikimin, V. V., Setenov, A. S. "The dynamics of emission from injection semiconductor lasers" UFN, 1969, V. 97, No. 4, p. 561-600.

19. Basov, N. G., Drozhbin, Yu. A., Zakharov, Yu. P. et al. "The influence of injection current on the temporal characteristics of emission from a GaAs laser" FTT, 1966, V. 8, No. 9, p. 2816-2818.

20. Basov, N. G., Bozdankevich, O. V., Devyatkov, A. G., "A fast electron-excited CdS laser" DAN SSSR, 1964, V. 155, No. 4, p. 78.

21. Basov, N. G., Grasyuk, A. Z., Katulin, V. A. "Induced emission in gallium arsenide under optical excitation" DAN SSSR, 1965, V. 161, p. 1306-1307.

22. Southgate, P. D. "Laser action in field-ionized bulk GaAs." APPL. PHYS. LETT., 1968, V. 123, No. 3, p. 61-63.

23. Basov, N. G., Molchanov, A. G., Nasibov, A. S., et al. "Solid state streamer lasers" ZHETF, 1966, V. 70, No. 5, p. 1751-1761.

24. Hinkley, E. D. "High resolution infrared spectroscopy with tunable diode laser." APPL. PHYS. LETT., 1970, V. 16, No. 9, p. 351-354.

25. Zasavitskiy, I. I., Matsonashvili, B. N., Shomov, A. P. "An investigation semiconductor lasers as emission sources for spectral measurements" ZHURN. PRIKL. SPEKTROSKOPII, 1971, V. 15, No. 2, p. 349-352.

26. Zasavitskiy, I. I., Matsonashvili, B. N., Shomov, A. P. "A continuous injection laser based on PbSe" PIS'MA V ZHETF, 1975, V. 1, No. 7, p. 341-343.

27. Alferov, Zh. I., Andreev, V. M., Korol'kov, V. I., et al. "Coherent emission in epitaxical structures with heterojunctions in an AlAs-GaAs system" FTP, 1968, V. 2, No. 10, p. 1545-1548.

28. Alferov, Zh. I., Andreev, V. M., Garbuzov, D. Z., et al. "An investigation of the influence of heterostructure parameters on the threshold current of lasers and continuous lasing at room temperature" FTP, 1970, V. 4, p. 1826-1830.

29. Hayashi, I., Panish, M. B., Foy, P. W., Sumski, S. "Junction lasers which operate continuously at room temperature" APPL. PHYS. LETT., 1970, V. 17, No. 3, p. 109-111.

30. Bronshmein, I. K., Dolginov, L. M., Druzhinina, L. V., et al. "A continuous injection laser operating at 300 K" KRAT, SOOBSHCH. GIREDMET V. 5, 1970, No. 21.

31. Dolginov, L. M., Eliseev, P. G., Mil'vidskiy, M. G. "Multicomponent semiconductor solid solutions and their applications in lasers" KVANTOVAYA ELEKTRON. 19076, V. 3, No. 7, p. 1381-1393.

32. Bogatov, A. P., Dolginov, L. M., Druzhinina, L. V., et al. "Heterolasers based on GaInAsP and AlGaSbAs solid solutions" KVANTOVAYA ELEKTRON., 1974, V. 1, No. 10, p. 2294-2295.

33. Bogatov, A. P., Dolginov, L. M., Eliseev, P. G., et al. "The emissive characteristics of InP-InGaAsP laser heterostructures" FTP, 1975, V. 9, No. 10, p. 1959-1961.

34. Hsieh, J. J. "Room temperature operation of GaInAsP/InP double-heterostructure diode lasers emitting at 1.1 μm" APPL. PHYS. LETT., 1976, V. 28, No. 5, p. 283-285.

35. Eliseev, P. G., Ismailov, I., Man'ko, M. A., Strakhov, V. P. "An injection semiconducting laser with a compound resonator" PIS'MA V ZHETF, 1969, V. 9, p. 594-595.

36. Akkerman, D., Eliseev, P. G., Kayper, A., et al. "Methods of mode selection in injection lasers" KVANTOVAYA ELEKTRON., 1971, No. 1, p. 85-90.

37. Eliseev, P. G., Strakhov, A. P., "Single mode lasing in injection lasers" PIS'MA V ZHETF, 1972, V. 16, p. 606-608.

38. Bazhenov, V. Yu., Bogatov, A. P., Gurov, Yu. V., et al. "Optical heterodyning of injection laser emission for a laser with an internal dispersion resonator" KVANTOVAYA ELEKTRON., 1980, V. 7, No. 11, p. 2510-2512.

39. Belichanskiy, V. L., Zibrov, A. S., Kargopol'tsev, V. S. et al. "The maximum lasing line width in an injection laser" PIS'MA V ZHETF, 1978, V. 4, No. 18, p. 1087-1090.

40. Molochev, V. I., Nikitin, V. V., Samoylov, V. D. "The optical interaction of injection lasers under uniform excitation" KVANTOVAYA ELEKTRON., 1971, V. 2, p. 25-31.

41. Goldobin, I. S., Dobkin, A. S., Kurnosov, V. D. et al. "Quantum-optical gallium arsenide integrated circuits" FTI, 1971, V. 5, No. 1, p. 170-171.

42. Bezomosnyy, V. V., Dolginov, L. M., Eliseev, P. G., et al. "GaInPAs/InP heterolasers based on a buried mesa stripe structure operating in a continuous mode at room temperature at 1.24-1.28 μm" KVANTOVAYA ELEKTRON., 1980, V. 7, No. 9, p. 1990-1992.

43. Bazhenov, V. Yu., Bogatov, A. P., Eliseev, P. G., et al. "Bistable modes and spectral tuning in an injection laser with an external dispersion resonator" KVANTOVAYA ELECTRON, 1981, V. 8, No. 4, p. 853-859.

44. Bogatov, A. P., Eliseev, P. G., Sverdlov, B. N., "Anomalous interaction of spectral modes in a semiconducting laser" IEEE J. QUANT. ELECTRON., 1975, V. 11, No. 7, p. 510-514.

45. Dolginov, L. M., Druzhinina, L. V., Eliseev, P. G., et al. "A new uncooled injection heterolaser for 1.5-1.8 μm" KVANTOVAYA ELEKTRON., 1976, V. 3, No. 2, p. 465-466.

46. Eliseev, P. G., "The optimum width of the active layer in a heterolaser" KVANTOVAYA ELEKTRON., 1971, No. 3, p. 120-121.

47. Eliseev, P. G., Vvedenie v fiziku inzhektsionnykh lazerov [An introduction to the physics of injection lasers] Moscow: Nauka, 1983, 294 p.

48. Tsang, W. T., Logan, R. A., Ditzenberger, J. A. "Ultralow threshold graded-index waveguide separate confinement CW buried-heterostructure lasers" ELECTRON. LETT., 1982, V. 18, No. 19, p. 845-847.

49. Tsang, W. T. "Extremely low threshold AlGaAs modified multiquantum well heterostructure lasers grown by molecular beam epitaxy" APPL. PHYS. LETT., 1981, V. 39, No. 10, p. 786-788.

50. Camras, M. D., Holonyak, N., Jr., Nixon, M. A., et al. "Photopumped low threshold AlGaAs-AlGaAs-AlGaAs single quantum well lasers" APPL. PHYS. LETT., 1983, V. 42, No. 9, p. 761-763.

51. Annenkov, D. M., Bogatov, A. P., Eliseev, P. G., et al. "The spectrally-matched modulation of emission from an injection laser in a traveling wave amplifier at frequencies up to 2 GHz" KVANTOVAYA ELEKTRON., 1984, V. 11, No. 2, p. 231-232.

52. Basov, N. G., Belenev, E. M., Letokhov, V. S. "Mode locking in a semiconductor laser with several *p-n*-junctions" FTT, 1065, V. 7, No. 1, p. 237-239.

53. Kobayashi, S., Kimura, T. "Injection locking characteristics of an AlGaAs semiconductor laser" IEEE J. QUANT. ELECTRON., 1980, V. 16, No. 9, p. 915-917.

54. Scifres, D. R., Burnham, R. D., Streifer, W. "Continuous wave high-power high temperature semiconducting laser phase-locked arrays" APPL. PHYS. LETT., 1982, V. 41, No. 11, p. 1030-1032.

55. Scifres, D. R., Sprague, R. A., Streifer, W., Burnham, R. D. "Focusing a high power phased array semiconducting laser" APPL. PHYS. LETT., 1982, V. 41, No. 12, p. 1121-1123.

56. Eliseev, P. G. "Degradation of injection lasers" J. LUMIN., 1973, V. 7, p. 338-356.

57. Bogatov, A. P., Eliseev, P. G., Man'ko, M. A., et al. "Reducing the emission divergence of emission from an injection laser by excitation of nonwaveguide modes" KVANTOVAYA ELEKTRON., 1979, V. 6, No. 12, p. 2639-2641.

58. Bogatov, A. P., Eliseev, P. G., Mikaelyan, L. T., Sverdlov, B. N. "A nonwaveguide mode injection GaInPAs-heterolaser with a divergence of 6-7°" KVANTOVAYA ELEKTRON., 1980, V. 7, No. 11, p. 2487-2488.

59. Soda, H., Iga, K., Kitahara, S., Suematsu, Y. "GaInAsP/InP surface emitting injection lasers" JAPAN J. APPL. PHYS., 1979, V. 18, No. 12, p. 2329-2330.

60. Eliseev, P. G., Ivanov, L. P., Logginov, A. S., Senatorov, K. Ya. "The influence of microwave modulation on the emission spectrum of an injection laser" KVANTOVAYA ELEKTRON., 1974, V. 1, p. 151-154.

61. Bogatov, A. P., Eliseev, P. G., Man'ko, M. A., Chan, Min' Tkhay "The influence of a compound resonator on the emission coherence of injection lasers" KRAT. SOOBSHCH. PO FIZIKE FIAN, 1971, No. 9, p. 60-65.

62. Eliseev, P. G., Khaydarov, A. V. "The electrooptical and piezoelectric tuning of a compound laser resonator" OPTIKA I SPEKTROSKOPIYA, 1973, V. 34, No. 2, p. 343-346.

63. Eliseev, P. G., Novikov, A. A., Fedorov, V. B. "Optical interaction effects between two diode lasers" PIS'MA V ZHETF, 1965, V. 2, p. 58-60.

PHENOMENA IN SEMICONDUCTOR LASERS ASSOCIATED WITH NON-LINEAR REFRACTION AND THE INFLUENCE OF CURRENT CARRIERS ON THE REFRACTIVE INDEX

P.G. Eliseev, A.P. Bogatov

Abstract: Semiconductor lasers have found many practical applications. The understanding of their optical properties is based on nonlinear models. Semiconductors are characterized by significant optical nonlinearity. The influence of free carriers on the refractive index is critical while the carrier concentration has an intensity dependence due either to photoelectric absorption or stimulated emission. In semiconductor lasers this produces a number of phenomena such as self-focusing, nonlinearity of optical losses with associated anomalies in emission dynamics, bistable modes in compound resonators and amplifiers, frequency self-modulation and nonlinear scattering by electron density waves. The same physical principles are used to achieve self-stabilization of single-frequency conditions which makes it possible to increase the coherent emission intensity of a semiconductor laser.

1. INTRODUCTION

An optical model of a semiconducting laser is based on dielectric waveguide theory, although an important feature of this model is the significant optical nonlinearity of the semiconductor medium, including nonlinear refraction (the intensity dependence of the refractive index). Among the physical mechanisms responsible for nonlinear refraction at the operating frequencies of laser emission, i.e., along the edges of fundamental absorption, a significant role is played by the powerful influence of the free carriers on the refractive index and the influence of emission intensity on their concentration. This fact has long been known and has always been borne in mind in discussing the properties of semiconductor lasers. In addition, nonlinear refraction in semiconductors has been the focus of recent attention in connection with studies on optical bistability in semiconductor resonators [1, 2]. It has become possible to develop new types of high speed optical devices based on the principle of bistability for emission control and optical data processing applications [3, 4].

As noted above, the influence of carriers on the refractive index is accounted for in semiconducting laser theory and in models of its behavior [5-7]. The dynamic phenomena caused by this influence represent an important chapter in the nonlinear optics of semiconductor lasers.

In this survey we will examine the physical principles of nonlinear refraction and the bistable phenomena in semiconductor resonators (Sections 2-4), the feature of nonlinear refraction in lasing (sign reversal of the nonlinearity, Section 5), self-focusing conditions and the single-dimensional case (Section 6, 7), nonlinear optical losses from nonlinear refraction (Section 8), bistable modes in lasers and amplifiers (Sections 9, 10), frequency modulation (Section 11), the nonlinear scattering phenomena by electron density waves and the autostabilization effect of the single-frequency mode (Sections 12, 13).

2. RESONANT NONLINEAR REFRACTION IN SEMICONDUCTORS

In phenomena associated with a "giant" nonlinearity, resonator nonlinear refraction occurs near the edge of fundamental absorption as a result of the influence of free carriers. In certain cases nonlinear refraction makes it possible to demonstrate optical switching or bistability with beams of only 1 mw (with an optical flux density of 10^3-10^4 w/cm^2). We will discuss frequencies near the fundamental absorption edge which includes the working frequencies of semiconductor lasers. We note that in other frequency ranges the nature of the nonlinearity may be quite different; for example, it may have a different sign, since it is caused by different mechanisms (interaction with plasmons, the Stark effect, striction etc.).

The third-order nonlinear susceptibility effects often are characterized by a refractive index dependent on the intensity Φ of emission

$$n(\Phi) = n_0 + n_2\Phi, \tag{2.1}$$

where n_2 is the coefficient of the "nonlinearity" period. Initial demonstrations of self-focusing, when $n_2>0$, as we know, employed dielectric media in which n_2 was 10^{-16} - 10^{-13} cm^2/watt. Values of $-4\cdot10^{-4}$ cm^2/watt (GaAs [1]) and $-3\cdot10^{-3}$ cm^2/watt (InSb [2]) were subsequently found in semiconductors; these values were ten times greater (H) in absolute value than for the case of classical nonlinear materials. Today other media with a "giant" nonlinearity have been discovered, including dyes, liquid crystals and alkaline metal pairs. Nonetheless, for practical purposes, semiconductors remain the most attractive media since they operate at a high speed and may be used in miniature (such as integrated-optic) devices [3].

If we assume an approximate linear relation between the change in the refractive index δn and the carrier concentration N,

$$\delta n = \sigma N, \tag{2.2}$$

where σ is the proportionality factor, then ignoring spatial relations we may write [8]

$$n_2 \approx \eta\sigma\alpha\tau/\hbar\omega, \tag{2.3}$$

where η is the coefficient representing the relative portion of photoelectric absorption, α is the absorption coefficient and τ is the lifetime of the excess carriers. Due to the mobility of the current carriers, the effect of the intensity Φ may be nonlocal; for example, the carriers may diffuse into the nonilluminated region etc. Hence, a precise theory must accurately account for spatial effects.

Thus, the subject of this discussion will include the phenomena of nonlinear refraction in semiconductors caused by a "giant" nonlinearity, which in turn is associated with the influence of free carriers on the refractive index. With external illumination, the effect results from photocarriers, while in semiconductor lasers it is produced by injected carriers. It will be demonstrated below that due to an opposite effect produced by the electromagnetic field on a normal and inverted medium, the sign of the nonlinearity n_2 will be opposite in these two cases.

We note that in experiments with external illumination, often we use emission with a longer wavelength than the fundamental absorption edge. This, however, does not mean that photoelectric absorption is not present. There always exists a photoelectric absorption tail near the absorption edge resulting from indirect transitions, the nonstationarity of quantum states in the bands, the influence of the chaotic Coulomb field of the dopants, etc. In order to observe nonlinear refraction, an absorption coefficient of ~1 cm^{-1} is sufficient. Moreover, nonlinear refraction is also observed rather far from the fundamental absorption edge, when photocarriers are generated due to the multiphoton process.

3. THE INFLUENCE OF CARRIERS ON THE REFRACTIVE INDEX

Electromagnetic wave dispersion theory for a medium with resonant absorption relates Re ε and Im ε by Kramers-Kronig relations. Changes in the Im ε spectrum result in predictable changes in the Re ε spectrum, so that if we have, for example, a variation in the refractive index $\Delta\alpha(\hbar\omega)$, we may calculate the changes in the refractive index $\Delta n(\hbar\omega)$ according to the formula [9]

$$\Delta n(\hbar\omega) = \frac{\hbar c}{\pi} \oint_0^\infty \frac{\Delta\alpha(\hbar\omega')}{(\hbar\omega')^2 - (\hbar\omega)^2}\, d(\hbar\omega') \tag{3.1}$$

(the integral is taken as the principal value). With changes in the carrier concentration, changes in the absorption spectrum are related primarily to the saturation of fundamental absorption at several frequencies near its edge, the frequency shift of plasma resonance as well as changes in the combination phonon-plasmon frequencies, shielding phenomena, absorption by the free carriers, etc.

The primary contribution is made by band-to-band transitions, and in certain cases by exciton transitions (for example, in pure GaAs at low temperatures). An approximate and accessible evaluation of the band-to-band effect is given in study [8]. Each transition from the valence band to the conduction band, characterized by the effective electrical dipole moment μ, produces a contribution Δn to the refractive index equal to

$$\Delta n \approx (2\pi/\hbar n_0)\ |\mu|^2/(\omega_0 - \omega), \tag{3.2}$$

where ω_0 is the central transition frequency (the influence of broadening is ignored for simplicity). If the upper state is filled, resonant absorption is excluded and the refractive index is reduced by Δn. In the vicinity $\omega<\omega_0$, the effect should be negative, i.e., the refractive index drops with an increase in the photocarrier concentration. Further calculations require accurate averaging of the contributions from carriers distributed in the band and interacting with various levels in the lower band. This results in the following formula for the band-to-band component σ:

$$\sigma = dn/dN \approx -(2\pi/\hbar n_0)\ |\bar{\mu}|^2/(\bar{\omega}_0 - \omega), \tag{3.3}$$

where $\bar{\mu}$ and $\bar{\omega}_0$ are the effective mean values of μ and ω_0, respectively. For InSb, such an estimate produced $\sigma \approx 5\cdot 10^{-18}$ cm^3 for emission 20 cm^{-1} from the fundamental absorption edge. With $\alpha = 1$ cm^{-1}, $\eta = 1$ and $\tau = 300$ ns, we derive $n_2 - 5\cdot 10^{-5}$ cm^2/watt, which is consistent with measurements at 5 K. At 77 K we obtain $\sigma \approx 2\cdot 10^{-18}$ cm^3 (estimates from study [8]). Study [9] derived a formula for A^3B^5 semiconductor compounds which may be reduced to the form

$$\sigma(\hbar\omega) \approx -3\cdot 10^{-17}\, T^{-1} (\hbar\omega)^{-2} J(\hbar\omega), \tag{3.4}$$

where $J(\hbar\omega)$ is the dimensionless resonance function equal to one along the nominal absorption edge and decreasing by a factor of approximately 5 when separated by an amount on the order of kT from the absorption edge [9]. Formula (3.4) gives σ in cm^3 when substituting T in degrees Kelvin and $\hbar\omega$ in electron volts.

The plasma transition effect, consistent with [10], may be expressed in the form

$$\delta\varepsilon = -4\pi Ne^2/m_{\text{эф}}\omega^2, \tag{3.5}$$

which is sufficiently simple in view of the condition that this frequency is greater than the plasma oscillation frequency. We may derive the following for the plasma component σ:

$$\sigma = -2\pi e2 / n_0 m_{э\Phi} \omega^2 . \tag{3.6}$$

It is clear that both fundamental components are negative and are inversely proportional to the square of the frequency.

Table 1

The quantities σ and R at near-threshold concentration in semiconductor lasers[1]

Material	*Wavelength,*	$\sigma \cdot 10^{21}$, cm^3	*R*	*Reference*
GaAs	0.9	-6.9	-	[12]
	0.9	-3.6	-	[13]
GaAlAs	0.88	-4.0	-2.5	[15]
	0.84	-4.9	-	[16]
InGaAsP	1.61	-7.2	-4.6	[15]
	1.57	-20*	-3.2*	[15]
	1.53	-40**	-2.4**	[15]
PbSe	4.8	-60	-	[17][2]
	7.0	-100***	-	[17][2]

[1] *These results apply to room temperature with the exception of figures marked by asterisks: * T = 190 K, ** T = 140 K, *** T = 80 K.*

[2] *Calculation results based on data from study [17].*

In GaAs near 880 nm at 300 K, the contribution of equal concentrations of electrons and holes is estimated [10] to be $-1.7 \cdot 10^{-21}$ cm^3. For comparison, an estimate of the band-to-band component using formula (3.4) gives a maximum effect of $-5 \cdot 10^{-20}$ cm^3, and accounting for the shift of the operating point from the nominal absorption edge, we have $-1.1 \cdot 10^{-20}$ cm^3. Thus, we may assume that the band-to-band component will also predominate in GaAs at room temperature. Theoretical calculations in study [11] produced σ of approximately -10^{-20} cm^3 for GaAs. A comparison to experimental values given in Table 1 is not too favorable for theory which gives noticeably higher values. Study [12] has indicated a range of possible values of σ from $-2.8 \cdot 10^{-21}$ to $-1.9 \cdot 10^{-20}$ cm^3 with the most probable value at $-6.9 \cdot 10^{-21}$ cm^3. Data obtained from spectral measurements with laser GaAs-diodes and a planar cavity were utilized. Bogatov et al. [13] measured changes in the optical length of a planar GaAs-cavity in a sensitive configuration of two optically-coupled diodes under the influence of carrier injection to one diode (the intracavity refractometry method [14]). In order to obtain σ a specific value τ was needed in calculating the carrier concentration

$$N = j\tau / ed \tag{3.7}$$

(j is the pumping current density, d is the thickness of the active layer). The derived value of $-3.6 \cdot 10^{-21}$ cm^3 corresponded to a pumping level of $j = 15$ kA/cm^2 in one-sided GaAlAs/GaAs-heterostructures with an active layer thickness of 2.2 μm. A value of τ of 2.5 ns from independent measurements was used. Values of σ lower than the theoretical estimates given above were derived in subsequent studies for GaAs and GaAlAs. Discrepancies in the experimental values may be explained by the effect of the pumping level, since $n(N)$ in the general case is a nonlinear function. It should also be noted that in spectral experiments to determine σ, it is necessary not only to know the value τ but also to carefully eliminate the temperature effect which produces a contribution of opposite sign with an increase in current.

Measurements on InGaAsP-diodes in the 1.5-1.65 μm range [15] were consistent with the theoretical estimate, producing $\sigma \approx -7 \cdot 10^{-21}$ cm^3 [15]. For PbSe longwave lasers Table 1 gives calculated data obtained from the curves in study [17]. These refer to the equilibrium carrier effect and are given for a reference estimate. A comparison of values of σ (300 K) at different wavelengths indicates a rather rapid increase in absolute value, although not precisely consistent with the relation $1/\omega^2$, as indicated from formula (3.4), (3.6). An increase in σ with cooling evidently confirms the predominance of the contribution of band-to-band transitions according to formula (3.4). The table also gives values of the coefficient =

$$R = -4\pi\sigma/\lambda a, \tag{3.8}$$

where a is the proportionality factor between the gain g (cm^{-1}) and the carrier concentration N (cm^{-3}). This quantity characterizes the relation

$$R = (d \operatorname{Re}\varepsilon/dN)/(d \operatorname{Im}\varepsilon/dN), \tag{3.9}$$

which figures in a number of problems in the nonlinear optics of semiconductor lasers.

Thus, we have a clear concept of a significant factor for an optical model of a semiconducting laser, specifically the dependence of n on the free carrier concentration. It is clear that theory correctly predicts the value of σ within an order of magnitude, although this value is greater in narrowband semiconductors. From here we see that the corresponding physical phenomena of this factor are expected to be stronger in lasers of longer wavelength, which have just begun to be investigated. The relation $n(N)$ results in a number of phenomena. The simplest observation is that the longitudinal mode frequencies of a planar GaAs-resonator in superluminescence conditions shift uniformly with an increase in the pumping current (this shift has an opposite sign from the temperature effect; hence, the observed effect corresponds to the algebraic sum of both shifts). Henceforth we will limit our consideration to the phenomena interpreted in the spirit of formula (2.1), i.e., optical nonlinear effects where n depends on Φ and Φ in turn influences N.

We will now consider certain properties of resonators containing a medium with nonlinear refraction.

4. OPTICAL BISTABILITY AND OTHER PHENOMENA IN NONLINEAR RESONATORS

Theoretical studies on nonlinear resonators have been performed for some time (see survey [18] and references). Initially, primary attention was focused on the Im ε nonlinearity (due to absorption saturation), then attention shifted to the Re ε nonlinear (i.e., nonlinear refraction). It has been noted that in resonance conditions, the emission intensity within the resonator may be significantly greater than the intensity of incident emission, which enhances the nonlinearity of the medium. A variety of conditions were predicted theoretically (emission limiting, bistability, optical hysteresis, amplification). The technology of feedback electrooptical resonators performing an analogous function has been developed together with the technology of other devices (see [19]).

We will examine in greater detail experimental results from the observation of nonlinear refraction in InSb and GaAs. Direct-band semiconductors are characterized by a sharp fundamental absorption edge, particularly at low temperatures. For example, in InSb the absorption coefficient increases by an order of magnitude greater than 3 within a few millielectronvolts from the absorption edge. A tunable CO-laser covering the wavelength range from 5-6.5 μm in steps was used in optical bistability experiments in InSb; in this case the absorption at 4 K is near 5.3 μm. In order to observe nonlinear refraction, the laser was tuned to the photon energy near the nominal fundamental absorption edge (within 10 mev), where the absorption coefficient amounted to only a few cm^{-1}. In experiments employing samples with a bleaching coating on the mirror facets, the propagation of emission without interference effects caused by multiple reflection was investigated. At a flux density of >10 w/cm^2, beam splitting into a number of rings was observed in the far field [20]. This behavior was interpreted as the result of selfdefocusing caused by the negative contribution of laser emission to the refractive index. The low intensity at which nonlinear refraction becomes evident indicates a very high third order nonlinear susceptibility χ. It was discovered that χ in InSb is approximately 4-6 times greater than was previously believed. This made it possible to conduct effective experiments on the transmittance of the InSb resonator [2]. The natural surface of InSb has a reflection coefficient for emission at the fundamental absorption edge of 36%, and a plane-parallel sample functions as a Fabry-Perot interferometer (resonator) with a relatively low Q. During the transmission of the laser beam, an increase in intensity reduces the optical length of the resonator and pulls the resonator to resonance with external emission, corresponding to the condition where a whole number of half waves are accumulated between the mirrors. As a result, the intensity within the resonator increases (compared to ambient light). The progression of resonances is reflected on the transmission graph in stages whose slope is much greater than 1, corresponding to differential amplification. Moreover, the calculated curve may have a negative slope (an S-nonlinearity). In this case, a hysteresis loop is formed, i.e., with a given intensity at the input, a double valued solution or bistability exist within the loop (two stable transmission states implemented depending on how (from above or from below) the intensity at the input is achieved). Regarding the flat sections of the graduated curve, these correspond to limiting conditions or stabilization of the intensity of the emission.

If two beams with a slightly different angle of incidence (for independent observation of the emission) are directed at the sample, changes in the intensity of one of the beams within the differential amplification region are accompanied by changes in the intensity of the second beam. Thus, a weak beam may be used to control a strong beam or to amplify a strong optical signal. By analogy to a transistor - a device in which resistance is transferred from one electrical circuit to another - an optical device of this type is called a "transphaser", since the change in phase on the optical path of one beam is transferred to the optical path of another [2]. We note that the "quality" of the nonlinear material for the resonator of a bistable element is characterized by the relation n_2/α. Optical bistability was observed in InSb at 77 K as well in polycrystalline and, consequently, inexpensive samples. The experiments were extended to the region of semiconductor magnetooptics. It is demonstrated that the Faraday rotation of the polarization plane depends on the emission intensity [21]. This leads to new possibilities for controlling light beams.

An experiment to observe optical bistability from nonlinear refraction in GaAs [1] was performed using a sample 4.2 μm in thickness covered by transparent GaAlAs layers up to 0.2 μm in thickness and dielectric mirrors with a reflection coefficient of 90%. The nonlinearity of the refractive index was caused by exciton absorption saturation, and the laser emission wavelength was 1 - 2.5 nm higher than the absorption resonance peak of the free excitons. Optical bistability was observed when focusing emission onto a spot ~10 μm in diameter in a temperature range of 5-120 K. It was determined that the critical intensity Φ_{Kp} for developing the hysteresis loop was ~1 mw/μm^2 (10^5 w/cm^2). In order to maintain an intensity on the order of 10^5 w/cm^2, 0.24 pJ/μm^2 are required for transparency. Here, the turn-on time was less than 1 ns, while the turn-off time was up to 40 ns [1].

Analogous light-controlled nonlinear elements have been tested and may be developed for converting the shape (waveform) of light pulses, for deflecting and scanning beams, for performing logic operations, for automatic monitoring of lumination levels, for display, etc. It should be noted that in view of the temperature dependence of the refractive index, possible practical devices based on nonlinear semiconductor resonators must be carefully temperature-controlled.

Table 2 gives a chart from study [3] containing the parameters of existing and promising future bistable devices. It is clear that the results achieved to date are far from the desirable characteristics that could be used as the basis for developing new optical systems. Since the beam cross-section in an ideal material is less than the wavelength, we use semiconductor-based integrated optical devices; for the wavelength this clearly is quaternary InGaAsP.

Table 2

A comparison of materials for optically bistable nonlinear resonators [3]

Material	Wave-length μm	Sample Diameter μm	Sample length μm	Time, ps Turn on	Time, ps Turn off	Switching energy fJ	Maintenance energy $mw/\mu m^2$	Maximum temperature K
InSb (Experiment)	5	180	560	-	-	-	10^{-2}	5
GaAs (Experiment)	0.8	10	5	10^2	$4\cdot10^4$	$6\cdot10^5$	1	120
GaAs (Extrapolation)	0.8	<1	<1	~1	$<10^2$	~1	10^{-2}	?
"Ideal" (Extrapolation)	1.3	<1	<1	~1	~1	~1	<1	300

The maintenance power (optical "shifting" of the operating point) and the similar critical power amount to the following power density values: 1 kw/cm^2 in InSb and 100 kw/cm^2 in GaAs; in both cases it is assumed that the medium remains absorption-linear. We note that the record critical density of optical flux, equal to 5 $nw/\mu m^2$ or 0.5 w/cm^2 still belongs to a nonlinear Na resonator; however, this device is slower than a semiconductor (in the μs range) and is also larger (~10 cm in size). An improvement in speed in the semiconductor may be anticipated by reducing the length and, consequently, the transit time of the resonator. Here, the turn-on time may be reduced by increasing intensity, although the turn off time may be much greater than the turn-on time since it is related to the lifetime of the excess carriers.

An interesting aspect of observations of optical and electrooptical bistability are photoelectric studies of semiconductor samples in the form of a Fabry-Perot resonator. Studies [22, 23] have derived wave-shaped volt-current characteristics (with decaying sections) in a nonlinear GaAs:Cr photoconducting Fabry-Perot resonator under external illumination. The cause of oscillations in the photocurrent were the resonances in the sample under the action of applied voltage from the electrooptic effect. A later study [24] also observed wave-like volt-current characteristics in a resonator with a *p-n*-junction along the axis (with a space charge layer thickness of 25-35 μm). Moreover, the dependence of the photocurrent on the illumination intensity had a hysteresis loop, indicating the existence of optical bistability. The experiment was conducted on a GaAs resonator 1 mm in length at 1.15 μm. The onset of photocurrent (1-2 μa at an illumination power of ~1 mw) resulted from two-stage optical transitions through deep, uncontrolled impurity centers (the photoelectric attenuation coefficient is estimated on the order of 10^{-2} cm^{-1}).

5. NONLINEAR REFRACTION IN PHOTOELECTRIC ABSORPTION AND OPTICAL AMPLIFICATION

Since our discussion involves the influence of free carriers on the refractive index and associated nonlinear phenomena, it is important to establish the intensity dependence of the carrier concentration. For the case of photoelectric absorption, an increase in intensity causes an increase in the concentration of excess carriers, approaching saturation at the high intensity limit. Here the absorption coefficient also drops (the Im ε optical nonlinearity).

Stimulated emission is an opposite effect of regular optical absorption, and hence it produces an internal photoeffect of opposite sign: a reduction in the concentration of excess carriers and an increase in intensity (concentration saturation may also occur in lasing). This "photoeffect" may be used in the electrical diagnostics of amplification modes, as recently demonstrated in study [25]. If we measure the electrical voltage across the *p-n*-junction in a diode (injection laser) during the propagation of resonant interacting emission, the absorption conditions produce regular photo-EMF or a photodiode effect with negative (reverse) or small forward biases, while amplification conditions produce an opposite effect.

Thus, in the conversion from absorption to amplification, the sign of the addition to the carrier concentration changes from negative to positive, thereby changing the sign of the nonlinear addition to the refraction index with all its attendant consequences. Above all self-defocusing becomes self-focusing.

For an elementary quantitative analysis we will consider kinetic equations for electrons accounting for the influence of external emission. Assuming that the spatial uniformity condition is satisfied, we may ignore the diffusion term and may represent the change in the concentration N of conduction electrons in the following form:

$$dN/dt = -N/\tau + G - KN_{\Phi}, \qquad (5.1)$$

where τ is the lifetime of the conduction electrons, G is the volume velocity of induced electron pumping; K is the coefficient describing the velocity of photon interaction with the medium and N_Φ is the photon density. We will limit our discussion to a relatively simple approximation in which K is a linear function of electron concentration

$$K = \beta\ (N - N_0), \tag{5.2}$$

where β is the proportionality factor (of differential gain), N_0 is the concentration at the inversion threshold, when the populations of the upper and lower working levels are compared and absorption (amplification) is absent. In steady-state conditions ($dN/dt = 0$), we obtain

$$N^0 = (G + \beta N_0 N_\Phi)\ (1/\tau + \beta N_\Phi)^{-1}, \tag{5.3}$$

$$dN^0/dN_\Phi = -\ \beta\ (N^0 - N_0)\ (1/\tau + \beta N_\Phi)^{-1} \tag{5.4}$$

(here the 0 superscript designates a fixed quantity). Since the intensity Φ is defined as

$$\Phi = (c\hbar\omega/n) N_\Phi, \tag{5.5}$$

it is possible to obtain the following formula for n_2

$$n_2 = \frac{dn}{dN}\frac{dN}{dN_\Phi}\frac{dN_\Phi}{d\Phi} = -\frac{dn}{dN}\ \frac{n}{c\hbar\omega}\ \frac{\beta\ (N^0 - N_0)}{1/\tau + \beta N_\Phi}\ . \tag{5.6}$$

Since $dN/dt < 0$, as demonstrated above,

$$n_2 = \left|\frac{dn}{dN}\right| \frac{n}{c\hbar\omega}\ \frac{\beta\ (N^0 - N_0)}{1/\tau + \beta N_\Phi} \tag{5.7}$$

and the sign of n_2 is determined by the quantity in the brackets in the last factor. At the inversion threshold n_2 vanishes. It follows from (3) that if $g < N_0\tau$, $N^0 < N_0$ and $n_2 < 0$ (including the case where $G = 0$). This corresponds to "defocusing." With $G > N_0/\tau$ we will have $n_2 > 0$ and the nonlinearity will correspond to the self-defocusing.

In laser conditions the photon density is not an independent quantity, since it is matched to the pumping rate. In order to account for this fact, it is necessary to incorporate an equation for the photon density

$$dN_\Phi/dt = -N_\Phi/\tau_\Phi + KN_\Phi, \tag{5.8}$$

where τ_Φ is the lifetime of photons in the laser resonator. In steady-state conditions

$$K = 1/\tau_\Phi, \tag{5.9}$$

and subject to (5.2) it is possible to derive the following constant for steady-state conditions

$$N_t = N_0 + 1/\beta\tau_\Phi, \tag{5.10}$$

which represents the concentration of carriers supporting the necessary (threshold) level of optical amplification in the laser. This means that in lasing, changes in Φ may occur without changes in N or in the refractive index.

This conclusion does not relate to nonsteady-state or local intensity deviations, nor to the situation when τ_Φ changes (or the resonator Q changes). An increase in τ_Φ serves to increase N_Φ and decrease N; these changes may be related by the same formula (5.4) if fixed quantities are involved. In nonsteady-state conditions it is necessary to solve complete velocity equations (5.1) and (5.8) from which we may derive the phase relations for oscillations in N and N_Φ.

We note that a quantity that is dependent on N_Φ may be found in the right factor in the right half of formula (5.4); this reflects the photon action on the effective lifetime of the electrons. When $1/\tau \gg \beta N_\Phi$ and $G = 0$, formula (5.6) may be written as

$$n_2 \approx -\ |dn/dN|\alpha\tau/\hbar\omega, \tag{5.11}$$

where

$$\alpha = (n/c)\beta N_0 \tag{5.12}$$

is the absorption coefficient in equilibrium (N_0 figures in here by virtue of the use of formula (5.2)). Formula (5.11) may easily be derived directly from general concepts for the case of the absorption of externally incident emission (see formula (2.3)).

Finally we see that the sign of the optical nonlinearity described by the coefficient n_2 depends on the pumping level and changes in the transition through the inversion threshold. In lasing conditions this optical nonlinearity factor continues to play a role since local and nonsteady-state deviations in N may occur; these are related to deviations in N_Φ. In the spatially-homogeneous steady-state case N is a constant equal to N_t. In practice the saturation condition in the laser affects only integral gain along the entire resonator and the mode "volume." This indicates the possibility of local deviations which in turn, with the activation of the optical nonlinearity, may change the lasing conditions (the mode volume, the value of τ_Φ, etc.

SELF FOCUSING AND ITS INFLUENCE ON THE OPTICAL STRUCTURE IN SEMICONDUCTOR LASERS

One feature of injection lasers is the waveguide nature of emission propagation. The waveguide effect in homostructures results from doping variations, i.e., in the final analysis it results from variations in the concentrations of free carriers. As early as 1961 the theoretical studies of Basov et al. [26] noted that the degenerate regions of a semiconductor may serve as a reflector to laser emission. Indeed, the waveguide effect is based on total internal reflection off of a barrier containing a medium of lower optical density. Since the high carrier concentration reduces the refractive index of the semiconductor, the heavily doped regions then become such a medium with a lower optical density compared to the region near the *p-n*-junction, where, due to the partial compensation of the impurities, an interlayer is formed with a relatively low concentration of free carriers. For emission propagating along the plane of the *p-n*-junction, this interlayer functions as a waveguide layer. In actual homostructures, the waveguide layer has a refractive index that differs from that of the adjacent layers by 0.1-1%.

In heterostructures, the differential in the refractive index Δn is caused by the significant difference in the electron energy spectrum of semiconductors of different chemical composition making contact. If the active layer consists of GaAs, and the broadband emitter layers use $Ga_{1-x}Al_xAs$, Δn between the active layer and the passive layers at the emission wavelength of GaAs (~900 nm) will amount to $\Delta n(x) \approx 0.6\ x$. The relative jump in the refractive index may exceed 10%. We note that a very large jump in the refractive index is occasionally undesirable due to the fact that it has a favorable influence on the formation of high order transverse modes. Thus, we may optimize a structure by the selection of x to provide desirable characteristics of laser emission ("unimodality", directivity, etc.). In InGaAsP lasers, we have analogous data for the refractive index. Here the wavelength is determined by the composition of the quaternary solid solution and varies from ~1 to 1.67 μm (at 300 K).

In analyzing the advantages of heterostructures over homostructures, specifically with regard to threshold current values, we normally use the so-called optical confinement effect: the compensation of the diffraction of laser emission propagating through the active layer. This waveguide effect, as we shall see, is characteristic of both homostructures and heterostructures, although it is expressed more strongly in the latter. This effect makes it possible to confine the laser emission flux near the active layer when the thickness of this layer is less than 0.1 μm.

We shall focus on cases where laser structures have, in spite of the capability for total control over the dielectric structure, a weak waveguide or are even characterized by an "antiwaveguide" (i.e., such structures have a refractive index at the optical axis of the resonator that is less than at the periphery). Such cases include significantly asymmetric structures; planar structures with a stripe contact; a structure with "emergence" modes and a InGaSbAs/GaSb heterostructure.

In an asymmetric structure, the sudden discontinuities in the refractive index differ on opposite boundaries of the waveguide layer. This means that the total internal reflection angles are different and may confine only beams incident on both boundaries at an angle greater than the largest total internal reflection angle. This angle is achieved on the boundary with the low value of Δn. Thus, the geometric optics approach demonstrates that the waveguide effect is determined by a small discontinuity (jump). An analysis of the wave equation indicates that in an asymmetric structure at this wavelength, a critical thickness exists below which the waveguide effect does not occur. The more powerful boundary pushes the wave field towards the weak boundary, and in the transition to a thickness below the critical thickness the field exits through the weak boundary to the passive layer. A standard example is a single-sided p-$Ga_{1-x}Al_xAs$/p-GaAs/n-GaAs heterostructure with a significant discontinuity in the refractive index at the heterojunction, but a small discontinuity at the p-n-junction. The waveguide model of such a heterostructure was given in study [7] using planar waveguide theory.

In stripe-contact planar lasers, we encounter another case; specifically, with a powerful symmetric waveguide in the vertical cross-section, the spatial laser channel has blurred boundaries in lateral directions. If no deliberate refractive index discontinuities are introduced here, the profile of laser emission distribution depends on the distribution of pumping current concentrated directly over the contact. This also depends on such factors as elastic stresses caused by the difference in the thermal expansion of the semiconductor and the contact material as well as that of the semiconductor and the material in the insulating mask as well as the temperature profile. Moreover, it should be noted that in the injection region, the refractive index is reduced compared to the neighboring nonpumped regions due to the increase in carrier concentrations. This indicates the probability of an "antiwaveguide" in the horizontal plane, if a negative carrier contribution predominates.

Nonetheless laser emission in such stripe structures is generated, although the lasers have very specific characteristic features that are undesirable for many applications in a specific, rather broad range of widths (10–20 μm). This applies above all to the shape of the watt-current characteristics, which often have a stepwise form, up to regions with a negative slope, and occasionally such characteristics contain hysteresis loops, which makes it possible to implement a bistable mode [27]. The undesirability of nonlinear and irregular volt-current characteristics is related to the distortions that arise in analog modulation as well as the noise accompanying the irregularities in the characteristic and, finally, with the variation in the directional pattern along the watt-current characteristic, which is also a factor in distortion and noise, such as when feeding emission to a fiber-optic light guide. Nonlinearities in the watt-current characteristics and accompanying phenomena are caused by an instability in the spatial laser channel in the waveguide plane, since it is not fixed in lateral directions (so-called lateral optical confinement does not exist). It is eliminated in structures with narrow active strips with a nonpumped material bounding on the lateral sides or one having a three-dimensional waveguide for fixing the laser mode.

In nonstripe lasers, the formation of spatial channels (filaments) is observed; this process influences the shape of the near field and emission directivity as well as the spectral characteristics and dynamics of lasing. Often it is this phenomenon that explains the degradation in the coherence properties of emission in large laser diodes. Several factors play a role in the formation mechanism of the filaments: inhomogeneities in the optical and electrical characteristics of the medium, mutual shunting of the filaments through the electrical circuit, self-focusing and other intrinsic nonlinear optical phenomena.

Initial discontinuities of the medium are caused by technological processes. Irregularities in the pumping current are caused by layer irregularities in electrical conduction in the substrates and in the epitaxical layers caused by the features of impurity distribution during the growing process. The segregation of impurities and the formation of extraneous phase inclusions near the active layer (in dislocations and other defects), incomplete metal coating of the contact surfaces and assembly defects also influence the pumping distribution. Variations in the concentration of impurities and defects, primarily deep centers, determine irregularities in the quantum efficiency of emission in the active region. Defects of this type include puncture dislocations (grown up from the substrate), mismatch dislocations in the vicinity of lattice spacing gradient and dislocations from the mechanical breakdown of surfaces from cleaving and assembly of the diodes. Finally, geometric structural irregularities (nonplanar boundaries, variations in the thicknesses of the active layer, etc.) create irregularities in optical losses. As a result, separate sections of the laser structure are more advantageous than others with respect to threshold lasing conditions. These types of inhomogeneities produce a more or less well expressed speckle structure of the near field, which is clearly visible at the lasing threshold and in imperfect samples when the threshold is significantly exceeded. Accelerated degradation sites ("dark line" defects) arise during laser aging; these are areas of rapid nonradiative recombination, which exacerbate the near field inhomogeneity. It should be noted that in the general case an optical inhomogeneity in the near field is quite sensitive to the pumping current near threshold and in investigating this process it is necessary to bear in mind that isolated transverse modes also produce a speckle image. Mode inhomogeneities, which are also characteristic of stripe lasers, should be differentiated from inhomogeneities caused by the formation of filament channels.

It has been established in special experiments with GaAlAs/GaAs nonstripe lasers [28], that at a certain high yet not excessive pumping level (say 10-30% above the threshold) a discrete (speckle) structure is formed in the near field even when an external cavity is used to control spatially-single mode conditions. At the same time, new side lobes in the directional pattern arise indicating, first, periodicity in the distribution of speckles and, second, coherence of the channels formed (by virtue of selection in the external portion of the cavity). Moreover, the appearance of the speckle structure was accompanied by the generation of a step, i.e., a nonlinearity in the watt-current characteristic. These phenomena were interpreted as the result of the self-focusing of laser flux which manifests a two-dimensional nature rather than a three-dimensional nature. Since there is a significant discontinuity in the refractive index in the normal direction of the heterostructure, the optical flux is already sufficiently compressed in one measurement, and the self-focusing process occurs in the other transverse direction, along the waveguide layer.

The influence of self-focusing on the behavior of semiconductor lasers may be characterized in the following manner:

1. The self-focusing instability is characteristic of a semiconductor active medium, although in a normal situation it acts in conjunction with other factors (a heterogeneity, multimodality, etc.) and may be masked by them.

2. For the case where a selective cavity is used to control spatially-single-mode lasing, self-focusing is manifest as a threshold phenomenon and produces deformation in the transverse mode profile, forming a speckle structure.

3. In stripe lasers of medium width (10-20 μm) in the absence of laser channel locking, self-focusing produces deformations in the profile of the dominant mode as well as a spatial instability of the laser channel in lateral directions, generating a nonlinearity in the watt-current characteristic and other undesirable phenomena.

4. Self-focusing is an important factor in the nonlinear behavior of a laser beyond the lasing threshold, when its optical structure in at least one measurement does not provide mode profile locking.

Reducing the width of the active layer increases the self-focusing threshold of the instability, and in narrow lasers with a good three-dimensional waveguide, its role may be reduced, since in the working range of the lasers, it may be ignored. This applies, for example, to lasers based on mesa-stripe heterostructures or other analogous laterally confined stripe lasers.

In addition to the optical nonlinearity described above, it is necessary to clearly identify phenomena associated with thermal effects and caused by temperature action on the refractive index of the semiconductor. For GaAs, for example, at the laser emission frequency [29]

$$dn/dT \approx 3 \cdot 10^{-3}\ K^{-1},$$

so that local temperature variations may have a significant effect on the transverse cross-section of the laser beam. Specifically, as demonstrated in study [30], a rather significant nonstationary thermal waveguide effect is observed in stripe lasers; this results from the pulsed adiabatic heating of the active region. This waveguide acts as a lateral optical confinement factor in planar stripe lasers, and helps to avoid discontinuities in the watt-current characteristic. Thermal effects are characterized by inertia, which makes it possible to identify such effects from coincident phenomena. Specifically, temperature action becomes manifest in the frequency modulation of laser emission due to modulation of the pumping current. At low modulation frequencies, a thermal effect predominates, at the same time that at high frequencies (above ~ 10 MHz) a free carrier effect predominates.

Thus, in the general case, a nonlinear optical model of a semiconductor laser includes the action of free carriers on the refractive index producing a self-focusing instability, together with the temperature action (which in turn depends on nonradiative electrical losses in the medium and on the nonresonant absorption of laser emission).

7. SINGLE DIMENSIONAL SELF-FOCUSING IN AN INJECTION LASER

The case of self-focusing in a single transverse direction admits an analytical solution and was examined theoretically as the first example of such phenomena [31]. We will represent the relative dielectric constant of a nonlinear medium in the form

$$\varepsilon(E) = \varepsilon_0 + \varepsilon_2 E^2, \tag{7.1}$$

where E is the field strength of the wave electrical field, ε_0 and ε_2 are the direct component and coefficient of nonlinearity, respectively ($\varepsilon_2 = (c/2\pi) n_0 n_2$). After substituting (7.1) into the Helmholtz wave equation, we obtain

$$\frac{d^2u}{dy^2} - u\left(\frac{\varepsilon_0\omega^2}{c^2} - \gamma^2\right) + \frac{\varepsilon_2\omega^2u^3}{c^2}\cos^2(\omega t - \gamma z) = 0, \tag{7.2}$$

where $u(y)$ is the transverse profile of the heterogeneous planar wave,

$$E\ (y,\ z,\ t) = u\ (y)\ \cos\ (\omega t - \gamma z), \tag{7.3}$$

γ is the propagation constant along the axis z. Assuming that the nonlinear effect is caused by the direct component of the oscillating term in the left half of equation (7.2), we will limit our analysis to an equation averaged over time

$$d^2u/dy^2 - \beta^2 u + (\varepsilon_2\omega^2/2c^2)u^3 = 0, \tag{7.4}$$

where $\beta = (\gamma^2 - \varepsilon_0\omega^2/c^2)^{1/2}$. This equation has the following solution:

$$u(y) = u_0/\mathrm{ch}\ \beta y, \tag{7.5}$$

where β and u_0 are interrelated,

$$\beta = u_0\omega\sqrt{\bar{\varepsilon}_2}/2c. \tag{7.6}$$

The width of the spatial channel w, defined by the half-intensity level compared to the channel axis, is equal to [32]

$$w = 1.76/\beta \tag{7.7}$$

and for the product $u_0 w$ we derive the universal formula

$$u_0 w = 3.52\,c/\omega\sqrt{\bar{\varepsilon}_2}. \tag{7.8}$$

For example, if we have a laser line of width w, given by the initial optical structure, as soon as the electrical field in the center exceeds u_0, the spatial channel begins to narrow to satisfy condition (7.8). This rule was valid with a certain error caused by the fact that the initial profile of the spatial channel differed from formula (7.5).

In order to obtain critical power we will integrate the optical flux density in terms of the beam cross-section. Then we obtain

$$P_{\text{кр}} = 0.178\ d_0\lambda_0^2/n_0 n_2 w, \tag{7.9}$$

where d_0 is the effective thickness of the channel in the second cross-sectional direction (along the x axis). Here for n_2 we use the formula

$$n_2 = (dn/dN)\ N_0\tau K n_0/c\hbar\omega. \tag{7.10}$$

Where $d_0 = 10^{-4}$ cm, $n_0 = 3.6$, $w = 10^{-3}$ cm and $n_2 = 10^{-8}$ cm^2/watt (GaAs at 300 K) we obtain $P_{\text{кр}} \approx 4.5$ mw. The critical threshold excess for self-focusing may be estimated by the formula

$$\frac{I_{\text{кр}}}{I_{\text{пор}}} = 1 + \frac{\lambda_0^2}{2n_0\omega^2 N_0\,|\,dn/dN\,|}. \tag{7.11}$$

For the same numerical parameters for GaAs and $N = 2\cdot10^{13}$ cm^{-3} we obtain $I_{\text{кр}} = 1.07\ I_{\text{пор}}$. The calculation yields a refractive index increment Δn of near $1.4\cdot10^{-3}$. It is clear that in actual conditions the self-focusing instability factor becomes significant at a very low excess over threshold (~7% and near similar values in lasers with an active region 10 μm in width). The output power level of 4.5 mw is typical for discontinuities in the watt-current characteristics of planar stripe lasers. Any method of narrowing the active region increases the relative threshold of self-focusing instability and increases the critical power.

The analysis was conducted assuming that n_2 is an actual quantity. In a more strict approach it is necessary to account for n_2 as a complex quantity. Indeed, a change in population N changes both the refractive index and the gain (absorption). The situation implemented in an injection laser is analogous to self-focusing in other resonant-absorption media, such as in Na vapors for the propagation of resonant laser emission [33, 34].

These studies carried out a calculation and study [34] performed an experiment which indicate that the real part of the quantity n_2 plays the primary role in the deformation of the transverse distribution. Here we may hope that in our case the error resulting from ignoring the imaginary part of n_2 in (7.1) will not be excessive.

The self-focusing effect in stripe lasers was first examined as the cause of the change (broadening) of the directional pattern with an increase in intensity [35]. Specific properties of the formation of spatial valleys of carrier concentration due to inversion "burning" in the laser channel were derived in studies [36, 37]. Naturally such "valleys" correspond to the refractive index maxima, i.e., the waveguide filaments, which may have a self-focusing nature. The study previously cited [28] has established by direct observation the formation of filaments in a nonstripe laser with internal control over single-mode conditions. The deformation in the mode profile associated with this mechanism imposes constraints on the luminosity in single-mode conditions, since it produces efficient broadening of the directional pattern of laser emission (including the appearance of sidebands with periodic distribution of "filaments"). A reflected flux similar to a planar wave is formed in the external cavity due to a spatial filter (slit). In an active element this flux is divided into channels formed in a single pass of the diode. A periodic speckle structure appears in the near field, and side lobes occur in the far field. Such a mechanism retards the increase in the luminosity of emission in the dominant mode and causes nonlinearity in the watt-current characteristic.

It is possible that the pattern of coherent channel formation in broad diodes is more complex, and includes, in addition to single-dimensional self-focusing channel compression, parametric interaction (noncollinear interaction of the degenerate type), caused by the same nonlinear refraction mechanism. This interaction may play a critical role in the onset of the transverse periodicity of channels, which is characteristic of speckles in the near field, and is responsible for the symmetric side lobes in the directional pattern.

8. NONLINEAR OPTICAL LOSSES IN WEAK AND ASYMMETRIC ACTIVE WAVEGUIDES

Optical losses in the cavity of an injection laser are highly dependent on the waveguide effect which eliminates emission diffraction from the thin layer or the narrow active stripe in lateral directions. The thickness of the active layer nearly always corresponds to significant diffraction losses, say on the order of 100-500 cm^{-1} depending on the thickness and properties of the structure layers (in the absence of optical confinement). The existence of the waveguide effect reduces this quantity to a value of ~10 cm^{-1} or less, depending on the structural loss coefficients. In a stripe structure the lateral loss factor becomes manifest at a width less than ~2-3 μm, i.e., in rather narrow stripes. The development of lateral optical confinement makes it possible to significantly reduce lateral diffraction to a width on the order of 0.2-0.3 μm (such as in TJS diodes), i.e., of the same order as the active layer thickness in a heterostructure. With moderate waveguide confinement, the optical confinement factor Γ provides a suitably accurate description; in the general case this factor depends on the dimensions and optical constants in both transverse directions with respect to the laser emission axis. We may use parameters independent for each of two transverse directions in different tasks. Initially we will consider certain anomalies in emission dynamics related to nonlinear optical losses from diffraction perpendicular to the plane of the active layer. Such anomalies occur when the waveguide effect (an "inbuilt waveguide") is insufficient to suppress diffraction. Such conditions occur primarily in homostructures or asymmetric heterostructures. The dielectric asymmetry in both homostructures and heterostructures generates a critical thickness d_{Kp} below which localized waveguide modes do not exist (the maximum of the transverse profile of the field is displaced to the region with a lower optical density of two adjacent regions). For example, in a single-sided heterostructure

$$\begin{array}{lll} n\text{-GaAs}/ & p\text{-GaAs}/ & p\text{-GaAlAs} \\ n=3{,}58 & n=3{,}61 & n=3{,}40 \end{array} \tag{8.1}$$

the field is displaced from the solid solution layer and, while its peak remains within the active *p*-GaAs-layer, the waveguide effect continues, although at a certain thickness $d < d_{Kp}$ the solution for the mode field reaches a maximum at some uncertain distance into the *n*-GaAs-region. This indicates, first, an increase in losses and, second, a reduction in mode amplification in accordance with the following simplified relations:

$$\alpha_m = \alpha_0\Gamma + \alpha_{pn}\,(1 = \Gamma),\ g_m = g_0\Gamma, \tag{8.2}$$

where α_m and g_m are the distributed losses and gain indicators for this lower order mode; α_0 and g_0 are local indicators of nonresonant absorption and amplification in the active layer (*p*-GaAs), α_{pn} is the effective refractive index in the passive regions.

For simplicity we may assume that the field does not penetrate into the p-GaAlAs-region; hence, α_{pn} corresponds to the refractive index in the *n*-GaAs-region. Let the gain g_0 be linearly related to the pumping current density

$$g_0 = b\,(j - j_0), \tag{8.3}$$

where b is the proportionality factor ("differential gain"), j_0 is the current density for the "inversion threshold" (at which the gain becomes greater than 0). Then we may show that the threshold current density $j_{\text{пор}}$ (corresponding to the equalization of gain and total loss in the mode) will be equal to

$$j_{\text{пор}} = j_0 + (\alpha_0/b) + [\alpha_{pn}(1 - \Gamma) + \alpha_{\text{ext}}]/b\Gamma, \tag{8.4}$$

where α_{ext} are external losses (through semitransparent mirrors), recalculated for the dimensions of the absorption coefficient. In an approximation of planar waves

$$\alpha_{\text{ext}} = L^{-1} \ln (1/R), \tag{8.5}$$

where R is the power reflection coefficient for perpendicular incidence. It should be remembered that in the general case α_{ext} for a specific mode depends on the transverse mode profile and may be one of the factors of the nonlinear behavior when the mode profile depends on its intensity.

We note now that if the refractive index of the active layer is reduced, the quantity $d_{\text{кр}}$ increases and may exceed the thickness of the active layer. In this case $\Gamma \to 0$ and, as we see from formula (8.4), the threshold current density becomes indeterminately large.

These discussions correspond (with some simplification) to waveguide theory of asymmetric heterostructure lasers as proposed in 1970 [7, 38] in order to explain the temperature breakdown of lasing in single-sided heterostructures. The temperature increase in the threshold current density and excess carrier concentration leads specifically to a reduction in the refractive index of the waveguide layer. The initial discontinuity of the index on the n-GaAs side is small, and reducing this value sharply increases the critical thickness. Hence, at a certain temperature $T_{\text{кр}}$ lasing breaks down. With $T > T_{\text{кр}}$ the laser structure, generally speaking, may remain in a lasing condition if the waveguide amplification effect is accounted for. Accurate experimental observations demonstrate the possibility of obtaining lasing although with significant anomalies in the dynamics, as discussed below. Study [7] has noted that accounting for the gain may produce a shift in the critical thickness. Since the gain has a self-consistent dynamic relation to the emission intensity, the probability of dynamic anomalies becomes clear, at least qualitatively. This general pattern became clear within the scope of the waveguide model after 1970, when alternative explanations were still advanced for lasing breakdown (the "through hole injection" model), and for the dynamic anomalies (the "slow trap" model). The final waveguide model, which explains from a single viewpoint all anomalies in the behavior of single-sided heterostructures observed in experiment was formulated in [39, 40]; these sources also provided confirmation experiments. Their conditions were enhanced by using an external cavity which made it possible to continuously tune the Q of the cavity and to observe virtually all characteristic anomalies.

The complex nature of lasing near T_{Kp} was discovered in study [41] where spectral chronography measurements of laser emission from single-sided heterostructures were carried out. It was discovered that in certain conditions, lasing bursts occur only at the beginning of the pumping pulse and at the end (with an emission spectrum characteristically broad for nonsteady-state conditions). During the pulse, a valley progressing through total breakdown was observed in the emission intensity. Such an unusual emission pulse shape provided the foundation for the name "H"-pulsing produced in a later study [42]. It should be stated that the peak conditions at the end of the pumping pulse was familiar from observations of certain homolasers [43], heterolasers [44, 45] and transverse electron pumped lasers [46] (this phenomenon is often called "internal Q-switching"). Regarding homolasers, their detailed dielectric structure depends on the doping gradients, the distribution of carrier concentrations, etc. Hence, it is difficult to identify the degree of dielectric asymmetry in specific cases. We note that "internal Q-switching" is observed in a certain temperature range and with thermal treatments of the laser structure, this interval may be varied. The single peak mode at the end of the pulse may be observed as a limiting case of an "anomalous" lasing delay, when the wavefront of the laser pulse is separated from the wavefront of the pumping pulse as the current increases (then the regular delay drops off with an increase in current). In homostructures, these anomalies may be related to the fact that the initial waveguide effect is weak and at the same time is suppressed due to possible dielectric asymmetry.

Henceforth we will limit our discussion to a qualitative examination of the first peak conditions at the wavefront of the pumping pulse, since here it is necessary to incorporate obscure arguments based on the development of the waveguide model [7]. Regarding the second peak and the "anomalous delay", these phenomena are discussed more persuasively in the literature by incorporating temperature effects [40, 47]. This latter fact introduces nothing new, but significantly complicates the qualitative description and makes it more cumbersome. The details in this process are given in study [40].

In order to explain why laser emission intensity first increases and then decays as the current increases (at the wavefront of the pumping pulse), it is necessary to incorporate the concept of spatially-heterogeneous (along the cavity length) gain saturation under laser emission action. Let the waveguide be at the edge of breakdown of the waveguide effect. Then conditions are generated sooner or later in a single-sided heterostructure when the temperature increases and reaches the nominal value T_{Kp}. Then when the pumping is increased over the threshold level, the concentration of excess carriers is redistributed along the resonator axis. Indeed, due to the inhomogeneity in the intensity distribution of traveling waves, they reduce gain to a greater degree at the semitransparent mirrors, and overall mode amplification is compensated by a certain increase in local gain and carrier concentration in the central portion of the diode, far from the mirrors. However, the increase in the carrier concentration in this portion of the waveguide soon begins to degrade the waveguide to its "breaking" point, which already produces the opposite result, and in the final analysis causes lasing breakdown. This process is valid in both nonsteady-state and steady-state conditions, as confirmed by corresponding experiments. The formation of the laser emission pulse at the beginning of the pumping pulse is determined by the passage of excess carrier concentrations through the region where lasing conditions are satisfied, although a further increase in pumping and inhomogeneous saturation cause a breakdown of the waveguide, a sharp increase in losses and lasing breakdown. Since the progression of these conditions depends on the slope of the wavefront as well as the waveform of the pumping current pulse, it is easy to control the duration of the leading lasing pulse. In experiments it was increased from a few nanoseconds to 50 ns, which corresponds to quasisteady-state lasing.

The steady-state lasing breakdown mechanism occurring with an increase in pumping (and laser emission intensity) plays an important role in explaining many behavioral anomalies of semiconductor lasers. These include above all the phenomena described above in single-sided heterostructures as well as in homostructures, where an analogous mechanism may produce anomalous delays and "internal Q-switching." Evidently an identical situation exists in transverse electron-pumped semiconductor lasers, where the dielectric asymmetry of the active region is caused by a semiconductor/vacuum boundary on one side of the active layer and a weak uncontrollable refractive index gradient within the semiconductor (if a special waveguide structure is not employed). Anomalous delays and other anomalies characteristic of the behavior of single-sided heterostructures have actually been observed in transverse electron-pumped lasers [46, 48].

In stripe lasers, this type of optical nonlinearity is manifest as deformations in the spatial profile of the laser mode, and accompanying discontinuities and jumps in the watt-current characteristic, changes in the directional pattern of emission and dynamic instabilities (excess noise, increased relaxation fluctuations, etc.). Since these phenomena, as a rule, are undesirable in applications, a number of publications have been devoted to their study (see, for example, [27, 49-53]). The phenomenology of nonlinearities in the watt-current characteristic may be described in the following manner. These are typical nonlinearities for planar stripe structures without lateral optical confinement; the power threshold of their appearance is inversely proportional to the width of the structures. As a result, such nonlinearities may be virtually eliminated in so-called narrow-stripe (less than 4 μm) planar structures. At a width of 10-15 μm, the nonlinearity in the shape of the discontinuities is more clearly expressed. Occasionally discontinuities and hysteresis loops in emission power are observed in addition [51]. In oxide insulated GaAlAs/-GaAs-lasers, a standard power level for this step is approximately 2 mw, which falls directly in the working range of the watt-current characteristic. At a width of greater than 20 μm, masking factors (spatial multimodality; the formation of several spatial lasing channels, etc.) play a role, since the watt-current characteristic begins to smooth out, and in samples with a broad contact, these types of nonlinearities may be completely absent. They have not been observed in various stripe structures with lateral optical confinement, in buried structures, in structures with a substrate channel, with a transverse *p-n*-junction (TJS), etc.

Another feature of these phenomena is that they are to some degree related to the spatial (transverse) instability of the laser mode; there is a noticeable spatial shift of the emission spot corresponding to the beginning of the nonlinearity (i.e., a reduction in the differential laser efficiency) together with a change in the spot shape. This shift is in the direction of the periphery of the active region. This may be related both to the parallel shift or the rotation of the laser filament within the pumped cross-section [50]. Indications of spatially-inhomogeneous inversion burning in the vicinity of the laser channel have been noted.

Thus, this is an approximate qualitative description of the phenomenon. As the superthreshold pumping increases together with the intensity of the mode arising at the threshold, the influence of transverse variations in the refractive index caused by inversion burning also increases, which in turn produces self-focusing mode deformation. As a result, the efficiency of inversion use (and the slope of the watt-current characteristic) is reduced. In the absence of an inbuilt waveguide, the compressed laser channel becomes unstable and it becomes more sensitive to small variations in the refractive index (from heating, mechanical stresses, etc.). Since the mode is formed under the action of the gain profile, both factors - gain and self-focusing - compete during the increase in pumping. This leads to a situation where sufficient gain develops outside the compressed channel to form a new mode profile, and the suppression of the first, deformed mode produces a sharp increase in the intensity of the new mode. The slope of the watt-current characteristic increases (occasionally in the form of a discontinuity). Thus, the beginning of the nonlinear may be identified with the self-focusing threshold.

The variety of individual features of nonlinearities in the watt-current characteristics and accompanying phenomena has been emphasized in a number of studies [50, 53]. Undoubtedly this reflects the powerful action of inhomogeneities of a medium in various samples. One aspect of this influence is critical. The argument is that self-focusing mode deformation, in conjunction with the inhomogeneities, converts to a new factor of nonlinear optical losses with an unpredictable magnitude and sign of the derivative of the mode intensity loss coefficient. Consequently one additional source of the individual nature of dynamic and spectral properties of lasers has been discovered (primarily the increment of self-modulation fluctuations which determines the dynamic modes). Such mechanisms may easily make a significant contribution and excite fluctuations in transient conditions or sustained pulsations (spiking). Certain observations have indicated the participation of self-focusing in relaxation pulsations [54, 54].

9. BISTABLE CONDITIONS IN LASERS WITH AN EXTERNAL RESONATOR AND IN A RESONANCE AMPLIFIER

As demonstrated in studies [56-59], bistable conditions are achieved in injection lasers with external reflection selectors if the diode resonator itself is not removed, i.e., in the compound ("three mirror") resonator. The diode plays the role of intracavity nonlinear Fabry-Perot etalon whose optical length depends on the emission intensity. Unlike a nonlinear resonator illuminated from external sources, in this case the source of laser emission coincides with the nonlinear element. In all other respects, the principle of operation is the same. A hysteresis loop is observed in the spectral tuning of the external resonator. The mechanism of its onset may be described in the following manner.

The external selector controls the wavelength generated in the laser, which may differ somewhat from resonance in the diode. With precise tuning of the external element (in principle discrete tuning over a large number of longitudinal modes of the external resonator) the Q of the compound resonator is changed, since the relative position of the resonances of the selector and the diode changes. The change in the Q produces a change in the threshold current; hence, at certain "fixed" pumping current values, tuning makes it possible to pass through the lasing threshold, i.e., to obtain laser emission with favorable tuning at the same time that spontaneous emission is produced with unfavorable tuning. We will now consider the nonlinearity of the diode resonator. An increase in the Q of the compound resonator causes an increase in emission intensity and a decrease in the concentration of free carriers and, consequently, a certain increase in the refractive index and the optical length of the resonator. Longitudinal resonance of a given order shifts towards the longer wavelengths. If the external element is tuned from the shorter wavelengths to the longer wavelengths, the process will be accompanied by "runaway" of nonlinear resonance; in the opposite direction tuning causes a convergent shift of longitudinal resonance. It is clear that the dependence of emission intensity on tuning (the tuning curve) will be spectrally asymmetric. This was noted previously in study [60]. When the Q of the compound resonator is increased to a sufficient level, the transition to lasing occurs in jumps, and the asymmetry of the tuning curve is expressed in the fact that the turn-on and turn-off of lasing will occur at different wavelengths. This produces a hysteresis loop. Analogous behavior is observed with a temperature drift in the diode, which changes the detuning of the diode resonator and the external selector. Since the thermal effect is highly inertial, it may be associated with the slow switching and low-frequency relaxation pulsations.

A theoretical analysis of this type of bistability is provided in study [56]. The model is as follows. The lasing threshold in a laser with an external resonator at a fixed wavelength is a periodic function of the phase shift δ in the active element (after two passes through the resonator). The emission power at the fixed point also has a periodic dependence on the phase shift δ, for which we derive the following in linearized form

$$\delta = \delta_0 + \mu\lambda + \theta N, \tag{9.1}$$

where δ_{0_2} is a certain constant characterizing the active element, $\mu = -4\pi Ln^*/\lambda_0$, L is the length of the active element, $n^* = n - \lambda dn/d\lambda$ is the group indicator, $\lambda = 4\pi L\sigma/\lambda$, where σ designates, as before, the quantity dn/dN. In this formula the second term on the right describes the dependence of the phase shift on wavelength, while the third term expresses the dependence on carrier concentration. Since N in turn depends on emission power P, we may derive (in linearized form)

$$\delta = \delta_0 + \mu\lambda + \gamma P, \tag{9.2}$$

where $\gamma = \theta\partial N/\partial P$. The location of the operating point on the plane of variables P and δ corresponds to a solution of the self-consistent problem. A quantitative relation for the bistability condition may be found in the form [56]

$$\frac{\sqrt{r_3\varkappa}}{\sqrt{r_2}(1+r_3\sqrt{\varkappa})} \geqslant R = \frac{\mathrm{Re}\,(\partial\varepsilon/\partial N)}{\mathrm{Im}\,(\partial\varepsilon/\partial N)}, \tag{9.3}$$

where r_2 is the reflection coefficient at the diode face opposite the selective element, r_3 is the effective reflection coefficient from the selective element, $\varkappa$ is the coefficient of emission coupling from the selective element to the active medium. It is clear that suppression of bistability may be achieved by bleaching of the face which faces the selective element. With $R = 3$, $r_3 = 0.5$ and $\varkappa = 0.8$, the bistability vanishes with $r_2 < 0.02$. Study [56] also gives an approximate solution for the spectral points of the hysteresis loop when exceeding the threshold slightly. An analogous model was examined in detail in study [58].

The effects of bistable tuning with stripe injection GaAlAs/GaAs CW lasers operating at room temperature have been observed in experiment; a total internal reflection holographic selector was used as the dispersion element of the external resonator. Its spectral selectivity is 4-10 times greater than that of a flat reflective diffraction grating. Optimum tuning produces single-frequency conditions, i.e., a single longitudinal model of the external resonator was excited, and tuning produced the serial excitation of these modes. The tuning range in the vicinity of the single longitudinal resonance of the diode was 80 GHz or ~160 modes of the external resonator. The maximum power of single-frequency emission was 10 mw.

In the vicinity of bistability, the laser has two power levels in spontaneous and laser modes; the power difference was 8 mw. An important experimental condition for the observation of bistability is high selectivity of the external resonator. The existence of two selective elements in a common resonator creates the necessary conditions for effects generally called "compound resonator" effects. Evidently these include the bistability observed for the case of a single-mode laser with an external heavy mirror forming an additional feedback loop [61].

This bistable laser represents a system with a memory. The laser may be switched from one stable state to another by introducing an optical signal from another laser. Unlike bistable optical schemes with electronic feedback, this case does not require electronic circuits. By virtue of the significant nonlinearity, the bistability is manifest at a very low power level (approximately 1 mw), which favorably differentiates this case from nonlinear Fabry-Perot etalons. From the viewpoint of emission dynamics, the fact that the steady-state conditions are stable across the entire tuning range, with the exception of the vicinity of the lasing breakdown point is important. Here it is possible to obtain sustained emission pulsations.

This type of bistability effect is characteristic of a resonance amplifier (or a laser locked to emission from an external source), based on a semiconductor laser. Here we refer to the mode arising from the optical injection of emission (from another single-frequency laser at ω_0) to the active region. If the pumping is near yet still below the threshold, the injected emission is amplified (such a mode is called regenerative). In pumping above the threshold, the emission conditions at ω_0 may be preserved, corresponding to the locking condition. Achieving such modes (essentially one above and one below the nominal lasing threshold) depends on the detuning of ω_0 with respect to the natural resonance of the diode ω_q, and the value $\omega_0 - \omega_q$ must be less than the half width of the locking band. The latter in turn depends on the pumping current and the intensity of the injected emission. A number of studies have been devoted to this mode [62-67]. It is of significant practical interest since it makes possible low noise laser transmitters and amplifiers for coherent optical transmission of information.

The onset of bistability in the regenerative mode is similar in its design to bistability in a nonlinear passive Fabry-Perot resonator, and is observed in the same manner on the basis of the hysteresis of the output optical power as a function of the input optical power. The difference is found in the sign of the change in the refractive index, resulting from the presence of pumping and inversion (this issue was discussed earlier in Section 5). By virtue of the amplification of emission, the sensitivity of the bistable amplifier is significantly greater than that of the passive etalon. A typical input power level for switching is in the order of fractions of a milliwatt.

The theory of nonlinear locking in a semiconductor laser was presented in study [67]. It is interesting that the boundaries on the stability region of locking depend on another effect, also caused by nonlinear refraction. This effect is nonlinear mode interaction and will be examined in Section 12. The stable locking bandwidth is rather narrow. With an input power of 0.01 mw at 1.3 μm when the threshold is exceeded by 50% in a standard InGaAsP/InP it is estimated at 1.4 GHz (0.008 nm). This requires temperature stabilization on the order of ±0.02°C [67].

Thus, it is possible to develop bistable optically- or electrically-controlled optical elements employing either internal or external emission and characterized by a rather high sensitivity and the capability to operate in a pulsed mode at room temperature based on the phenomenon of nonlinear refraction in semiconductor lasers. Evidently they may compete with bistable semiconducting laser circuits based on bleaching media (i.e., the nonlinear absorption effect), since they require lower energy. On the other hand, their tie to specific resonant frequencies requires temperature, current, etc. stabilization.

10. C^3-LASERS AND OTHER COUPLED CAVITY DESIGNS

The phenomena of injection laser interaction has been the focus of attention since the 1960's [68-73] since these phenomena underlie the physics of superhigh speed optical logic. Initially the idea of using the nonlinearity of absorption-amplification in lasers to obtain optical switching and perform logic operations predominated. These studies evolved into a more promising direction (see [74-76]) encompassing research on the dynamics of switching, bistable and multistable states, and various circuit designs employing two or more active elements, including integrated optical circuits [76] based on injection lasers.

Recently so-called C^3-lasers (cleaved-coupled cavity lasers) have been the focus of attention. These devices consist of a pair of laser diodes connected in series along the axis of the cavity and fabricated by cleaving one diode wafer. In assembly the diodes are not completely separated, but rather are left in close proximity to provide strong optical coupling. Separate electrical pumping makes it possible to control lasing by two independent parameters: the pumping currents in both sections of the laser device. The configuration of a cleaved-coupled cavity laser was first implemented in study [71] in order to increase optical coupling between interacting diodes. The gap between diodes was less than 5 μm with section lengths of 1450 and 425 μm. A nonstripe laser operated in a pulsed condition at 77 K between 842 and 848 μm. The gain and lasing quenching in coupled diodes caused by the interaction of amplification emission in the sections (effects associated with Im ε) were investigated. We note that in fabricating coupled cavity lasers a reinforced deposition technique for the initial diode wafer was used where the metallic Ag film is not separate during the cleaving process, and this made it possible to maintain the close proximity of the laser sections.

The next stage in investigating coupled cavity lasers involved investigating the spectral effects of interaction [78]. Identical laser diodes with differing cavity lengths L_1 and L_2 were used in the experiments. Unlike the C^3-laser, the sections were placed in parallel and optical coupling was provided by means of an external spherical mirror. Since the intermode distances in the longitudinal mode spectra differed, "beating" arose in the common spectrum, i.e., the superluminescence intensity was frequency modulated with a period of

$$\Delta\lambda = \eta^2/2n^*|L_1 - L_2|, \tag{10.1}$$

where n^* is the effective (group) refractive index. This spectral modulation is explained by the near-periodic functions of the spectral dependence of the Q of longitudinal modes in the two cavities (the "vernier" effect). Its influence on the longitudinal mode envelope produces additional narrowing of the superluminescence spectrum, if the modes at its center have the same wavelength in both sections. Noncoincident modes have significant Q losses and are suppressed. Thus, the introduction of optical coupling between laser diodes of various length is one method of spectral selection in injection lasers [79, 80] and was used to obtain the single longitudinal mode.

Since pumping variations in one of the sections changes the concentration of carriers, the refractive index also changes and the resonances are shifted in this section. This produces an even more noticeable shift of the envelope of longitudinal mode "beating" in the presence of optical coupling, which was used in studies [13, 14] to determine dn/dN, as was discussed earlier in Section 3. We will discuss the tuning of compound resonators and C^3-lasers below.

Spectral phenomena in compound or multimirror (multicavity) lasers became the focus of common interest not only with regard to selection and tuning but also as a widespread phenomenon in coupled devices (a laser/fiber-optic light guide) and in integrated optical schemes. In its pure form, the compound resonator of an injection laser is investigated in studies [81-86] and later in studies [87-90], etc. Longitudinal mode selection as well as piezoelectric and electrooptical tuning of a compound resonator injection laser are described in [84]. The advantages of electrooptical tuning and alignment of a compound resonator include the solid design of the device which reduces the influence of temperature drift and vibrations as well as the capability to easily implement automatic tuning. In view of the high sensitivity of laser conditions to the optical length of the external portion of the compound resonator, special displacement transducers have been developed recently (for measuring acceleration, magnetic field, for recording oscillations, for applications as microphones, hydrophones, etc.) [91].

For the case of electrooptical tuning, the modulated portion of the compound resonator consisted of an electrooptical material: lithium niobate or barium-sodium niobate [84]. The voltage to the diode was applied to the silver-coated external face of the electrooptical wafer. The tuning acceleration resulting from the "vernier" effect was studied and the possibility for using this effect for discrete frequency modulation of emission was discussed.

If the optical lengths of both sections of the compound resonator are approximately equal, $n_1^* L_1 \simeq n_2^* L_2$, the mode intervals differ by only a relatively small amount,

$$\delta\lambda_1 - \delta\lambda_2 = (\lambda^2/2)(1/n_1^* L_1 - 1/n_2^* L_2). \tag{10.2}$$

If one of the mode systems, due to a change in n_2^*, approaches this difference quantity, the point of correlation of the two modes from the different systems shifts by $\delta\lambda \approx \delta\lambda_1$, and also produces the "vernier" tuning acceleration. The tuning coefficient is equal to

$$d\lambda/dn = (\lambda/n_2)(n_2^* L_2/n_1^* L_1 - 1)^{-1}, \tag{10.3}$$

where the second multiplier on the right gives the magnitude of tuning acceleration.

We will now return to C^3-lasers. The new stage of investigation of these devices began with studies [92, 93]. Practical interest in these devices has derived primarily from their application in fiber-optic communications links (specifically, at ~1.55 μm where dynamic single-mode emission is required, i.e., a single longitudinal mode with high speed pumping current modulation). One feature of C^3-lasers is that significant spectral selectivity is achieved with a comparatively simple fabrication method, together with the capability to control the emission wavelength and a mode system in each section that in principle depends on pumping. The laser frequency is also discretely tunable by virtue of the vernier effect described above. The action of excess carriers on the refractive index of the semiconductor medium is used in place of the electrooptical effect; this eliminates the use of high voltage in the device and maintains the electrical control method.

Normally a C^3-laser consists of two sections of approximately equal length, one of which functions as the active element and is pumped above the threshold, while the second is the modulator with pumping below the natural threshold. In the general case the laser is a four-mirror system, if the gap between the diodes is significantly less than a quarter wavelength. The gap L_g plays an important role in mode selection and its optimum value corresponds to a few half waves [93]. The possibility for obtaining stable emission in a single longitudinal mode under RF modulation is an important feature of such lasers, but is not their only advantage. Another feature is the capability for electrical tuning of the laser emission; for example, for spectral channel multiplexing. Only a small change in the direct current flowing through the modulator section of the laser is needed for this process, since the tuning rate is significantly enhanced by the "vernier" effect. This method may be used to switch a laser to several working wavelengths (for example, 12). In conjunction with amplitude modulation, this method makes it possible to expand information transmission capabilities and to achieve channel multiplexing [92]. Study [94] has demonstrated the use of a C^3-laser to execute logic operations using an optoelectronic method.

The rapid switching of the current in the modulator section makes it possible to achieve frequency (discrete) modulation with a broad frequency deviation (one or several mode intervals). Demodulation in this case may be achieved by a simple interferential filter. Two electrical inputs make it possible to simplify the conditions for independent control over emission intensity and frequency, and, for example, for frequency modulation it is possible to eliminate parasitic amplitude modulation which has been observed in regular laser diodes.

The next important feature of C^3-lasers is the variety of possible operating conditions, particularly with bistable conditions used for specific modulation variants and to perform logic operations.

Spectral bistability was observed in a C^3-laser in study [95] by varying the current in the modulator section, and this phenomenon was expressed in the coupled power hysteresis at two modes (at 1.5122 and 1.5102 μm) with full power (±5%); the width of the current hysteresis loop was 10 ma with a current of 60 ma in the laser section (threefold above the threshold). Switching from one state to another (i.e., switching the laser wavelength) may be achieved by means of ultranarrow (0.2 ns) electrical pulses superimposed on the d.c. bias current of the modulator section; the switching time was less than 1 ns. The product of the switching signal power and the switching delay time was 10^{-12} J. We note that the spectral switching speed is higher than the power switching speed (in hard conditions), since it is less closely related to the accumulation of excess carriers. Spectral switching is not related to absorption saturation, and may be observed with pumping currents in both sections exceeding the lasing thresholds.

The origin of spectral bistability, as for the case of bistability in the external cavity, is based on accounting for the dependence of the refractive index on the carrier concentration. Bistability is observed in a comparatively narrow range of current values through both sections, covering the stability regions of each of the spectral modes considered. The span in turn is possible due to the satisfaction of threshold conditions for each of the two modes with different concentrations of excess carriers in the sections, corresponding to tuning to different modes. Study [96] has demonstrated that the bistable behavior is sensitive to the size of the gap L_g between the sections. If $L_g = m\lambda/2$, where m is a small integer, the stability regions of the separate modes do not overlap. If L_g is somewhat greater than the total number of half waves, the stability regions are separated by single frequency instability regions; in the other case, when L_g is somewhat less than $m\lambda/2$, the stability regions overlap and, consequently, bistability regions arise.

Another model of optical bistability is considered in study [97]. It is assumed that both sections are pumped above the threshold (so that the absorption saturation factor is eliminated), and lasing is considered at one (the common) longitudinal mode with a narrow gap between the sections (its width is less than $\lambda/4$ or the gap is reduced to a single semitransparent surface). As is normally done in experiments with C^3-lasers, the lengths of both sections are similar; in order to simplify the calculations, the internal parameters of the medium in both sections are identical. The steady-states of the laser system with various pumping levels in the sections correspond to predominant photon injection from one section to another. As the two pumping levels converge, two steady states are realized with an asymmetric intensity distribution through the sections, and power hysteresis regions appear in the watt-current characteristics. We note that the influence of the carrier concentration on the refractive index is not incorporated in the laser model; consequently, the physical bistability mechanism is caused by the nonlinearity from gain saturation with supercritical pumping. Numerical calculations run for the case of an InGaAsP-laser (wavelength of 1.3 μm) have revealed bistability in this model, and a comparison to experimental data has produced satisfactory agreement.

In concluding this section we will focus on one form of optical bistability in injection lasers caused by the interaction of the parallel spatial lasing channels [98]. Sensitivity of the spatial distribution of laser emission in a twin-stripe laser structure to emission injection in the gap between the active regions has been observed experimentally. This makes it possible in principle to achieve optically-controlled switching of the directional pattern of an injection laser. One theoretical variant of bistability of this type is given in study [99].

C^3-lasers, as discussed above, represent a very promising design of a comparatively simple, compact and heavy-duty laser device with good spectral selectivity providing dynamic single-mode lasing with the capability for electrical tuning or direct frequency modulation. Compared to another laser design with dynamic single mode lasing - a Bragg reflector laser - the C^3-laser has the advantage of a simpler technology and more extensive controllability. It is interesting that recently in the course of the development of Bragg reflector lasers it was noted that the diode wafer (without the Bragg lattice), which has three or more concentrated reflection regions within the cavity, manifests sufficient spectral selectivity to proceed without the Bragg reflector. This selectivity is caused by the compound resonator effect (see [100]).

11. FREQUENCY SELF-MODULATION AND MODULATION

In this section we will discuss the inevitable interrelationship between amplitude and frequency variations in semiconductor lasers, which is manifest both in self-modulation instabilities and with a given modulation of the pumping current. We should differentiate the thermal current modulation effect that produces frequency modulation of emission at low frequencies (in practice up to ~10 MHz), and the higher frequency effect (up to several Gigahertz) resulting from the action of carriers on the refractive index.

Observations of high-speed spectral chronograms in a number of earlier studies [101-104, 86] have shown that the frequencies of the separate laser modes often experience significant deviations whose rates may be as high as 60 GHz/ns, while the amplitude is comparable to the mode interval. This serves to expand the modes in the averaged spectrum. It was discovered in examining stripe injection lasers [86, 103, 104] that amplitude self-modulation is accompanied by frequency self-modulation, and for the case of the periodic process, both forms of self-modulation oscillations are synchronized and are approximately $\pi/2$ out of phase (the frequency self-modulation leads amplitude self-modulation). Simultaneous frequency self-modulation has been observed in stripe lasers in the entire family of observed modes (including superluminescent modes), indicating a change in the optical length of the cavity. In broad (nonstripe) diodes this property is masked by the superimposition of patterns from asynchronous spatial channels. Study [103] identifies variations in the refractive index of the medium as possible causes of frequency self-modulation, while study [105] produces quantitative estimates demonstrating the suitability of such an explanation.

A physical model of frequency self-modulation is given in study [106]. The essential nature of this effect is that the oscillations in laser emission intensity and photon density in the cavity are accompanied by oscillations in the electron concentration, as may easily be found from rate equations. The electron concentration is related to the refractive index and thus influences the optical length of the cavity and the resonant frequencies of the laser modes.

If, for example, radiated power P fluctuates at a frequency Ω obeying a sinusoidal law

$$P|(t) = P_0 + P_1 \sin\Omega t,$$

with fixed pumping, the electron concentration N will obey the following law over time [76]

$$N = N_0 + (1/KP)\, dP/dt,$$

where K is the proportionality factor between the velocity of the stimulated transitions and the product PN (this interrelationship is linearized for simplicity). From here we have

$$N|(t) = N_0 + (P_1/KP_0)\Omega\cos\Omega t/[1 + (P_1/P_0)\sin\Omega t],$$

where N_0 is the constant component of the electron concentration. It is clear that at a low modulation percentage ($P_1/P_0 \ll 1$) oscillations in electron concentration have a cosinusoidal shape, i.e., they lead intensity oscillations by $\pi/2$. The variable portion of the refractive index (and the optical length of the cavity) obey the same law. As the intensity modulation percentage increases, inversion oscillations assume a clearly nonsinusoidal shape; wavelength deviations of the laser mode amount to 0.1-0.2 nm in a GaAlAs laser at 870-890 nm with a discontinuity in N at $2 \cdot 10^{17}$ cm^{-3}. The latter quantity represents 10-20% of the threshold concentration. Thus, variations in the carrier concentration under intensity fluctuations or under the joint action of these fluctuations and modulation of the pumping current causes a frequency deviation (oscillations with regular fluctuations) or rapid scanning ("chirping") in pulsed lasing. In order to eliminate these deviations or to reduce their influence, it is advisable to use an external cavity in which the longitudinal mode frequencies are less dependent on the refractive index of the semiconductor active element.

Moreover, study [106] has provided an estimation of line broadening (in a time-averaged spectrum), resulting from the low laser intensity oscillations (with fixed pumping). This broadening may be expressed in the relative expression

$$\Delta\omega/\omega_0 = \sigma(\Omega/4nK)(P_1/P_0),$$

where $\sigma = dn/dN$; the remaining designations are given earlier in this section. Broadening of this order or much greater order (with significant pulsations) is quite characteristic of the nonsteady-state lasing conditions of semiconductor lasers, although a thermal effect is often involved in practice.

Studies [105, 107] have demonstrated that since the wavelength deviation may serve as a measure for oscillations in carrier concentration, one interesting possibility in semiconductor lasers is to employ spectral chronograms to simultaneously trace two important dynamic variables characterizing lasing conditions: inversion intensity and density. Studies [104, 86] have reported the form of fluctuations on the boundary of steady-state single-mode lasing and self-modulation pulsation conditions. Then similar spectral chronograms were analyzed and phase trajectories of the operating point were drafted to demonstrate the spiral coil of two-dimensional oscillations as the critical cycle is approached.

The possibility of frequency or phase modulation of emission by current modulation is of practical significance. The contribution of the thermal effect and the contribution of free carriers may be differentiated here. The generation of FM by the thermal effect was described in study [108]. Recently interest in these modulation techniques has increased sharply due to the expanding prospects of coherent optical communications. Measurements in study [109] have demonstrated that the frequency deviation $\delta\nu/\delta I$ (where δI is the amplitude of the a. c. pumping current) amounts to 1-10 GHz/ma in various GaAlAs/GaAs stripe lasers. In the low modulation frequency range the thermal effect predominates; this makes it possible to achieve deviation on the order of greater than 1 GHz/ma at a frequency of ~0.1 MHz. At higher frequencies the contribution of free carriers increases, reaching a peak at a frequency of above 1 GHz. This peak corresponds to relaxation resonance which is observed in amplitude modulation as well. The behavior of the carriers in the active region influences the frequency response of frequency modulation; specifically, in order to obtain a flatter response, lateral carrier leakage structures are preferable. In buried mesa-stripe ("BH") lasers with significant lateral confinement, a deeper valley has been observed at frequencies of approximately 0.1 GHz (a reduction of the induced deviation by more than an order of magnitude).

12. NONLINEAR SCATTERING BY ELECTRON DENSITY WAVES AND ASYMMETRIC MODE INTERACTION

Study [10] focuses on the mechanism of nonlinear interaction of spectral modes in a semiconductor active medium which may explain experimental observations of anomalous (asymmetric) mode competition in a two-mode injection laser [110, 111]. As a result, indicators of asymmetric competition (more effective suppression of the neighboring shortwave mode) were observed in research on laterally optically confined stripe structures [112, 113].

We will examine the physical mechanism of nonlinear mode interaction which may be defined as stimulated scattering by electron density oscillations (waves). Below we will demonstrate that this mechanism may exist in both an externally-pumped medium (a laser) and in an unexcited medium, if the incident emission generates photoelectrons. It will also be demonstrated that gain is produced at new frequencies due to the nonlinear refraction of the medium and that the position of the auxiliary amplification band changes with respect to the initial spectral line in the transition from absorption to amplification in the medium.

We will limit our examination to the case of noncoherent interaction of emission and a medium, which for simplicity will be a set of bilevel systems. The noncoherent interaction condition may be represented $T_1 \gg T_2$, where T_1 and T_2 are the longitudinal and transverse relaxation time, respectively. Assume two waves are propagating in the medium. One wave is assumed to be "strong", the other "weak." These are given as

$$E_L = E_{L0} \exp(-i\omega_0 t + ik_0 \mathrm{r}), \; E_S = E_{S0} \exp(-i\omega_1 t| + ik_1 r), \quad (12.1)$$

where $E_{L0} \gg E_{S0}$ (wave amplitudes). Here ω_0 and ω_1 are the angular frequencies, k_0 and k_1 are the wave vectors of both waves, respectively. The superposition of waves produces a space-time lattice; the electromagnetic wave intensity is

$$|E|^2 = |E_{L0}|^2 + |E_{S0}|^2 + E^*_{L0} E_{S0} \exp[i\Omega t - i(k_0 - k_1) r], \tag{12.2}$$

where $\Omega = \omega_0 - \omega_1$ is the difference angular frequency. The following analysis will focus on the cross term (the third term in the right half) which is often improperly ignored in "multimode" problems. The dynamic intensity lattice has an amplitude

$$E^*_{L0} E_{S0} \exp[-i(k_0 - k_1)r], \tag{12.3}$$

where a dynamic lattice of corresponding geometry arises in the population distribution by virtue of the influence of emission intensity on the population of the bilevel particles. In order to find the value N_2 of the concentration of bilevel systems in the upper state, it is necessary to solve the equation

$$\dot{N}_2 + N_2/T_1 + B(N_2 - N/2)\,|E|^2 - G = 0, \tag{12.4}$$

where N is the concentration of particles, B is the proportionality factor describing the rate of stimulated emission; G is the external pumping rate (if external pumping is used). It is valid to ignore the diffusion processes when the diffusion coefficient D satisfies the condition

$$D \leqslant \Omega/|k_0 - k_1|^2, \tag{12.5}$$

i.e., is sufficiently small. We note that it is easier to satisfy condition (12.5) when the vectors k_0 and k_1 are similar; in other words, it is most appropriate to ignore diffusion for the case of interaction of waves propagating in the same or similar directions.

Substituting (12.3) into (12.4) makes it possible to calculate the oscillating portion δN_2 of the concentration of active particles:

$$\delta N_2 = -\frac{B(N_2 - N_2/2)\, E^*_{L0} E_{S0} \exp[i\Omega t - i(k_0 - k_1)r]}{i\Omega + 1/T_1 + B|E_{L0}|^2}. \tag{12.6}$$

Population oscillations cause oscillations in the optical characteristics of the medium: absorption (amplification) and the refractive index. We will represent corresponding variations in the dielectric constant $\delta\varepsilon$ in linearized form

$$\delta\varepsilon = (\partial\varepsilon/\partial N_2)\delta N_2 \tag{12.7}$$

and will represent the dielectric constant in simplified form

$$\varepsilon = \varepsilon_0 + (\partial\varepsilon/\partial N_2) N_2, \tag{12.8}$$

Thus, the addition δD arises in the electrical induction D of the electromagnetic field; this is related to the scattering of the strong field by the phase dynamic lattice $\delta\varepsilon$:

$$\delta D = E_{L0} e^{-i\omega_0 t + ik_0 r} \delta\varepsilon(r, t) = \\ = \left[|E_{L0}|^2 \frac{\partial\varepsilon}{\partial N_2} B\left(\frac{N}{2} - N_2\right) E_{S0} e^{-i\omega_1 t + ik_1 r} \right] \left(i\Omega + \frac{1}{T_1} + B|E_{L0}|^2 \right)^{-1}. \tag{12.9}$$

This addition occurs at the frequency of the weak signal ω_1. This makes it possible to obtain an expression for the additional complex dielectric constant $\delta\varepsilon_1$ at ω_1 induced by the strong field:

$$\delta\varepsilon_1 = \frac{|E_{L0}|^2 (\partial\varepsilon/\partial N_2) B (N/2 - N_2)}{i\Omega + 1/T_1 + B|E_{L0}|^2}. \tag{12.10}$$

From here we will have the following for the additional (induced) gain factor δg (or $\delta\alpha$: the induced absorption):

$$\delta g = -\frac{\omega}{c\sqrt{\varepsilon_0}} \operatorname{Im}(\delta\varepsilon_1) = \\ = \frac{\omega}{c\sqrt{\varepsilon_0}} |E_{L0}|^2 \operatorname{Re}\left(\frac{\partial\varepsilon}{\partial N_2}\right) B\left(\frac{N}{2} - N_2\right) \frac{\Omega - \Gamma\zeta}{\Omega^2 + \Gamma^2}, \tag{12.11}$$

where

$$\zeta = \operatorname{Im}(\delta\varepsilon/\partial N_2)/\operatorname{Re}(\partial\varepsilon/\partial N_2) \quad \text{and} \quad \Gamma = 1/T_1 + B|E_{L0}|^2.$$

It is clear from this result that either amplification or absorption may result depending on the symbol Ω, i.e., on the position of the frequency of the weak field with respect to the frequency of the strong field. This also depends on the symbols ζ and $N/2 - N_2$. The additional amplification (absorption) is the result of the scattering of the strong wave at the phase lattice induced by population discontinuities ("waves"), which in turn arise from the beating of both waves. This process is analogous to the well-known stimulated nonlinear scattering processes. The parameter of the medium responsible for the optical nonlinearity is the population of the working level. The equation for the population N_2 manifests a relaxation nature; hence, the spectral profile of the gain addition manifests a nonresonant nature, and appears as

$$\delta g \sim (\Omega - \Gamma\zeta)/\Omega^2 + \Gamma^2). \tag{12.12}$$

In this respect the mechanism is analogous to stimulated temperature scattering or stimulated scattering of the Rayleigh wing. The form of (12.12) was derived in study [111]. The frequency scale characterizing the distance at which the strong field acts on the weak fields is determined by Γ: the probability of optical transition between the levels. The profile δg consists of the even Lorentz form

$$-\Gamma\zeta/(\Omega^2 + \Gamma^2), \tag{12.13a}$$

from the Im ε lattice, and the odd form

$$\Omega/(\Omega 2 + \Gamma^2),\tag{12.13b}$$

from the Re ε lattice, and describing asymmetric interaction. We note that the value of (12.13a) is negative, i.e., it corresponds to additional absorption, at the same time that the value of (12.13b) changes sign and may yield amplification [111, 114].

We will now consider the particular case of a semiconductor such as GaAs interacting with emission near the absorption edge. The time T_2 at room temperature is 12^{-12} - 10^{-13} seconds, which is significantly less than the radiative recombination time $T_1 \approx 10^{-9}$ seconds. Hence for the range of difference frequencies $|\Omega| \ll T_2^{-1}$ we may assume that the initial assumptions of the properties of the medium are valid. The concentration of excess conduction electrons may be assigned to the quantity N_2, while the waves δN_2 are the electron density waves; the quantity $N/2$ corresponds to the concentration at the inversion threshold at which the difference of the Fermi quasilevels is $\hbar\omega_0$. From the experimental data, the constant ζ is 0.2-0.3 (at T = 300 K). Since $\mathrm{Re}(\partial\varepsilon/\partial N)$ is negative for GaAs, while the quantity $N/2 - N_2$ without pumping is positive, gain, according to (12.11) will be observed in the anti-stokes region ($\Omega < 0$). The spectral extrema of profile (12.12) are at frequencies

$$\Omega_{1,2} = \Gamma\,(\zeta \pm \sqrt{\zeta^2 + 1}),\tag{12.14}$$

where the minus sign corresponds to the amplification peak at which the value of (12.12) takes the form

$$-\sqrt{\zeta^2 + 1}/2\,[\zeta^2 - \zeta\sqrt{\zeta^2 + 1} + 1].\tag{12.15}$$

Accounting for the interrelationship between the quantities N_2, Γ, and $B|E_{L0}|^2$, we may derive

$$\delta g = -\frac{\omega}{c\sqrt{\varepsilon_0}}\,\mathrm{Re}\left(\frac{\partial\varepsilon}{\partial N_2}\right) N_2\left(1 - \frac{2N_2}{N}\right)\frac{\sqrt{\zeta^2+1}}{2\,(\zeta^2 - \zeta\sqrt{\zeta^2+1}+1)}\,.\tag{12.16}$$

We have absorption at the frequency of the additional amplification peak

$$\alpha = \frac{\omega}{c\sqrt{\varepsilon_0}}\,\mathrm{Im}\left(\frac{\partial\varepsilon}{\partial N_2}\right)\left(N_2 - \frac{N}{2}\right),\tag{12.17}$$

so the resulting amplification quantity is

$$\delta g' = \delta g - \alpha = -\frac{2\omega}{cN\sqrt{\varepsilon_0}}\,\mathrm{Re}\left(\frac{\partial\varepsilon}{\partial N_2}\right)\left(\frac{N}{2} - N_2\right)[N_2 - \varphi(\zeta)\,N],\tag{12.18}$$

where

$$\varphi(\zeta) = \zeta\,(\zeta^2 - \zeta\sqrt{\zeta^2 + 1} + 1)(\zeta^2 + 1)^{-1/2}.\tag{12.19}$$

with ζ = 1/3 this quantity is $\varphi \approx 0.24$.

We see that with $N_2 = N/2$ the amplification (gain) is equal to 0, $\delta g = 0$, which corresponds to the inversion threshold (which is not attainable by single resonant pumping). A positive value of δg is possible if $\varphi(\zeta) < 1/2$ and δg has a maximum with

$$N_2 = (N/2)(1/2 + \varphi), \tag{12.20}$$

equal to

$$\delta g_{\max} = -\frac{\omega}{c\sqrt{\varepsilon_0}} \operatorname{Re}\left(\frac{\partial \varepsilon}{\partial N_2}\right) \frac{N}{2} \left(\frac{1}{2} - \varphi\right)^2. \tag{12.21}$$

A numerical estimate for emission at ω_0, corresponding to absorption saturation at a concentration $N/2 \approx 10^{18}$ cm^{-3} produces a maximum amplification (gain) $\delta g_{max} \approx 40$ cm^{-1} (we have $\operatorname{Re}(\partial\varepsilon/\partial N_2) = -3 \cdot 10^{-20}$ cm^3, $\lambda = 0.85$ μm). In spite of the fact that total absorption saturation did not occur at this working transition ($N_2 = 0.4\ N < N/2$), nonetheless not only absorption compensation but optical amplification occur at the frequency $\omega = \omega_0 + \Omega_{max}$ (where Ω_{max} is determined by formula (12.14) with the minus sign). With an electron lifetime of 2 ns, Ω_{max} is approximately 50 MHz. Sufficiently monochromatic emission and high resolution spectral measurements are needed to observe this effect. The most convenient observation method involves directing both waves at a small relative angle sufficient for spatial separation of the beams.

This effect may be observed in other media such as metal vapors (K, Na, Cs, etc.), if the emission frequency falls at the edge of the resonant absorption line. In this case α may be sufficiently small to observe amplification of emission.

In a medium with population inversion, the additional gain peak falls in the Stokes region. According to calculations from study [111] the maximum additional gain may be estimated as

$$\delta g \approx 2 \cdot 10^{-16} \bar{N}, \tag{12.22}$$

where $\bar{N}$ is the average concentration of excess carriers. We see that this estimate gives a very high gain of ~200 cm^{-1} at $\bar{N} = 10^{18}$ cm^{-3} at a short spectral distance from the strong mode. When the difference frequency is increased, the additional gain diminishes as $1/\Omega$, so that for the neighboring mode in a standard laser diode, it is 0.1 cm^{-1}. Nonetheless such a value may be sufficient to significantly alter the laser emission spectrum. The fact is that the gain shortage in neighboring modes is quite small, say 0.01-0.1 cm^{-1}, due to the significant width of the amplification band. If the additional gain reaches this value, it may compensate for the gain shortage and cause a switch to the neighboring mode. We will apply the spectral mode interaction mechanism to the case of modes with different transverse indices, i.e., energy transfer may also take place between modes of different orders. In calculating this interaction we need an accurate account of the transverse profiles of these modes, since in the general case they are significantly different. In other words, the calculation must incorporate the corresponding overlap integrals requiring rather cumbersome calculations (an approximation of planar waves here is clearly unsuitable).

13. THE AUTOSTABILIZATION EFFECT OF SINGLE-FREQUENCY EMISSION

The preceding section demonstrated that accounting for the cross term in the electromagnetic field intensity (in the two-mode case) leads to additional amplification and absorption in the vicinity of the spectral line of a strong electromagnetic field. The profile is described by formula (12.12). Thus, when lasing occurs in a laser at a specific frequency, such as in the center of the amplification band, as the electromagnetic field intensity increases, a certain increasing deformation of the spectral amplification profile occurs for the second wave. An additional effect of amplification suppression occurs in the immediate vicinity of the strong mode and on the shortwave end of the strong mode. Thus, the strong mode not only saturates gain (by homogeneous broadening), but even suppresses the amplification process in a certain spectral range, and the extent of this suppression is proportional to the intensity of the strong wave as described by formula (12.11). The longwave boundary of the suppression region (ignoring the influence of the slope of the initial amplification profile) is at a spectral distance of $\Gamma\zeta$ from the strong line; in GaAs this is approximately 100-300 MHz. Here if we use selection to avoid excitation of modes greater than $\Gamma\zeta$ from the longwave end, it is possible to use the automatic suppression of all such modes by the nonlinear scattering process described in the preceding section. For example, if we construct a resonator with a resonant band half width of less than $\Gamma\zeta$, we may expect to achieve autostabilized single-frequency emission, since the other frequencies in this band would be suppressed by the powerful field.

It should be remembered that these concepts follow from an examination of two-wave interaction. If we examine actual conditions in a laser, it is necessary to remember the large number of modes with comparable Q. For the case of weak waves (aside from the single laser wave) we may use an independent analysis for each difference frequency Ω, although it is necessary to remember that there cannot be nonindependent waves whose frequencies are symmetric with respect to the strong line. Indeed, with interaction of two waves with angular frequencies of ω_0 and ω_1 a dynamic lattice arises at the beat frequency $\Omega = \omega_0 - \omega_1$, together with induced polarization of the medium at $\omega_0 \pm \Omega$. Analogously a wave with an angular frequency $\omega_2 = \omega_0 + \Omega$ generates induced polarization of the medium at $\omega_0 \pm \Omega$. This indicates energy exchange between symmetric modes. Such an optical field interaction mechanism in the inverted medium by saturation of the working transition was introduced in study [116] for a ruby laser. This mechanism was, however, considered insignificant and was eliminated from the consideration of the stability of single-frequency lasing. This is related to the comparatively long time period T_1 that determines the characteristic frequency scale ($\sim 1/T_1$) of this mechanism. It turns out that it is limited in ruby to frequencies of less than 10^3-10^4 Hz, which is much less than the standard mode interval.

A detailed analysis in study [115] has demonstrated that this phenomenon may be used successfully to suppress neighboring side modes, including those on the longwave end by reversing the phase of the shortwave mode and equalizing the action of the powerful field on both symmetric waves. Thus, the electromagnetic field suppresses neighboring weak modes in the spectral vicinity; this process increases with intensity of single-frequency emission, which is the source of the term autostabilization.

In order to experimentally observe this effect, studies [115, 117] investigated the emission spectra of an ejection laser with an external selective cavity with resolution of 100 MHz. It was determined that the emission was stable single-frequency emission (a single longitudinal mode of the external cavity was excited). Microwave pumping to resonance with photon transit in the cavity was employed to increase intensity at the neighboring longitudinal modes (i.e., active mode-locking was achieved). Further the relationship between the lateral mode intensity and the central mode intensity was observed with a fixed microwave pumping level, and it was demonstrated that an increase in the central mode intensity is accompanied by suppression of side modes.

Deformation of the amplification profile is identified experimentally by observation of phonon superluminescence in single-mode lasers in which the transverse mode profile is well stabilized, and hence it becomes possible to obtain a significant intensity in the single longitudinal mode. The characteristic shape of the superluminescent mode envelope in single-frequency emission corresponds to a certain additional suppression from higher frequencies.The calculation in study [113] has indicated that it is satisfactorily described by formula (12.12); however, the intraband relaxation time ($\sim 10^{-12}$ sec) is used as the characteristic relaxation time that determines Γ. This corresponds to a change in the model and assumes uniform energy saturation of the transitions within the line. Within the scope of the initial model, a reduction in the relaxation time, (i.e., an increase in Γ) does not increase the effect at greater frequency distances, since in addition to broadening of the band there is a reduction in its amplitude (this is evident from formula (12.12)). The initial model accounts for the contribution of all excess carriers, and not solely those occupying the working states. Thus, the observed asymmetric behavior in the spectrum of superluminescent modes may be explained by the mechanism described above (without assuming violation of quasiequilibrium in the bands).

Nonetheless, some care must be taken in analyzing the deformation of the amplification profile with an increase in intensity in the dominant mode. Above all it is necessary to differentiate the spectral mode amplification profile from the local amplification profile corresponding to specific pumping and field intensities. The modal profile is obtained by integration with respect to space with variable local amplification parameters. This is the primary essence of the problem of spatially-inhomogeneous inversion "burning" and the action of spatial pumping irregularities. An analysis of these problems is covered in [118] to some degree where it is demonstrated that the corresponding deformation in the modal profile is clearly comparable to the amplification deficit in the neighboring modes. This fact, in conjunction with the temperature action significantly complicates the behavior of the separate superluminescent and laser lines.

14. CONCLUSION

The study of nonlinear refraction effects in semiconductor lasers has enhanced the understanding of laser emission properties and has made it possible to identify a wide range of anomalies that are not understood from the viewpoint of the linear optical model. A wide range of empirical results is classified and a more or less adequate explanation from a unified viewpoint is obtained. More comprehensive research may produce new phenomena, particularly those characteristic of multiwave interactions producing phase conjugation, etc. This class of phenomena includes nonlinear mode interaction and autostabilization of single-frequency emission (based on the phase reversal of neighboring modes). This interaction increases the output power of highly-coherent emission when there is sufficient selection in the cavity (in order to suppress remote modes and eliminate transverse modal instability). Undoubtedly nonlinear refraction plays an important role in the nonregular nonsteady-state mechanism that arises at points significantly over threshold [104] and is the cause of limitation on the coherence of laser emission at high power (see also [78]). This process has not yet been completely investigated. Another important practical problem is the investigation and application of RF FM and PM based on nonlinear refraction.

A comprehensive optical model of a semiconductor laser will include these nonlinear refraction effects together with the thermal nonlinearity effect and the directional gain effect characteristic of such semiconductor lasers.

BIBLIOGRAPHY

1. Gibbs, H. M., McCall, S. L., Venkatesan, T. N. C., et al. "Optical bistability in semiconductors" APPL. PHYS. LETT., 1979, V. 35, p. 451-453.

2. Miller, D. A. B., Smith, S. D., Johnston, A. "Optical bistability and signal amplification in a semiconductor crystal: application of new low power nonlinear effect in InSb" APPL. PHYS. LETT., 1979, V. 35, p. 658-660.

3. Gibbs, H. M., McCall, S. L., Venkatesan, T. N. C. "Optical bistable devices; the basic components of all-optical systems?" OPT. ENGINEER., 1980, V. 19, No. 4, p. 463-468.

4. Abraham, E., Seaton, C. T., Smith, S. D. "The Optical computer" SCI. AMER., 1983, V. 248, No. 2, p. 85-93.

5. Stern, F. "Radiation confinement in semiconductor lasers" In: Radiative recombination in semiconductors. P., 1965, p. 165-169.

6. Anderson, W. W. "Mode confinement and gain in injection lasers" IEEE J. QUANT. ELECTRON., 1965, V. 1, No. 6, p. 228-238.

7. Eliseev, P. G., "O volnovodnoy strukture PKG na *p-n*-perekhodakh i geteroperekhodakh" [The waveguide structure of a laser at *p-n*-junction and heterojunction lasers] Preprint. FIAN No. 33, Moscow, 1970.

8. Miller, D. A. B., Smith, S. D., Seaton, C. T., "Optical bistability in semiconductors" IEEE J. QUANT. ENECTRON., 1981, V. 17, No. 3, p. 312-317.

9. Miller, D. A. B., Seaton, C. T., Prise, M. E., Smith, S. D. "Band-gap-resonant nonlinear refraction in III-V-semiconductors" PHYS. REV. LETT., 1981, V. 47, No. 3, p. 197-200.

10. Thompson, G. H. B. "The physics of semiconductor laser devices" N.Y.; London: J. Wiley and Sons, 1980, p. 185-186.

11. Mendoza-Alvarez, J. G., Nunes, F. D., Patel, N. B. "Refractive index dependence on free carriers for GaAs" J. APPL. PHYS., 1980, V. 51, No. 8, p. 4365-4367.

12. Selway, P. R., Thompson, G. H. B., Henshall, G. D., Whiteaway, J. E. A. "Measurement of the effect of injected carriers on the *p-n* refractive index step in single heterostructure diode lasers" ELECTRON. LETT., 1974, V. 10, p. 453-455.

13. Bogatov, A. P., Eliseev, P. G., Mamutin, V.V. "The influence of excess current carriers on the dielectric constant of GaAs at the radiative transition frequency in injection lasers" KVANTOVAYA ELEKTRON, 1976, V. 3, No. 7, p. 1609-1610.

14. Bogatov, A. P., Eliseev, P. G., Sverdlov, B. N. "Isledovanie inzhektsionnykh kvantovykh generatorov. II" [An investigation of injection lasers. II] Tr. FIAN, 1977, V. 91, p. 75-114.

15. Stubkjaer, K., Asada, M., Arai, S., Suematsu, Y. "Spontaneous recombination, gain and refractive index variation for 1.6 μm wavelength InGaAsP/InP lasers" JAPAN J. APPL. PHYS., 1981, V. 20, No. 8, p. 1499-1505.

16. Olsson, A., Tang, C. L. "Injected-carrier induced refractive index change in semiconductor lasers" APPL. PHYS. LETT., 1981, V. 39, No. 1, p. 24-26.

17. Vyatkin, K. V., Shotov, A. P. "The optical properties of PbSe epitaxial films" FTP, 1980, V. 14, No. 7, p. 1331-1334.

18. Lugovoy, V. N. "Nonlinear optical cavities" KVANTOVAYA ELEKTRON, 1979, V. 6, No. 10, p. 2053-2077.

19. Collins, S. A., Jr., Wasmundt, M. "Optical feedback and bistability: a review. OPT. ENGINEER., 1980, V. 19, No. 4, p. 478-487.

20. Miller, D. A. B., Mozolowski, M. H., Miller, A., Smith, S. D. "Nonlinear optical effects in InSb with CW CO laser" OPT. COMMUNS, 1978, V. 27, p. 133-136.

21. Mackenzie, H. A., Dennis, R. B., Voge, D., Smith, S. D. "Intensity dependent Faraday rotation in indium antimonide" OPT. COMMUNS, 1980, V. 34, p. 205.

22. Ryvkin, B. S. "The wave-shaped volt-current characteristic of a photoconducting plane-parallel wafer under laser bombardment" PIS'MA V ZHTF, 1979, V. 5, p. 65-69.

23. Ryvkin, B. S. "Current and light intensity oscillations with laser emission bombardment of a plane-parallel photoconducting wafer" PIS'MA V ZHTF, 1979, V. 5, p. 586-589.

24. Ryvkin, B. S. Stepanova, M. N. "Bistable optical characteristics of a resonant photoelement with two-stage optical transitions" ZHETF, 1982, V. 8, p. 951-954.

25. Vu Van Lyk, Eliseev, P. G., Man'ko, M. A., Mikaelyan, G. T. "The electrical diagnostics of a laser diode-based amplifier-monitor" KVANTOVAYA ELEKTRON, 1982, V. 9, No. 9, p. 1851-1853.

26. Basov, N. G., Krokhin, O. N., Popov, Yu. M. "Obtaining negative temperature states at the *p-n*-junctions of degenerate semiconductors" ZHETF, 1961, V. 40, p. 1879-1880.

27. Bogatov, A. P., Eliseev, P. G., Okhotnikov, O. G., Pak, G. T. "The properties of planar stripe heterolasers. Vol. 1. Nonlinear and discontinuous watt-current characteristics" KVANTOVAYA ELEKTRON, 1980, V. 7, No. 8, p. 1664-1669.

28. Bakhert, Kh.-Yu., Bogatov, A. P., Eliseev, P. G. "Mode deformation in an injection laser due to self-focusing and its relation to the nonlinearity of the output characteristic" KVANTOVAYA ELEKTRON, 1978, V. 5, No. 3, p. 603-608.

29. Turley, S. E. H., Thompson, G. H. B., Lovelace, D. F. "Effect in injection current on the dielectric constant of an inbuilt waveguide in twin transverse junction stripe lasers" ELECTRON. LETT., 1979, V. 15, No. 9, p. 256-257.

30. Vu Van Lyk, Eliseev, P. G., Man'ko, M. A., Mikaelyan, G. T. "Bliyanie nestatsionarnogo temperaturnogo volnovoda na kharakteristiki izlucheniya geterolazerov" [The influence of a nonsteady-state temperature waveguide on the emission characteristics of heterolasers] Preprint, FIAN, No. 224, Moscow, 1982, 28 p.

31. Chiao, R. Y., Garmire, E., Townes, C. H. Self-trapping of optical beams. PHYS. REV. LETT., 1964, V. 13, Noi. 15, p. 479-482.

32. Eliseev, P. G. "Vvedenie v fiziku inzhektsionnykh lazerov" [An introduction to the physics of injection lasers] Moscow: Nauka, 1983, p. 103-104.

32. Yabuzaki, T., Hory, H., Kitano, M. "Theoretical study on optical-pumping self-focusing" JAPAN. J. APPL. PHYS., 1982, V. 21, No. 3, p. 504-512.

33. Boshier, M. C., Sandle, W. J. "Self-focusing in a vapor of two-state atoms" OPT.COMMUNS, 1982, V. 42, No. 5, p. 371-376.

34. Yabuzaki, T., Hory, H., Kitano, M. "Theoretical study on optical-pumping self-focusing" JAPAN. J. APPL. PHYS. 1982, V. 21, No. 3, p. 504-512.

35. Thompson, G. H. G. "A theory for filamentation in semiconductor lasers including the dependence of dielectric constant on injected carrier density" OPTO-ELECTRON., 1972, V. 4, p. 257-310.

36. Eliseev, P. G., Krasil'nikov, A. I., Man'ko, M. A., Strakhov, V. P. "Issledovanie inzhektsionnykh lazerov na postoyannom toke" [An investigation of d.c. injection lasers] In: Fizika *p-n*-perekhodov i poluprovodnikovykh priborov [The physics of *p-n*-junctions and semiconductor devices] Leningrad: Nauka, 1969, p. 131-141.

37. Kawaguchi, H., Takahei, K. "Direct observation of the spatial hole burning and the saturation of the spontaneous emission in InGaAsP/InP lasers" IEEE J. QUANT. ELECTRON., 1980, V. 16, No. 7, p. 706-708.

38. Eliseev, P. G. "An optimum waveguide structure for an injection laser" KVAT, SOOBSHCH. PO FIZIKE FIAN, 1970, No. 4, p. 3-7.

39. Bogatov, A. P., Gurov, Yu. V., Eliseev, P. G., Khayretdinov, K. A. "Anomalous lasing dynamics in semiconductor lasers with an asymmetric waveguide structure. Part 1. Experimental investigation with an external cavity" KVANTOVAYA ELEKTRON, 1978, V. 5, No. 11, p. 2402-2407.

40. Bogatov, A. P., Eliseev, P. G., Popov, Yu. M., et al. "Anomalous lasing dynamics in semiconductor lasers with an asymmetric waveguide structure. Part II. Theory" KVANTOVAYA ELEKTRON, 1978, V. 5, No. 11, p. 2408-2415.

41. Eliseev, P. G., Ivanov, L. P., Logginov, A. S., Nikitin, Ye. P. "Internal Q switching in injection lasers with a single-side heterojunction" KVANTOVAYA ELEKTRON, 1973, No. 5 (17), p. 116-117.

42. Grundorfer, S., Adams, M. J., Thomas, B. "*H*-pulsing: a transient effect in GaAs/GaAlAs injection lasers" IEEE J. QUANT. ENECTRON., 1975, V. 11, No. 7, p. 532-537.

43. Ripper, J. E., Dyment, J. C. "Internal Q-switching in GaAs junction lasers" APPL. PHYS. LETT., 1968, V. 12, No. 11, p. 365-367.

44. Uno, Y., Yamamoto, M., Jida, S. "Internal Q-switching in GaAlAs-GaAs heterojunction lasers" JAPAN. J. APPL. PHYS. 1970, V. 9, p. 1181-1182.

45. Ulmer, E. A., Hayashi, I. "Internal Q-switching in GaAs-GaAlAs heterostructure lasers" IEEE J. QUANT. ELECTRON., 1970, V. 6, No. 6, p. 297-299.

46. Masuyama, A., Kawabe, M., Masuda, K., Namba, S. "Internal Q-switching in electron beam excited GaAs lasers" JAPAN. J. APPL. PHYS. 1971, V. 10, No. 2, p. 161-165.

47. Nunes, F. D., Patel, N. B., Ripper, J. E. "Theory on long time delay Q-switching and phenomena in GaAs junction lasers" ELECTRON. LETT., 1976, V. 12, No. 22, p. 574-575.

48. Aritome, H., Nasuda, K., Namba, S. "Time delay in CdS laser pumped by an electron beam" OPTO-ELECTRON., 1970, V. 2, No. 4, p. 239-244.

49. Dixon, R. W., Nash, F. R., Hartman, R. L., Hepplewhite, R. T. "Improved light-output linearity in stripe-geometry DH (AlGa)As lasers" APPL. PHYS. LETT., 1967, V. 29, No. 6, p. 372-375.

50. Campos, M. D., Hwang, C. J., Bossi, R. I., Ripper, J. E. "Cavity competition in anomalous emission intensity in DH lasers" IEEE J. QUANT. ELECTRON., 1977, V. 13, No. 8, p. 687-691.

51. Bogatov, A. P., Eliseev, P. G., Okhotnikov, O. G., Pak, G. T. "Output emission power hysteresis in CW AlGaAs heterolasers" KVANTOVAYA ELEKTRON., 1978, V. 5, No. 11, p. 2493-2495.

52. Boykachev, A. I., Gribkovskiy, V. P., Kononenko, V. K., et al. "The features of the watt-current characteristics of stripe-contact lasers" KVANTOVAYA ELEKTRON 1979, V. 6, No. 5, p. 972-978.

53. Eliseev, P. G. "Stripe-geometry lasers and their properties" CZECH J. PHYS. B, 1980, V. 30, No. 3, p. 300-317.

54. Selway, P. R., Kirkby, P. A., Goodwin, A. R., Thompson, G. H. B. "Dynamics of self-focusing in stripe-geometry semiconductor lasers" IEE PROC. SOLID STATE ELECTRON. DEVICES, 1978, V. 2, No. 1, p. 38-40.

55. Van der Ziel, J. P. "Self-focusing effects in pulsing AlGaAs DH lasers" IEEE J. QUANT. ELECTRON., 1981, V. 17, No. 1, p. 60-68.

56. Bazhenov, V. Yu., Bogatov, A. P., Eliseev, P. G., et al. "Bistable conditions and spectral tuning in an injection laser with an external dispersive cavity" KVANTOVAYA ELEKTRON., 1981, V. 8, No. 4, p. 853-859.

57. Bazhenov, V. Yu., Bogatov, A. P., Eliseev, P. G., et al. "Bistable operation and spectral tuning of injection laser with external dispersive cavity" PROC. INST. ELEC. ENGINEER., 1982, V. 129, Pt. 1, No. 2, p. 77-82.

58. Glas, P., Muller, R. "Bistable operation of a GaAs-AlGaAs diode laser coupled to an external resonator of narrow spectral bandwidth" OPT. AND QUANT. ELECTRON., 1982, V. 14, No. 5, p. 375-389.

59. Stallard, W. A., Bradley, D. J. "Bistability and slow oscillation in an external cavity semiconducting laser" APPL. PHYS. LETT., 1983, V. 42, No. 10, P. 858-859.

60. Velichanskiy, V. L., Zibrov, A. S., Molochev, V. I., et al. "The asymmetry of select characteristics of tunable lasers with an external dispersive cavity" KVANTOVAYA ELEKTRON., 1980, V. 7, No. 11, p. 2510-2512.

61. Kawaguchi, H. "Bistable operation of semiconductor laser by optical injection" ELECTRON. LETT., 1981, V. 17, No. 20, p. 741-742.

62. Kawaguchi, H. "All-optical bistable operation in semiconductor lasers" IEEE TRANS. ELECTRON. REV., 1982, V. 29, No. 10, p. 1676.

63. Nakai, T., Ito, R., Ogasawara, N. "Asymmetric frequency response of semiconducting laser amplifier" JAPAN. J. APPL. PHYS., 1982, V. 21, Pt. 2, No. 11, p. 680-682.

64. Nakai, T., Ogasawara, N., Ito, R. "Optical bistability in a semiconductor laser amplifier" JAPAN. J. APPL. PHYS., 1983, V. 22, No. 5, p. 310-312.

65. Otsuka, K., Kobayashi, S. "Optical bistability and nonlinear resonance in a resonant-type semiconductor laser amplifier" ELECTRON. LETT., 1983, V. 19, No. 7, p. 262-263.

66. Goldberg, L., Taylor, H. F., Weller, J. F. "Locking bandwidth asymmetry in injection-locked GaAlAs lasers" ELECTRON. LETT., 1982, V. 18, No. 23, p. 986-987.

67. Lang, R. "Injection locking properties of a semiconductor laser" IEEE J. QUANT. ELECTRON., 1982, V. 18, No. 6, p. 976-983.

68. Lasher, G. J. "Analysis of a proposed bistable injection laser" SOLID STATE ELECTRON., 1964, V. 7, No. 10, p. 707.

69. Lasher, G. J., Fowler, A. B. "Mutually quenched injection lasers and bistable devices" IBM J. RES. AND DEVELOP., 1964, V. 8, No. 4, p. 471.

70. Fowler, A. B. "Cooperative effect in GaAs lasers" J. APPL. PHYS., 1964, V. 35, p. 2275-2276.

71. Eliseev, P. G., Novikov, A. A., Fedorov, V. B. "Optical interaction effects of two diode lasers" PIS'MA V ZHETF, 1965, V. 2, No. 2, p. 58-62.

72. Nathan, M. I., Marinace, J. C., Rutz, R. F., et al. "A GaAs injection laser with novel mode control and switching properties" J. APPL. PHYS., 1965, V. 36, No. 2, p. 473.

73. Basov, N. G., Zakharov, Yu. P., Nikitin, V. V., Sheronov, A. A. "The interaction of *p-n*-junction GaAs optically-coupled lasers" FTT, 1965, V. 7, No. 11, p. 3460.

74. Basov, N. G., Culver, W. H., Shah, B. "Application of lasers to computers. In: Laser handbook. N. Y.: Amer. Elsevier, 1970, V. 2, p. 1649.

75. Basov, N. G., Nikitin, V. V., Semenov, A. S. "The dynamics of emission from semiconductor injection lasers" UFN, 1969, V. 97, No. 4, p. 561-600.

76. Riblin, L. A. "Dinamika izlucheniya poluprovodnikovykh kvantovykh generatorov" [The dynamics of emission from semiconductor lasers] Moscow: Sov. radio, 1976.

77. Goldobin, I. S., Dobkin, A. S., Kurnosov, V. D., et al. "Quantum-optical gallium arsenide integrated circuits" FTP, 1971, V. 5, No. 1, p. 170-171.

78. Bakhert, Kh.-Yu., Eliseev, P. G., Man'ko, M. A., Raab, Z. "A spectral investigation of oscillation synchronization in injection lasers" ZHURN. PRIKL. SPEKTROSKOPII, 1970, V. 13, No. 2, p. 232-237.

79. Akkerman, D., Eliseev, P. G., Kayper, A., et al. "Methods of mode selection in injection lasers" KVANTOVAYA ELEKTRON., 1971, No. 1, p. 85-90.

80. Eliseev, P. G., Shuikin, N. N., "Single-mode and single-frequency semiconductor lasers" KVANTOVAYA ELEKTRON., 1973, No. 3, (15), p. 5-26.

81. Eliseev, P. G., Ismailov, I., Man'ko, M. A., Strakhov, V. P. "An injection laser with a compound resonator" PIS'MA V ZHETF, 1969, V. 9, p. 594-595.

82. Eliseev, P. G., Man'ko, M. A. "An injection laser with a tunable compound resonator" KVAT. SOOBSHCH PO FIZIKE FIAN, 1970, No. 4, p. 3-7.

83. Bakhert, Kh.-Yu., Eliseev, P. G., Raab, Z. "An investigation of the characteristics of injection lasers with a tunable compound resonator" ZHURN. PRIKL. SPEKTROSKOPII, 1972, V. 16, No. 5, p. 810-813.

84. Eliseev, P. G., Khaydarov, A. B. "Electrooptic and piezoelectric tuning of a compound laser resonator" OPTIKA I SPEKTROSKOPIYA, 1973, V. 34, No. 2, p. 343-346.

85. Eliseev, P. G., Ismailov, I., Fedorov, Yu. F. "Injection lasers for multichannel optical communications" IEEE. J. QUANT. ELECTRON., 1970, V. 6, No. 1, p. 38-41.

86. Bogatov, A. P., Eliseev, P. G., Manko, M. A., et al. "A study of single-mode injection laser" IEEE J. QUANT. ELECTRON., 1973, V. 9, No. 2, p. 392-394.

87. Kurnosov, V. D., Semenov, A. T. "A two-cavity injection laser with strong optical coupling" KVANTOVAYA ELEKTRON, 1974, V. 1, No. 1, p. 35.

88. Renner, D., Caroll, J. E. "Simple system for broad-band single mode tuning of DH GaAlAs lasers" ELECTRON. LETT., 1979, V. 15, p. 72-74.

89. Lin, C., Burrus, C. A., Linke, R. A., et al. "Short-coupled-cavity (SEMICONDUCTOR) InGaAsP injection lasers for CW and high speed single-longitudinal mode operation" ELECTRON. LETT., 1983, V. 19, p. 561-562.

90. Van der Ziel, J. P., Mikulyak, R. M. "Single-mode operation of 1.3 μm InGaAsP/InP buried crescent lasers using a short external optical cavity" IEEE J. QUANT. ELECTRON., 1984, V. 20, N. 3, p. 223-229.

91. Miles, R. O., Dandridge, A., Tveten, A. B., Giallorenzi, T. G. "An external cavity diode laser sensor." J. LIGHTWAVE TECHNOL., 1983, V. 1, No. 1, p. 81-93.

92. Tsang, W. T., Olsson, N. A., Logan, R. A. "High speed direct-single-frequency modulation with large tuning rate and frequency excursion in cleaved-coupled-cavity semiconductor lasers" APPL. PHYS. LETT., 1983, V. 42, No. 8, p. 650-652.

93. Tsang, W. T., Olsson, N. A., Linke, R. A., Logan, R. A. "1.5 μm wavelength GaInAsP C^3 lasers: single frequency operation and wideband frequency tuning" ELECTRON. LETT., 1983, V. 11, p. 415-417.

94. Tsang, W. T., Olsson, N. A., Logan, R. A. "Optoelectronic logic operations by cleaved coupled cavity semiconductor lasers" IEEE J. QUANT. ELECTRON., 1983, V. 19, No. 11, p. 1621-1625.

95. Olsson, N. A., Tsang, W. T., Logan, R. A., et al. "Spectral bi-stability in coupled cavity semiconductor lasers" APPL. PHYS. LETT., 1984, V. 44, No. 4, p. 375-577.

96. Henry, C. H., Kazarinov, R. F. "Stabilization of single frequency operation of coupled-cavity lasers. IEEE J. QUANT. ELECTRON., 1984, V. 20, No. 7, p. 733-744.

97. Agrawal, G. P., Dutta, N. K. "Optical bistability in coupled-cavity-semiconductor lasers" J. APPL. PHYS., 1984, V. 56, No. 3, p. 664-669.

98. White, I. H., Carroll, J. E. "New mechanism for bistable operation of closely coupled twin stripe lasers" ELECTRON. LETT., 1983, V. 19, No. 9, p. 337-339.

99. Shore, K. A., Rozzi, T. E. "Transverse switching due to Hopf bifurcation in semiconductor lasers" IEEE J. QUANT. ELECTRON., 1984, V. 20, No. 3, p. 246-255.

100. Garmire, E., Evans, G., Niesen, J. "Longitudinal mode control in GaAs lasers using a three mirror active-passive cavity. APPL. PHYS. LETT., 1981, V. 39, No. 10, p. 789-791.

101. Vargashkin, A. I., Pleshkov, A. A., Trukhan, V. G., Tsvetkov, V. V. "Spontaneous transient processes in the emission spectrum of a semiconductor laser" PIS'MA V ZHETF, 19709, V. 12, No. 1, p. 3-6.

102. Bogatov, A. P., Eliseev, P. G., Panteleev, V. I., Shevchenko, Ye. G. "A comparison of the instantaneous and average emission spectrum of an injection laser in spontaneous pulsation conditions" KVANTOVAYA ELEKTRON., 1971, V. 5, p. 93-95.

103. Eliseev, P. G., Ivanov, L. P., Logginov, A. S., Senatorov, K. Ya. "Frequency self-modulation of emission in an injection laser" KVAT. SOOBSHCH PO FIZIKE FIAN, 1972, No. 6, p. 53-55.

104. Bogatov, A. P., Eliseev, P. G., Ivanov, L. P., et al. "The kinetics of the emission spectrum of an injection laser and the phenomena of single-mode lasing breakdown" KVANTOVAYA ELEKTRON., 1973, No. 5, (17), p. 14-20.

105. Eliseev, P. G. "Poluprovodnikovye lazery i preobrazovateli" [Semiconductor lasers and converters] In: Itogi nauki i tekhniki [Achievements of science and technology] RADIOTEKHNIKA. Moscow: VINITI, 1978, V. 14, No. 1, p. 195-198.

106. Eliseev, P. G., "An interpretation of the frequency self modulation effect in a semiconductor laser" KVANTOVAYA ELEKTRON., 1979, V. 6, No. 10, p. 2243-2245.

107. Eliseev, P. G., Morozov, V. N., Pashko, S. A., et al. "Spectral mode broadening of a semiconductor laser with intensity fluctuations" KVANTOVAYA ELEKTRON., 1980, V. 7, No. 10, p. 2197-2201.

108. Bykovskiy, Yu. A., Velichanskiy, V. L., Maslov, V. A., Smirnov, V. L. "Frequency modulation of a semiconductor laser by injection current" KVANTOVAYA ELEKTRON., 1971, No. 3, p. 90-92.

109. Yamamoto, Y., Kobayashi, S., Kimura, T. "Direct frequency modulation and FM noise in semiconductor lasers" REV. ELEC. COMMUN. LAB., 1983, V. 31, No. 3, p. 349-360.

110. Bogatov, A. P., Eliseev, P. G., Sverdlov, B. N., "The anomalous interaction of spectral modes in a semiconductor laser" KVANTOVAYA ELEKTRON, 1974, V. 1, No. 10, p. 2286.

111. Bogatov, A. P., Eliseev, P. G., Sverdlov, V. N. "Anomalous interaction of spectral modes in a semiconductor laser" IEEE J. QUANT. ELECTRON., 1975, V. 11, No. 7, p. 510-514.

112. Ishikawa, H., Imai, H., Tonahashi, T., Takusagawa, M. "Longitudinal mode behavior of transverse-mode-stabilized DH laser. APPL. PHYS. LETT., 1981, V. 38, p. 962.

113. Ishikawa, H., Yano, M., Takusagawa, M. "Mechanism of asymmetric longitudinal mode competition in InGaPAs/InP lasers" APPL. PHYS. LETT., 1982, V. 40, No. 7, p. 553-555.

114. Bogatov, A. P., Eliseev, P. G. "Stimulated scattering of emission by the population waves of the excited state" KVANTOVAYA ELEKTRON., 1983, V. 10, No. 4, p. 826-828.

115. Bogatov, A. P., Eliseev, P. G., Okhotnikov, O. G., et al. "Mode interaction and autostabilization of single-frequency lasing in

injection lasers" KVANTOVAYA ELEKTRON., 1983, V. 10, No. 9, p. 1851-1865.

116. Kuznetsova, T. I. "The interaction of electromagnetic fields and active media" TR. FIAN, 1966, V. 43, p. 116-159.

117. Bogatov, A. P., Eliseev, P. G., Okhotnikov, O. G., et al. "Investigation of injection lasers with an external dispersive cavity" TR. FIAN, 1983, V. 141, p. 62-68.

118. Seki, K., Kamiya, T., Yanai, H. "Effect of waveguiding properties on the axial mode competition in stripe geometry semiconductor lasers" IEEE. J. QUANT. ELECTRON., 1981, V. 17, No. 5, p. 706-713.

MODE INTERACTION AND AUTOSTABILIZATION OF SINGLE-FREQUENCY LASING IN INJECTION LASERS

A.P. Bogatov, P.G. Eliseev, O.G. Okhotnikov, M.P. Rakhval'skiy, K.A. Khayretdinov

Abstract: Mode interaction in an injection laser is investigated both theoretically and experimentally, including the case of an external dispersive cavity. A theory of nonlinear mode interaction is presented and it is demonstrated that such interaction does not disappear when the intraband relaxation time is reduced, and serves to suppress amplification at frequency distances of approximately $1/\tau$ (τ is carrier lifetime) from the strong mode. A "self-stabilization" effect of single-frequency lasing has been observed experimentally with selectivity in a range of approximately $1/\tau$.

1. INTRODUCTION

Single-frequency lasing with a low background noise emission level is needed for a number of practical applications such as coherent optical communications, and laser spectroscopy. An important parameter is the stability of single-frequency lasing with respect to the excitation of other modes at frequencies near the fundamental line.

It is necessary to identify the nature and characteristic of the interaction of a laser field at the fundamental frequency and weak (with respect to the laser field) noise emission at other frequencies. Obviously for a semiconductor medium the primary interaction mechanism is the gain saturation effect along the line profile which occurs due to the laser field stabilization of the Fermi level [1]. This is the result of quasihomogeneous amplification line broadening of the semiconductor. This mechanism is activated when the threshold is achieved in the laser and hinders mode excitation at other frequencies. As pumping and laser emission intensity increase, other field interaction mechanisms may arise as a result of the limited relaxation time both in the bands and between the bands. The situation where the laser field is so strong that the Rabi frequency Ω_R exceeds or is on the order of the inverse relaxation time in the bands $1/\tau_B$, is examined in studies [2, 3]. In this case, the line profile of the working transition is deformed by the coherent interaction of the field and active particles (an energy gap is formed) near the strong field with a frequency scale in the order of $\Omega = 2Ed/\hbar$, where E is the field intensity of the light wave, d is the dipole moment for the working transition and $\hbar$ is Planck's constant. One characteristic of this mechanism is the suppression of gain not only in the vicinity of the laser field but also at the frequency of this field. As a result, when this interaction occurs, the power in single-frequency emission becomes saturated, and a further increase in pumping will produce lasing at other frequencies at a distance exceeding Ω.

Studies [4-8] have examined the intermediate case when the field, on the one hand, is sufficiently weak to form the energy gap, yet on the other hand is sufficiently strong so that the relaxation processes in the band do not completely equalize the valley in the energy spectrum of the active particles. These studies report that as the monochromatic field intensity increases, the amplification profile is deformed, which may both result in a breakdown in single-frequency lasing [4, 5] and may increase the stability of single-frequency lasing [7, 8]. For example, it is stated in studies [7, 8] that the strong field suppresses weak fields in its spectral vicinity $\pm 1/\Gamma_B$.

Unlike the mechanisms discussed above, which are related to the finite relaxation time of the carriers in the bands, this study will examine the field interaction mechanism related to the optical nonlinearity of the working transition itself (due to the saturation effect). The characteristic time of this mechanism is the radiative band-to-band transition time τ. Hence its action is most intensive in the spectral interval $\pm 1/\tau$ and diminishes as $1/\Omega\tau$. In our preceding study [9] we demonstrated that when only two fields are present, one "strong" field and one "weak", this mechanism has the strong field generating for the weak field either additional gain or absorption depending on the spectral position of the weak field with respect to the strong field. This analysis was carried out for an experiment with a laser with a special dispersive cavity allowing excitation of only two spectral lines. Clearly in the more general case when it is possible to excite a large number of spectral frequencies in the laser, it is necessary to perform the analysis accounting for at least three frequencies: a strong field and two weak fields at a distance of Ω and positioned symmetrically with respect to the strong field on both the short and long wave sides. Indeed, with a strong field and only one weak field at a distance of Ω, optical nonlinearity will clearly generate one additional field symmetrically detuned with respect to the strong field. Hence, the optical nonlinearity creates an interrelationship between all amplitudes of weak fields symmetric with respect to the strong field.

Accounting for this discussion, our study analyzes the interaction mechanism between the strong field and weak fields in the active region of an injection laser; expressions are derived for the additive absorption (gain) coefficients of weak fields induced by the strong field. Experiments are given that illustrate this interaction.

2. THEORETICAL ANALYSIS

2.1. Field interaction in a laser with a cavity formed by natural facets

We will examine the case where the field in the laser $\mathscr{E}(x, y, z, t)$ is the sum of three fields corresponding to the three axial modes of a laser diode with a single transverse index. For simplicity we will also assume that the spatial field distribution along the cavity axis is similar to standing waves, while the transverse distribution is homogeneous:

$$\mathscr{E} = E_0(t)\, e^{-i\omega_0 t} \sin q_0 z + E_1(t)\, e^{-i(\omega_0-\Omega)t} \sin q_1 z + E_2(t)\, e^{-i(\omega_0+\Omega)t} \sin q_2 z \qquad (1)$$

where E_0 is the slow amplitude of the "strong" field, $E_1(t)$, $E_2(t)$ are the slow amplitudes of the "weak" fields, symmetrically detuned by $\pm\Omega$ from the frequency of the strong field ω_0; q_0, q_1, q_2 are the longitudinal wave vectors corresponding to the three closest axial modes,

where l is the length of the laser diode, m is an integer (the longitudinal mode index). It may be demonstrated that the approximation of planar waves, which is permitted here, results in a different solution from a more strict solution accounting for the actual spatial distribution with a factor in the order of one. Hence, such an approximation does not produce significant error, although it significantly simplifies calculation.

In order to establish the behavior of the field amplitudes over time, it is necessary to solve the wave equation:

$$\Delta\mathscr{E} - c^{-2}\partial^2 D/\partial t^2 = 0. \qquad (2)$$

We may represent the relationship between the electrical field $\mathscr{E}$ and induction $\mathscr{D}$ in the linear case (such as below the lasing threshold or at the lasing threshold) by means of the complex dielectric constant ε or (which is the same thing) the refractive index n and gain g:

$$n = \mathrm{Re}\,\sqrt{\varepsilon}, \quad g = -(\omega/cn)\ \mathrm{Im}\ \varepsilon, \qquad (3)$$

where c is the speed of light and ω is frequency.

Normally the following relation is used for the refractive index of the active medium of an injection laser:

$$n = n_0 + (\partial n/\partial N)\, N, \qquad (4a)$$

where n_0 is the refractive index of an unexcited semiconductor, N is the concentration of injected nonequilibrium electrons and the coefficient $\partial n/\partial N$ characterizes the dependence of the refractive index on N.

The following relation is used for the local value of the gain g_0:

$$g_0 = b(\omega)[N - N_0(\omega)],$$

where b is the proportionality factor, N_0 is the concentration of electrons at the inversion threshold. In principle it is possible to use experimental data to find the quantities b and N_0. However we should remember that since we are not using boundary conditions in our case, mirror losses α_m and nonresonant losses α may be incorporated into the final gain g:

$$g_0 = b\ (N - N_0) = \alpha_m - \alpha.$$

The value of N with excitation levels below the threshold or at threshold is determined only by pumping and is independent of the field. Hence the complex ε will also be independent of the field, and wave equation (2) will be linear. Consequently the amplitudes E_0, E_1, and E_2 will be independent. The situation changes significantly when the laser excitation level exceeds the threshold level. In this case N will no longer be determined only by pumping but also by the intensity of the laser field. As a result ε also becomes field-dependent, i.e., it ceases to be a constant, and equation (2) becomes significantly nonlinear. Two types of physical processes derive as a consequence of this nonlinearity. The first corresponds to the case where the solutions of equation (2) are found as a function of the coordinate but in the form of stationary amplitude over time. In this case, the solutions will describe the self-focusing phenomenon [10]. The second case, which is analyzed in this study, corresponds to spatially-homogeneous solutions that are time-dependent. We will examine this nonlinearity mechanism.

The concentrations of nonequilibrium electrons N in the active region obey the equation

$$\dot{N} + N/\tau_s + \beta\ (N - N_0)\ |\mathscr{E}_t|^2 - P = 0, \tag{5}$$

where $\beta = cnb/8\pi\hbar\omega$ is the stimulated recombination coefficient, τ_s is the lifetime of a nonequilibrium electron with respect to spontaneous band-to-band recombination; P is the pumping rate.

If we substitute field (1) into equation (5), we may rewrite this equation in the form of equations for a time-averaged value of N_{th} and a time- and coordinate-dependent addition δN:

$$N_{th}/\tau_s + (\beta/2)\ (N_{th} - N_0)\ |E_0|^2 - P = 0, \tag{6a}$$

$$\delta\dot{N} + \delta N\Gamma = -(\beta/2)\cos(\Delta qz)\ (N_{th} - N_0)\ [E_0^* E_1 + E_0 E_2^*]\ e^{i\Omega t} + \text{к. с.} \tag{6b}$$

The following designations are used in these equations: $\Delta q = q_0 - q_1$, $N = N_{th} + \delta N$, $\Gamma = 1/\tau_s + \beta|E_0|^2/2$ is the sum of probabilities for the spontaneous and stimulated band-to-band transitions.

In making the transition from equation (5) and (1) to system (6) we used the condition $|E_0| \gg |E_1|$, $|E_0| \gg |E_2|$; hence, terms of only the first order of smallness for δN, E_1, E_2 are preserved. In relations (6) N_{th} and δN are quantities averaged over a length in the order of the diffusion length. Hence the fact that no diffusion term is found in equation (5) is irrelevant for this calculation.

Equation (6b) is easily solved and for δN we have

$$\delta N = -\frac{\beta}{2}(N_{th} - N_0)\frac{E_0^* E_1 + E_0 E_2^*}{i\Omega + \Gamma} e^{i\Omega t}\cos\Delta q z + \text{к. с.} \tag{7}$$

Thus, we have derived an expression for the oscillating portion of the concentration of injected electrons. These oscillations have resulted from the beating between the strong field and the weak fields. Such oscillations also cause oscillations in the dielectric constant $\delta\varepsilon$ in accordance with the relations (4), (3)

$$\delta\varepsilon = (\partial\varepsilon/\partial N)\delta N, \tag{8}$$

where

$$\frac{\partial\varepsilon}{\partial N} = \frac{cnb}{\omega}(R + i), \qquad R = \frac{2n\partial n/\partial N}{cnb/\omega} = \frac{\operatorname{Re}(\partial\varepsilon/\partial N)}{\operatorname{Im}(\partial\varepsilon/\partial N)}.$$

Accounting for relations (8), (7) as well as (4) and (3), the expressions for the dielectric constant may be written in the following form:

$$\begin{aligned}
&\varepsilon(\omega_0, N) = \varepsilon_0 - i\frac{cn}{\omega} g(\omega_0, N_{th}) + \frac{\partial\varepsilon}{\partial N}\delta N, \\
&\varepsilon(\omega_0 - \Omega, N) = \varepsilon_0 - \left.\frac{\partial\varepsilon}{\partial\omega}\right|_{\omega_0}\cdot\Omega - i\frac{cn}{\omega} g(\omega_0 - \Omega, N_{th}) + \frac{\partial\varepsilon}{\partial N}\delta N, \\
&\varepsilon(\omega_0 + \Omega, N) = \varepsilon_0 + \left.\frac{\partial\varepsilon}{\partial\omega}\right|_{\omega_0}\cdot\Omega - i\frac{cn}{\omega} g(\omega_0 + \Omega, N_{th}) + \frac{\partial\varepsilon}{\partial N}\delta N, \\
&\varepsilon_0 = n^2 = n_0^2 + 2n_0\frac{\partial n}{\partial N}N_{th}, \qquad g(\omega_0, N_{th}) = b(\omega_0)(N_{th} - N_0) - \alpha_m - \alpha.
\end{aligned} \tag{9}$$

We will represent the quantities $g(\omega_0 - \Omega, N_{th})$ and $g(\omega_0 + \Omega, N_{th})$ in the form

$$\begin{aligned}
g(\omega_0 - \Omega, N_{th}) &= g(\omega_0, N_{th}) - \Delta g_1, \\
g(\omega_0 + \Omega, N_{th}) &= g(\omega_0, N_{th}) - \Delta g_2.
\end{aligned} \tag{10}$$

The quantities Δg_1 and Δg_2 are the initial (threshold) gain deficit in the neighboring modes.

Further, substituting expression for ε from (9) and for the field (1) into equation (2) and equating termwise the coefficients with identical frequencies, we derive the following expressions for the slow field amplitudes:

$$\dot{E}_0 - E_0\, cg\,(\omega_0, N_{th})/2n = 0, \tag{11a}$$

$$\dot{E}_1 - E_1 \frac{c}{2n}\left[g(\omega_0, N_{th}) - \Delta g_1\right] + i\frac{\omega_0\beta}{8\varepsilon_0}\frac{\partial\varepsilon}{\partial N}\frac{(N_{th} - N_0)(E_1 + E_2^*)E_0^2}{i\Omega + \Gamma} = 0, \tag{11b}$$

$$\dot{E}_2 - E_2 \frac{c}{2n}\left[g(\omega_0, N_{th}) - \Delta g_2\right] + i\frac{\omega_0\beta}{8\varepsilon_0}\frac{\partial\varepsilon}{\partial N}\frac{(N_{th} - N_0)(E_1^* + E_2)E_0^2}{-i\Omega + \Gamma} = 0. \tag{11c}$$

In deriving equations (11) we employed the fact that $q_0 = \omega_0 n/c$ and Ω are equal to the unperturbed mode interval, i.e., $\pi c/ln^*$, where $n^* = n + (\partial n/\partial\omega)\omega_0$. Moreover, we consider the amplitude E_0 to be real. This may always be done by selection of the beginning of the measurement period.

Since we assume that the strong field E_0 is stationary, i.e., $\dot{E}_0 \equiv 0$, then from equation (11a) we have

$$g(\omega_0, N_{th}) = b(\omega_0)(N_{th} - N_0) - \alpha_m - \alpha = 0. \tag{12}$$

This is the normal threshold condition indicating that for a field at ω_0, the gain will equal the losses. This relation makes no new contribution compared to the case of results from a regular intensity analysis utilizing rate equations. However, for fields at $\omega_0 \pm\Omega$, new terms proportional to E_0^2 without corresponding analogs in the intensity analysis appear in the equations. Carrying out complex conjugation of (11c) and using (11b) and (12), we may derive a linear system for E_1 and E_2^*, which is easily solvable, and for the corresponding increments we derive the values

$$\lambda_{1,2} = \frac{c}{2n}\left\{-\frac{\Delta g_1 + \Delta g_2}{2} - i\frac{aE_0^2}{i\Gamma - \Omega} \mp \left[\frac{\Delta g^2}{4} + \frac{\Delta g Ra}{\Gamma i - \Omega}E_0^2 - \frac{a^2E_0^4}{(i\Gamma - \Omega)^2}\right]^{1/2}\right\}, \tag{13}$$

where

$$\Delta g = \Delta g_1 - \Delta g_2, \qquad a = (\beta/4)\, b(\omega_0)\,[N_{th} - N_0(\omega_0)].$$

It is easy to see from equation (12) that when $E_0 \to 0$ we find that $\lambda_1 = -(c/2n)\Delta g_1$ and $\lambda_2 = -(c/2n)\Delta g_2$, i.e., the fields E_1 and E_2 decrease independently with decrements proportional to the threshold gain deficits for each mode. This is fully consistent with the normal result obtained from an intensity analysis. If $E_0 \neq 0$, the result is significantly different from that given by an analysis of rate equations.

We note that due to the fact that the fields E_1 and E_2 are coupled, the increments λ_1 and λ_2 are no longer increments for the separate fields E_1 and E_2, but rather correspond to a combination of these fields with different phase and amplitude relations between them.

We will examine two particular cases. The first is when the initial gain deficits for both modes are significantly different compared to the saturation factor $|aE_0^2/(i\Gamma - \Omega)|$, specifically:

$$| \Delta g | = | \Delta g_1 - \Delta g_2 | \gg | aE_0^2/(i\Gamma - \Omega) |. \tag{14}$$

In this case, relation (13) may be simplified to the form:

$$\lambda_1 \simeq \frac{c}{2n}\left[-\Delta g_1 + \frac{a(R\Omega - \Gamma)}{\Omega^2 + \Gamma^2} E_0^2 + i\frac{a(\Omega + R\Gamma)}{\Omega^2 + \Gamma^2} E_0^2\right], \tag{15a}$$

$$\lambda_2 \simeq \frac{c}{2n}\left[-\Delta g_2 - \frac{a(R\Omega + \Gamma)}{\Omega^2 + \Gamma^2} E_0^2 + i\frac{a(\Omega - R\Gamma)}{\Omega^2 + \Gamma^2} E_0^2\right]. \tag{15b}$$

It is clear from relations (15) that as the intensity of the laser field (E_0^2) increases, the decrements of the weak fields begin to change from their initial values. This change may be characterized by a certain additional gain δg (absorption) generated by the strong field. For one of the combinations of fields E_1, E_2 corresponding to excitation in one of the longwave components E_1, for the quantity δg we have

$$\delta g_1 \simeq \frac{\beta E_0^2}{4} b(N_{th} - N_0)\frac{R\Omega - \Gamma}{\Omega^2 + \Gamma^2}, \tag{16a}$$

while the corresponding correction to the intermodal interval will be

$$\delta\Omega_1 \simeq \frac{\beta E_0^2}{8}\frac{cb}{n}(N_{th} - N_0)\frac{R\Gamma + \Omega}{\Omega^2 + \Gamma^2}. \tag{16b}$$

For the relations between E_1 and E_2, which are similar to the excitation of the short-wave component E_2, we will have the following, respectively:

$$\delta g_2 = -\frac{\beta E_0^2}{4} b(N_{th} - N_0)\frac{R\Omega + \Gamma}{\Omega^2 + \Gamma^2} \tag{17a}$$

and

$$\delta\Omega_2 = \frac{\beta E_0^2}{8}\frac{cb}{n}(N_{th} - N_0)\frac{\Omega - R\Gamma}{\Omega^2 + \Gamma^2}. \tag{17b}$$

Further we will examine the second case when the laser field intensity is sufficiently great that

$$| aE_0^2/(i\Gamma - \Omega) | \gg | \Delta g_1 - \Delta g_2 |. \tag{18}$$

In this case we have from relation (13) for the roots λ_1 and λ_2

$$\lambda_1 = \frac{c}{2n}\left[-\frac{\Delta g_1 + \Delta g_2}{2} - i(\Delta g_1 - \Delta g_2)R\right], \tag{19a}$$

$$\lambda_2 = \frac{c}{2n}\left[-\frac{\Delta g_1 + \Delta g_2}{2} - \frac{2a\Gamma}{\Omega^2 + \Gamma^2} E_0^2 + i\frac{2a\Omega}{\Omega^2 + \Gamma^2} E_0^2\right]. \tag{19b}$$

The root λ_1 corresponds to the case where the phase shift between the amplitudes E_1 and E_2 is near π (phase modulation of the lasing frequency), and in this case the decrement of the weak fields is independent of the intensity of the strong field and is determined by the arithmetic mean of the threshold values of the decrements of each field. The effective absorption $\delta\alpha$ for the E_1 and E_2 fields will be

$$\delta\alpha_1 = (\Delta g_1 + \Delta g_2)/2. \tag{20a}$$

For the correction to the intermodal frequency $\delta\Omega_1$ we have

$$\delta\Omega_1 = -(c/2n)(\Delta g_1 - \Delta g_2)R. \tag{20b}$$

The root λ_2 corresponds to the case where the phase shift between E_1 and E_2 is near 0 (amplitude modulation of lasing). In this case, the action of the strong field is reduced to suppression of the weak fields by inducement of additional absorption $\delta\alpha$,

$$\delta\alpha_2 = (\beta E_0^2/2)\, b\, (N_{th} - N_0)\, \Gamma/(\Omega^2 + \Gamma^2). \tag{21a}$$

The change in the mode distance $\delta\alpha_2$ will be

$$\delta\Omega_2 = \frac{\beta E_0^2}{4}\frac{cb}{n}\frac{(N_{th} - N_0)\,\Omega}{\Omega^2 + \Gamma^2}. \tag{21b}$$

We note that relations (16), (17), (20) and (21) for δg and $\delta\alpha$ may also be rewritten in another form if we account for the fact that consistent with (6a) the following equality holds

$$\frac{\beta E_0^2}{2}(N_{th} - N_0) = P - P_{th} = \eta\frac{N_{th}}{\tau_s}, \qquad \eta = \frac{P - P_{th}}{P_{th}}, \tag{22}$$

where $P_{th} = N_{th}/\tau_s$ is threshold pumping rate, and η is the relative level of pumping over the threshold.

We will perform certain estimates. First, we will establish the criterion of strong or weak saturation (relations (14) and (18)). For the saturation factor consistent with (12) we have

$$\left|\frac{aE_0^2}{i\Gamma - \Omega}\right| = \frac{\beta E_0^2}{4}\frac{(N_{th} - N_0)\, b}{\sqrt{\Gamma^2 + \Omega^2}} = \frac{\eta N_{th} b}{2\tau_s\sqrt{\Gamma^2 + \Omega^2}}. \tag{23}$$

The product $N_{th}b$ may be estimated as

$$N_{th}b \simeq g. \tag{24}$$

Hence, for the saturation factor $|aE_0^2/(i\Gamma - \Omega)|$ we have the estimate

$$\left|\frac{aE_0^2}{i\Gamma - \Omega}\right| \simeq \frac{\eta g_0}{2\tau_s\sqrt{\Gamma^2 + \Omega^2}}. \tag{25}$$

In order to estimate the gain deficits Δg_1 and Δg_2 we will assume that the spectral gain profile $g_0(\omega)$ has a parabolic peak, i.e.,

$$g_0(\omega) = g_0(\omega_{\max}) - 2g_0(\omega_{\max})\,[(\omega - \omega_{\max})/\sigma]^2, \tag{26}$$

where σ is the width of the spectral gain profile at half amplitude and ω_{max} is the frequency corresponding to the spectral gain maximum. Then for Δg_1 and Δg_2, according to (10) and (26), we will have

$$\Delta g_1 = \frac{2g_0(\omega_{max})}{\sigma^2}(\Omega^2 + 2\Delta\omega\Omega),$$
$$\Delta g_2 = \frac{2g_0(\omega_{max})}{\sigma^2}(\Omega^2 - 2\Delta\Omega), \qquad \Delta\omega = \omega_{max} - \omega_0. \tag{27}$$

Here $\Delta\omega$ designates cavity mode detuning with respect to ω_{max}. Then for Δg we will have

$$\Delta\lambda = \Delta g_1 - \Delta g_2 = 8g_0(\omega_{max})\Delta\omega\Omega/\sigma^2. \tag{28}$$

Accounting for this we find that the boundary pumping level over lasing threshold η_0 which divides the weak and strong saturation regions will be

$$\eta_0 \simeq 8\frac{\Omega\Delta\omega}{\sigma^2}\tau_s\sqrt{\Gamma^2+\Omega^2} \simeq 8\frac{\Omega\Delta\omega}{\sigma^2}\Omega\tau_s. \tag{29}$$

For an AlGaAs-laser with a diode length l = 300 μm, n^* = 4.5 and an amplification width of σ corresponding to 30 nm, $\tau_s = 4\cdot10^{-9}$ seconds and with detuning of $\Delta\omega = \Omega/2$, we find that

$$\eta_0 = 2.$$

From here we see that in order for the strong saturation conditions to be satisfied for any detuning $\Delta\omega$ (we assume $|\Delta\omega| \leqslant \Omega/2$), it is necessary for the pumping level over threshold to be significantly greater than 2, i.e.,

$$\eta \gg \eta_0 \simeq 2.$$

For standard injection lasers this is a difficult level to achieve. Hence the strong saturation conditions will be satisfied only for very minor detuning $\Delta\omega$. For example, if $|\Delta\omega| \leqslant \Omega/100$, the strong saturation condition will be satisfied by exceeding the lasing threshold by only 10%.

We will now estimate the additional gain δg and frequency shift $\delta\Omega$ for the case corresponding to relations (16). Using relations (22) and (24), we have

$$\delta g = \frac{\beta E_0^2}{4} b(N_{th} - N_0)\frac{R\Omega - \Gamma}{\Omega^2 + \Gamma^2} \simeq \frac{\eta g_0(\omega_{max})R}{2\tau_s\Omega}. \tag{30}$$

For η = 0.5 and the laser parameters outlined above and also assuming that $N_{th}b \simeq g_0$ = 50 cm^{-1} and R = 3, we find that $\delta g \simeq 3\cdot10^{-4}\ g_0 \simeq 1.5\cdot10^{-2}$ cm^{-1}. It is clear from this estimate that δg is comparable to the initial gain deficit Δg. For a relative frequency shift $\delta\Omega/\Omega$, analogously we derive the estimate

Thus, we may make the following conclusion from the analysis.

When the lasing threshold is achieved in one of the modes (initially the mode with the higher Q) and pumping is increased further (together with the intensity of this mode) the spectral gain profile near this mode changes. The nature of this change is such that if the Q of the modes near the laser frequency differ, the change will produce additional gain for the longwave mode (if $\Omega > \Gamma/R$), and will produce additional absorption for the shortwave mode in accordance with relations (16a) (17a). We note that within the interval $\Omega < \Gamma/R$ the strong field produces only absorption for both modes. At the same time the strong field will produce a certain change in the mode interval in accordance with the relations (16b), (17b).

The amplification profile will be deformed until the lasing frequency switches to the longwave mode. Then the intensity of the laser wave will be quite high (inequality (18) is satisfied), and conditions in which all other fields will be suppressed by the laser field consistent with (20a) and (21a) will be observed. Indeed $\delta\alpha$, in accordance with (20a) and (21a), will always be greater than zero. As a result, with a significant laser field intensity, stable single-frequency lasing will be observed. Moreover, switching of the excited mode may occur with a certain hysteresis, because this mode creates a gain valley around itself.

2.2. Field interaction in an injection laser with an external dispersive cavity

A diagram of the laser with an external dispersive cavity is given in Fig. 1. There are two aspects to the characteristic feature of this cavity. First, the spectral distance between the longitudinal modes of such a cavity is two orders of magnitude less than in a regular laser diode. Hence the parameter $1/\Omega\tau$ will be in the order of one.

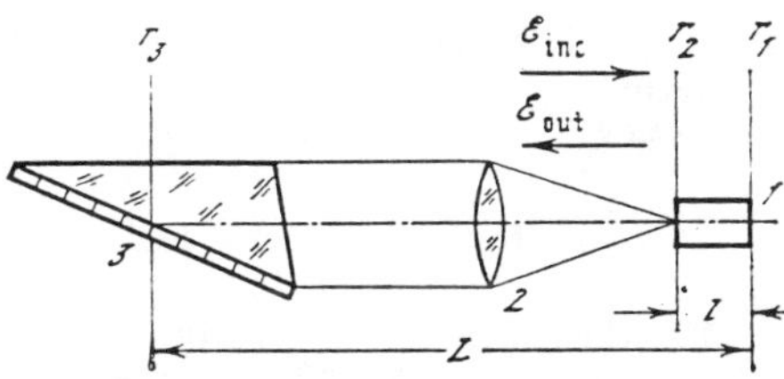

Fig. 1. Optical scheme of an injection laser with an external dispersive cavity.
1 - laser diode;
2 - collimating lens;
3 - dispersing element (in this case a total internal reflection selector)

From here the intensity of mode interaction in the external cavity will be significantly greater than for the case of modes in the natural cavity of the diode. Second, the spatial configurations of the fields of these closest modes in the cavity of the diode may be considered to be coincident. Hence, the natural cavity of the diode in this configuration plays the role of a Fabry-Perot etalon, where the etalon has nonlinear properties. Inversion oscillations arising from field beating cause a periodic change in the properties of this Fabry-Perot etalon (in the transmission function). As a result, mutual energy exchange occurs in the waves passing through this dynamic etalon and thus their amplitudes become coupled.

We will examine the field Re $\mathscr{E}(z, t)$, representing the sum of three traveling waves

$$\mathscr{E}_{\text{int}} = E_0(t)\, e^{ik_0 z - i\omega_0 t} + E_1(t)\, e^{ik_1 z - i(\omega_0 - \Omega)t} + E_2(t)\, e^{ik_2 z - i(\omega_0 + \Omega)t}, \tag{31}$$

incident on a laser diode on the side of the dispersing element. We will also assume that $|E_0| \gg |E_1|$, $|E_2|$, while k_0, k_1, k_2 are the wave vectors in the vacuum. After entering the laser diode, the waves experience multiple reflection off its faces as well as amplification. As may be easily shown, the exit field $\mathscr{E}_{out}$ will take the following form:

$$\begin{aligned}\mathscr{E}_{\text{out}} = \mathscr{K}(\omega_0, N)\, E_0 e^{-ik_0 z - i\omega_0 t} + \mathscr{K}(\omega_0 - \Omega, N)\, E_1 e^{-ik_1 z - i(\omega_0 - \Omega)t} + \\ + \mathscr{K}(\omega_0 + \Omega, N)\, E_2 e^{-ik_2 z - i(\omega_0 + \Omega)t},\end{aligned} \tag{32}$$

where

$$\begin{aligned}&\mathscr{K}(\omega, N) = \sqrt{r^2}\left(\frac{\xi j e^{i2ql}}{1 - je^{i2qj}} - 1\right), \\ &\xi = T\,(1 - r_2)/r_2, \quad j = \sqrt{r_1 r_2}\, e^{[g_0(\omega,\, N) - \alpha]l}, \quad q = (\omega/c)\, n\,(\omega, N),\end{aligned} \tag{33}$$

T is the emission coupling coefficient from the external portion of the cavity within the active region of the laser (in intensity), r_1, r_2 are the reflection coefficients of the external and internal faces of the diode, respectively, j is the underexcitation parameter of the laser diode, which is always less than 1, l is the length of the diode and q is the wave vector in the active region. The function $\mathscr{K}(\omega)$ is a periodic function of frequency with a period corresponding to the load distance of the diode cavity. We note further that the function $\mathscr{K}(\omega, N(t))$ also oscillates over time:

$$\mathscr{K}(\omega, N(t)) = \mathscr{K}(\omega, N_{th}) + \left.\frac{\partial \mathscr{K}}{\partial N}\right|_{N=N_{th}} \delta N(t). \tag{34}$$

This results from the fact that the beating of the fields at three frequencies causes oscillations in the inversion time N with an amplitude $\delta N(t)$.

In order to establish the value of $\delta N(t)$ we will use the fact that the spatial configurations of the mode fields of the external cavity are nearly coincident within the diode cavity. Hence there will be no spatial variable in δN in this case. Carrying out calculations analogous to those we performed earlier for equations (5), (6a) and (6b), we will have

$$N_{th}/\tau_s + \beta\,(N_{th} - N_0)\,|\,E_0^{\text{int}}\,|^2 - P = 0, \tag{35a}$$

$$\delta N = -\beta\,(N_{th} - N_0)\left(\frac{E_0^{\text{int}} E_1^{\text{int}} + E_0^{\text{int}} E_2^{*\,\text{int}}}{i\Omega + \Gamma}\, e^{i\Omega t} + \text{к. с.}\right), \tag{35b}$$

where E_0^{int}, E_1^{int}, E_2^{int} are the amplitudes of fields of frequencies ω, $\omega_2 - \Omega$ and $\omega_0 + \Omega$ within the laser diode, while the value of Γ is

$$\Gamma = \tau_s^{-1} + \beta \overline{|E_0^{int}|^2}, \tag{36}$$

The line designates averaging over the length of the laser diode. Below we will consider the case where the frequency ω_0 is near one of the maxima of the function $\mathcal{K}(\omega, N)$. In other words, we are interested in the case where the dispersing element of the external cavity is tuned to a frequency near one of the longitudinal resonances (modes) of the diode ω_M.

In this case it is easy to show that the following relations hold:

$$E_0^{int}(E_1^{int} + E_2^{*int}) \simeq FE_0(E_1 + E_2^*), \tag{37}$$

$$\overline{|E_0^{int}|^2} \simeq FE_0^2, \tag{38}$$

where

$$F = T(1 - r_2)\,\overline{(e^{gz} + (j^2/r_2)\,e^{-gz})}/(1 - j)^2\, n.$$

With standard coupling coefficient values of $T = 0.25$ and for the case of an unbleached diode $r_1 = r_2 = 0.3$ for a factor $j = 0.9$ we may derive the estimate $F = 8$. The coefficient of F characterizes the relation between fields in the external portion of the cavity and within the diode. We note further that the quantity δN is of the first order of smallness for E_1 and E_2; hence, expression (32) subject to (34) and accurate to the first order of smallness for E_1 and E_2, may be rewritten in the following form:

$$\mathscr{E}_{out} = \mathcal{K}(\omega_0, N_{th})\, E_0 e^{-ik_0z - i\omega_0 t} + \left.\frac{\partial \mathcal{K}}{\partial N}\right|_{\omega_0, N_{th}} \delta N(t)\, E_0 e^{-ik_0z - i\omega_0 t} + \\ + \mathcal{K}(\omega_0 - \Omega, N_{th})\, E_1 e^{-ik_1z - i(\omega_0 - \Omega)t} + \mathcal{K}(\omega_0 + \Omega, N_{th})\, E_2 e^{-ik_2z - i(\omega_0 + \Omega)t}. \tag{39}$$

The second term in the right half of equality (39) corresponds to field interaction. This term is an addition of the strong field at ω_0 to the amplitude of the weak fields E_1 and E_2. The mode excitation conditions may be found from the boundary conditions for the field component at ω_0, $\omega_0 - \Omega$ and $\omega_0 + \Omega$. Subject to relations (31), (35), these conditions have the following form (39):

- for the frequency ω_0

$$\mathcal{K}(\omega_0, N_{th})\, e^{i2k_0L}\sqrt{r_3} = 1; \tag{40a}$$

- for the frequency $\omega_0 - \Omega$

$$\left[\mathcal{K}(\omega_0 - \Omega, N_{th})\, E_1 - \frac{\partial \mathcal{K}}{\partial N}\beta\frac{(N_{th} - N_0)\, FE_0^2}{i\Omega + \Gamma}(E_1 + E_2^*)\right] e^{i2k_1L}\sqrt{r_3} = E_1; \tag{40b}$$

- for the frequency $\omega_0 + \Omega$

$$\left[\mathcal{K}(\omega_0+\Omega, N_{th}) E_2 - \frac{\partial \mathcal{K}}{\partial N} \beta \frac{(N_{th}-N_0) F E_0^2}{-i\Omega+\Gamma} (E_1^* + E_2)\right] e^{i2k_2L} \sqrt{r_3} = E_2, \tag{40c}$$

where L is the length of the external portion of the cavity, r_3 is the efficiency of the dispersing element (the reflection coefficient of the external mirror).

Expanding expression (33) for the function $\mathcal{K}$ (ω, N_{th}) into a series in the vicinity of its spectral maximum ω_M in powers of $\omega - \omega_M$ and preserving all terms up through the second order of smallness inclusively, we may rewrite the threshold condition (48) in the form

$$\left\{\frac{\xi j}{1-j}\left[1+\frac{i\varphi_0-(\varphi_0^2/2)(j+1)/(j-1)}{1-j}\right]-1\right\}\sqrt{r_2r_3}\, e^{i2k_0L}=1. \tag{41a}$$

Here it was assumed that $2q_0$ (ω_0, N_{th}) $l = 2\pi m + \varphi_0$, where m is an integer and $\varphi_0 \simeq 2(\omega_0 - \omega_M) ln^*/c$.

Analogously we may transform equations (40b) and (40c) so that, subject to (40a), they take the form

$$E_1 e^{-i2\Omega L/c}[1-\rho u(u+i)-\psi]-\frac{\theta E_0^2(1-iR)}{i\Omega+\Gamma}(E_1+E_2^*)e^{-i2\Omega L/c}=E_1, \tag{41b}$$

$$E_2^* e^{-i2\Omega L/c}[1-\rho u(u+i)+\varphi]-\frac{\theta E_0^2(1+iR)}{i\Omega+\Gamma}(E_1+E_2^*)e^{-i2\Omega L/c}=E_2^*, \tag{41c}$$

where the following designations are used:

$$\rho=\frac{2(1+\sqrt{r_2r_3})}{j-\sqrt{r_2r_3}}, \qquad u=\left(j-\sqrt{r_2r_3}\right)\frac{ln^*\pi}{L},$$

$$\varphi = (1+\sqrt{r_2r_3})(j-\sqrt{r_2r_3})\,\varphi_0, \qquad \theta=(1+\sqrt{r_2r_3})\, blF\beta\,(N_{th}-N_0).$$

We derive the following equation for frequencies Ω based on the compatibility condition of system of equations (41b) and (41c):

$$e^{i2L\Omega_{1,2}/c}=1-\rho u(u+i)-\frac{\theta E_0^2}{i\Omega+\Gamma}\pm\left[\varphi^2-2i\frac{\theta E_0^2}{i\Omega+\Gamma}R\varphi+\frac{\theta^2E_0^4}{(i\Omega+\Gamma)^2}\right]^{1/2}. \tag{42}$$

The two roots of system (41) as in the case of root system (11), designate two types of solutions which have different amplitude and phase relations for the fields E_1 and E_2.

We will consider the case where the intensity of the laser field E_0^2 is sufficiently low so that the saturation parameter is less than the difference in the losses of the fields E_1 and E_2, i.e.,

$$|\varphi| \gg |\theta E_0^2/(i\Omega+\Gamma)|. \tag{43}$$

In this case, the radicand in the right half of (42) may be expanded into a series in terms of the parameter $\theta E_0^2/\varphi$. As a result we have

$$e^{i2L\Omega_1/c}=1-\rho u(u+i)-\varphi-\theta E_0^2\left(\frac{\Gamma-\Omega R}{\Gamma^2+\Omega^2}-i\frac{R\Gamma+\Omega}{\Gamma^2+\Omega^2}\right), \tag{44a}$$

$$e^{i2L\Omega_2/c}=1-\rho u(u+i)+\varphi-\theta E_0^2\left(\frac{\Gamma+R\Omega}{\Gamma^2+\Omega^2}+i\frac{R\Gamma-\Omega}{\Gamma^2+\Omega^2}\right). \tag{44b}$$

Consequently for the case of strong saturation, i.e., when the inequality

$$|\sigma E_0^2/(i\Omega+\Gamma)| \gg |\varphi|, \tag{45}$$

is satisfied, for the eigenfrequencies Ω_1 and Ω_2 we will have

$$e^{i2L\Omega_1/c} = 1 - \rho u (u + i) - iR\varphi, \tag{46a}$$

$$e^{i2L\Omega_2/c} = 1 - \rho u (u + i) - 2\theta E_0^2\left(\frac{\Gamma}{\Gamma^2+\Omega^2} - i\frac{\Omega}{\Gamma^2+\Omega^2}\right). \tag{46b}$$

In order to find the decrements of the fields, we will represent Ω_1 and Ω_2 in the form

$$\Omega_{1,2} = \pi c/\omega + \delta\Omega' + i\delta\Omega''. \tag{47}$$

Here $\delta\Omega'$ is the correction to the nonperturbed mode interval of the external cavity $\Omega = \pi c/L$ related to the action of the external cavity of the diode. The decrement of the fields is characterized by $\delta\Omega''$.

It is easy to demonstrate from relations (44) that for $\delta\Omega'$ and $\delta\Omega''$ the following relations are valid:

$$\delta\Omega'' = \frac{c}{2L}\left(\rho u^2 + \varphi + \theta E_0^2\frac{\Gamma-\Omega R}{\Gamma^2+\Omega^2}\right), \qquad \delta\Omega' = \frac{c}{2L}\left(-\rho u + \theta E_0^2\frac{R\Gamma+\Omega}{\Gamma^2+\Omega^2}\right)$$

- (first root, the E_1 field predominates); (48)

$$\delta\Omega'' = \frac{c}{2L}\left(\rho u^2 - \varphi + \theta E_0^2\frac{\Gamma+R\Omega}{\Gamma^2+\Omega^2}\right), \qquad \delta\Omega' = \frac{c}{2L}\left(-\rho u - \theta E_0^2\frac{R\Gamma-\Omega}{\Gamma^2+\Omega^2}\right)$$

- (second root, the E_2 field predominates). (49)

Qualitatively the image is the same as observed for the case of natural mode interaction in a laser diode (relations (16), (17)). The laser field induces a positive addition to the gain on the longwave side and induces absorption on the shortwave side.

Consequently, the values for $\delta\Omega'$ and $\delta\Omega''$ for the case of strong saturation may be found from relations (46):

$$\delta\Omega'' = \frac{c}{2L}\rho u^2, \qquad \delta\Omega' = \frac{c}{2L}(-\rho u - R\varphi)$$

- (first root, frequency modulation); (50)

$$\delta\Omega'' = \frac{c}{2L}\left(\rho u^2 + 2\theta E_0^2\frac{\Gamma}{\Omega^2+\Gamma^2}\right), \qquad \delta\Omega' = \frac{c}{2L}\left(-\rho u + 2\theta E_0^2\frac{\Omega}{\Gamma^2+\Omega^2} + R\varphi\right)$$

- (second root, amplitude modulation). (51)

This case is also analogous to that for the modes of a laser diode with a high laser field intensity. The laser field, independent of its frequency position with respect to the threshold effective gain maximum (maximum of the function $\mathcal{K}(\omega)$), deformed the effective gain profile so that for the phase relations corresponding to frequency modulation, it had a maximum at the laser field frequency, while all other frequencies are effectively suppressed. For amplitude modulation it became symmetric with respect to the laser frequency, although still possessing additional absorption with the factor $\Gamma/\Omega^2 + \Gamma^2)$.

We also note that by using balance equation (35a) we may derive the relation

$$\beta\,(N_{th} - N_0)\,|\,\overline{E_0^{\text{int}}}\,|^2 = \beta F E_0^2\,(N_{th} - N_0) = \eta N_{th}/\tau_s, \tag{52}$$

where η is the relative level of pumping over the threshold. This relation is convenient for calculating δg and $\delta\Omega$ by formulae (48)-(51).

We will now estimate the strong or weak saturation conditions in a laser with an external cavity (inequalities (43), (45)). Consistent with (52) $|\theta E_0^2/ i\Omega + \Gamma)|$ may be rewritten as

$$\left|\frac{\theta E_0^2}{i\Omega + \Gamma}\right| = \frac{\beta b F E_0^2\,(N_{th} - N_0)\,l\,(1 + \sqrt{r_2 r_3})}{\sqrt{\Omega^2 + \Gamma^2}} \simeq \frac{\eta N_{th} b l\,(1 + \sqrt{r_2 r_3})}{\tau_s \sqrt{\Omega^2 + \Gamma^2}}. \tag{53}$$

Equating the expressions for the saturation factor (53) to the difference in losses φ for the boundary excess of pumping over the threshold η_0, which separates weak saturation and strong saturation, we have

$$\eta_0 = \frac{4\,|\,\omega_M - \omega_0\,|\,\Omega\,(n^* l)^2\,\tau_s \sqrt{\Omega^2 + \Gamma^2}\,(j - \sqrt{r_2 r_3})}{c^2 N_{th} b l}. \tag{54}$$

For detuning $\omega_M - \omega_0$ of three mode intervals ($|\omega_M - \omega_0| \approx 3\Omega$) and for an external cavity length of $L = 30$ cm and for the laser parameters outlined above, we derive, consistent with (55):

$$\eta_0 \simeq 3 \cdot 10^{-3}.$$

We see here that even with a 1% level of pumping current over the threshold, strong saturation occurs. Hence in a laser with an external cavity relations (50) and (51) are nearly always valid.

Henceforth we will estimate the effective absorption $\delta\alpha$ induced by the strong field. From relation (51) we find that for $\eta \simeq 0.5$ $\delta\alpha$:

$$\delta\alpha = (1n/l)\theta E_0^2 \Gamma/(\Omega^2 + \Gamma^2) \simeq 2\ \text{cm}^{-1}.$$

The laser parameters are assumed to be the same as indicated above.

Thus, by virtue of the fact that the parameter $(\Omega\tau)^{-1}$ for a laser with an external cavity is more than two orders of magnitude greater than the case of a diode cavity, the magnitude of this effect is also two orders of magnitude greater. In this case, field interaction through inversion causes additional absorption only for weak fields from the strong field, at the same time that for the case of a diode cavity, the strong field may induce both absorption for the shortwave mode and amplification for the longwave mode.

3. EXPERIMENT

The experimental set-up is shown in Fig. 2. Mode interaction was investigated in an injection laser with an external dispersive cavity. The external cavity had a length of L = 30 cm and a lens with numerical aperture of 0.8 and a focal distance of 6.3 mm was used to establish optical coupling between the active region of the laser and the external portion of the cavity. A holographic total internal reflection selector [11] was used as the dispersing element; this selector has a density of 3100 lines/mm. An estimate of the spectral selectivity of this cavity produces a value on the order of approximately 10 GHz. CW AlGaAs-diodes at room temperature are used as the active elements of the laser. The laser is powered simultaneously by d.c. and a.c. at a frequency near the mode beating frequency (approximately 540 MHz). In order to eliminate feedback from scattered emission from the recording equipment, a Fresnel rhomb was mounted immediately behind the outcoupling lens of the laser.

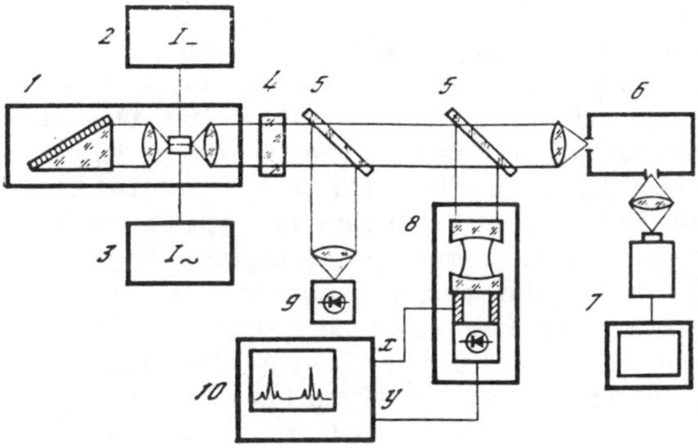

Fig. 2. Experimental set-up.
1 - laser with external cavity; 2 - d.c. power supply, 3 - a.c. power supply, 4 - optical isolation (Fresnel rhomb), 5 - semitransparent mirrors, 6- MDR-4 monochromator; 7- MTU-1 television; 8- scanning confocal interferometer with a free dispersion range of 3.7 GHz; 9 - optical power meter; 10- oscillograph

This study investigated the spectral composition of laser emission at various excitation levels. A scanning confocal interferometer was used for this purpose; the output signal from the interferometer was injected to the vertical deflection plates of the oscillograph. A grating monochromator in conjunction with a television monitor was used to observe the laser emission spectrum.

Numerous experiments to investigate the spectra of emission from a laser with an external cavity in which highly-selective dispersing elements are used such as a holographic total internal reflection selector or a reflection diffraction grating in conjunction with a Fabry-Perot etalon, have demonstrated that single-frequency lasing is maintained over a wide range of excitation levels of the laser [12]. It was also concluded on the basis of observations of such single-frequency lasing that the stability of single-frequency lasing is highly dependent on the power level. Near the lasing threshold (excess over threshold of less than 1%), when the output emission power is low, near single-frequency lasing may still be observed although, as a rule, it is unstable. Due to mechanical perturbations of the cavity or due to instability of the power supply, laser frequency jumping from one external mode of the external cavity to another is observed. Moreover, several longitudinal modes may exist simultaneously.

As a result, an averaged spectral pattern may consist of several modes (3-5) with a certain unstable intensity distribution among such modes over time. However this pattern changes significantly as the laser power is increased. Here, a dominant mode emerges from the overall spectrum at the same time that other modes begin to be suppressed. As a result, at a sufficiently high output power (in the order of a few milliwatts) stable single-frequency lasing is observed. The lasing frequency jumps from mode to mode cease, or become much rarer. In other words, a certain "autostabilization" of single-frequency lasing occurs. Such behavior is easily reproduced for an optimally-tuned cavity or for laser prototypes with a near-linear watt-current characteristic. Normally in this case bistable lasing with spectral tuning of the cavity is also observed [12].

A detailed pattern of side mode suppression with an increase in laser intensity at the central mode, of course, depends on the initial laser spectrum at the lasing threshold and, evidently, on the nature of mechanical and electrical instabilities in the laser. Hence, we introduced a special controlled laser perturbation by modulating the pumping current to create the initial multifrequency spectrum at the lasing threshold in order to conduct a comprehensive investigation of the side mode suppression effect. The pumping current was modulated at the mode beating frequency, and it was easy to excite several longitudinal modes in the external cavity at the lasing threshold. The number of such modes depended on the modulation amplitude and normally we limited the modulation amplitude to the level at which two symmetric additional modes could be excited in addition to the central mode.

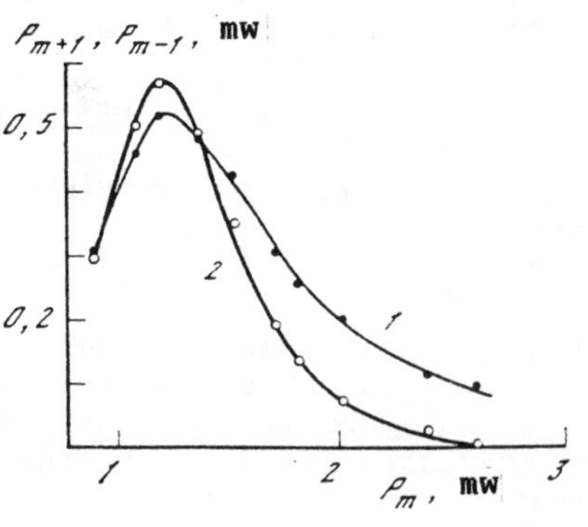

Fig. 4. The power of the shortwave mode P_{m+1} (1) and the longwave mode P_{m-1} (2) plotted as a function of central mode power P_m.

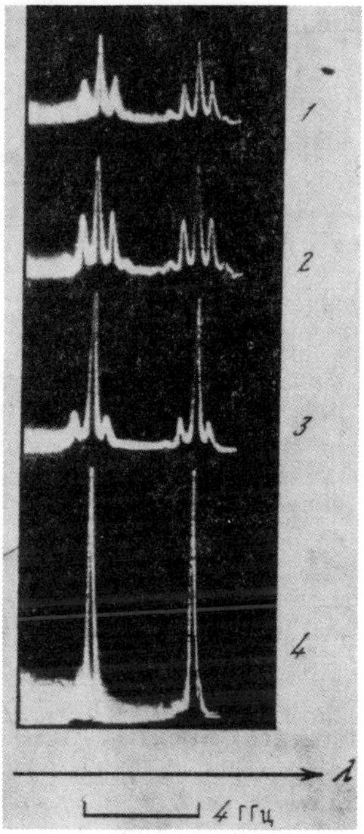

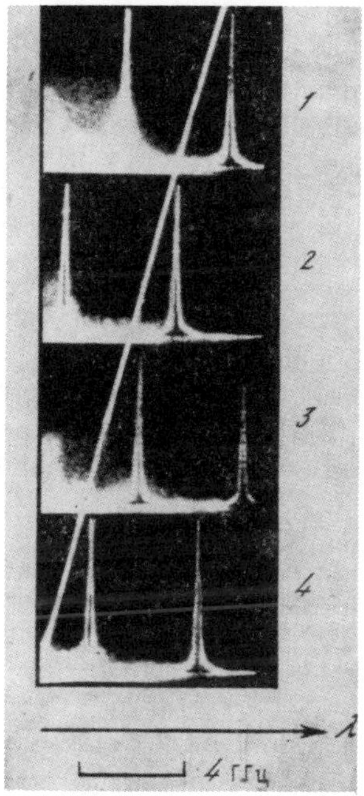

Fig. 3. The laser spectrum for various output power levels P_m. P_m = 1.1 mw (1); 1.4 (2); 1.6 (3); 2.6 (4). The a.c. pumping current amplitude was fixed at ~4 ma (two transmission levels of the interferometer are shown).

Fig. 5. Single-frequency lasing spectrum for the laser with alteration of resonator length L by ΔL by means of piezoceramics. ΔL = 0 (1); 1.2 μm (2); 2.1 (3); 3.2 (4). Pumping current: 120 ma, threshold current: 104 ma, output power ~2.4 mw.

An example of such spectra is given in Fig. 3. This diagram reveals the variation in the lasing spectrum with an increase in power at the central mode. Overall this pattern is similar to when modulation is absent, and the initial spectrum is formed from random perturbations.

Fig. 4 gives results from the processing of such spectra. We see that in the initial section (near the lasing threshold) the power at the side modes P_{m-1} and P_{m+1} increases and then becomes saturated, reaching a maximum and then falls to near zero. The initial increase in P_{m-1} and P_{m+1}, evidently, is related to the fact that the introduction of modulation to the pumping current produces a certain "blurring" of the threshold on the watt-current characteristic by a value on the order of twice the modulation amplitude. Hence, until the pumping current exceeds this interval, the attainment of the threshold in the modes will be continuous. The magnitude of the pumping current interval corresponding to the rising section for P_{m-1} and P_{m+1} and P_{m+1} is near twice the modulation amplitude, which supports this explanation. With a further increase in pumping and the intensity of the central mode P_m, a reduction in P_{m-1} and P_{m+1} is observed, which is characteristic for autostabilization of single-frequency lasing.

Tuning of the lasing frequency by rotating a holographic selector has produced jumps in the lasing frequency among the modes of the external cavity with a range of 1.5 Å in the vicinity of each mode of the diode. Such a tuning arrangement has been described in detail previously in study [12]. Experiments were carried out to investigate continuous lasing tuning by varying the length of the external cavity by means of piezoceramics as a supplement to such experiments. It was discovered that it is possible to achieve continuous tuning of lasing within a few mode intervals (2-3) in such a laser. Moreover, here lasing frequency jumps may occur from one mode to another at a distance of a few mode intervals. Hysteresis of the dependence of the lasing frequency on cavity length is evident here. As a result the use of piezoceramics to produce a single continuous change in cavity length may be used to tune the lasing frequency within 3 GHz. This is illustrated in Fig. 5. This behavior indicates that the laser field creates a certain spectral gain valley around itself that hinders the excitation of other modes that would be excited in the absence of this valley as the lasing frequency is tuned away from the gain maximum.

Thus, the following conclusions may be drawn on the basis of these experiments.

1. When the laser emission power is increased, the stability of single-frequency lasing is improved and weak "noise" fields are suppressed in the spectral vicinity of this laser field.

2. Tuning of lasing by employing piezoceramics to alter cavity length may occur over anomalously large distances of up to several gigahertz, which significantly expands the spectral range between two neighboring modes.

4. DISCUSSION

The problem of obtaining high-intensity single-frequency emission remains an important problem for injection lasers. The autostabilization effect of single-frequency emission described in this study makes it possible to optimize the tuning conditions of the laser cavity and, in the final analysis, simplifies the solution to this problem. When using active elements to provide CW emission at 10-15 mw, we were able to reliably obtain single-frequency emission over the entire range of emission power using an external dispersive cavity. The following conditions are necessary: 1) continuous lasing; 2) improved selectivity of the external cavity (i.e., the use of a holographic selector or a diffraction grating in conjunction with a Fabry-Perot etalon).

An analysis of the interaction of spectral modes applied to the case of three modes (one of which is a "strong" mode), has demonstrated that with certain conditions autostabilization of single-frequency emission will occur. This is expressed as an increase in the stability of such emission with an increase in intensity, i.e., an increase in the threshold values of external perturbations which disrupt emission. Such an autostabilization effect is confirmed experimentally in experiments where the perturbation appeared as resonant pumping at the intermode frequency (active mode-locking). Moreover, such an effect is revealed by the experience of tuning lasers, i.e., when an external perturbation is caused by changing the laser detuning levels.

Underlying the autostabilization effect is the generation of a spectral valley near the strong mode, which suppresses the excitation of neighboring spectral modes. The "strong saturation" condition (for the appearance of this valley) is easily satisfied in a laser with an external dispersive cavity, when the threshold is exceeded by more than 1%. An important feature is the elimination of modes at spectral distances of $\Omega > 1/\tau$, where this mechanism ceases to operate. This explains the weak manifestation of this effect with insufficient selectivity of external elements.

This mode interaction mechanism, unlike those discussed in the introduction, is insensitive to a reduction in the intraband relaxation time, which occurs with an increase in temperature or doping (but does not disappear as the intraband relaxation time approaches zero). Moreover, it acts before the power saturation conditions are achieved in the dominant mode (due to distortion of the band spectrum in the semiconductor), and, thus, may be effectively utilized to increase power in single-frequency emission. As is the case for other mechanisms, this mechanism may be masked by spatial inhomogeneities ("spatial burning of the valley"); hence, it is most clearly manifest in stripe lasers with a stabilized spatial position of the single laser channel, as well as in the absence of lasing at the high order transverse modes.

BIBLIOGRAPHY

1. Eliseev, P. G., "An investigation of injection lasers" TR. FIAN, 1970, V. 52, p. 3-117.

2. Goreslavskiy, S. P., Galitskiy, V. M., Elesin, V. F. "The electromagnetic and magnetic properties of a semiconductor in a powerful magnetic field" ZHETF, 1969, V. 57, No. 1, p. 207-215.

3. Elesin, V. F. "The theory of the electron spectrum of a semiconductor in a powerful electromagnetic field" FTP, 1970, V. 4, No. 8, p. 1524-1528.

4. Galitskiy, V. M., Elesin, V. F. "The kinetics of electrons and steady-state lasing in semiconductor lasers" ZHETF, 1973, V. 64, No. 2, p. 691-702.

5. Galitskiy, V. M., Elesin, V. F. "Kinetic theory of strong field lasing in semiconductor lasers" ZHETF, 1975, V. 68, No. 1, p. 216-222.

6. Zee, B. "Broadening mechanism in semiconductor (GaAs) lasers limitations of single mode power emission" IEEE. J. QUANT. ELECTRON., 1978, V. 14, No. 10, p. 727-736.

7. Yamada, M., Suematsu, Y. "A condition of single mode longitudinal mode operation in injection lasers with index-guiding structure" IEEE J. QUANT. ELECTRON., 1979, V. 15, No. 8, p. 743-749.

8. Yamada, M., Suematsu, Y. "Analysis of gain suppression in undoped injection lasers" J. APPL. PHYS., 1981, V. 52, No. 4, p. 2653-2664.

9. Bogatov, A. P., Eliseev, P. G., Sverdlov, V. N. "Anomalous interaction of spectral modes in a semiconductor laser" IEEE J. QUANT. ELECTRON., 1975, V. 11, No. 7, p. 510-515.

10. Bakhert, Kh. Yu., Bogatov, A. P., Eliseev, P. G. "Mode deformation in an injection laser due to self-focusing and its relation to nonlinearity of the output characteristic" KVANTOVAYA ELEKTRON., 1978, V. 5, No. 3, p. 603-608.

11. Soskin, M. S., Taranenko, V. B. "A holographic total internal reflection selector for tunable lasers" KVANTOVAYA ELEKTRON., 1977, V. 4, No. 3, p. 536-543.

12. Bazhenov, V. Yu., Bogatov, A. P., Eliseev, P. G., et al. "Bistable emission and spectral tuning in an injection laser with an external dispersive cavity" KVANTOVAYA ELEKTRON., 1981, V. 8, No. 4, p. 853-859.

ANALYSIS OF A LASER DIODE AMPLIFIER

A.P. Bogatov

Abstract: A travelling wave amplifier based on an injection laser is investigated theoretically. This analysis accounts for the dependence of the refractive index of the medium on the concentration of injected electrons. Analytical expressions are derived for the dependence of the amplitude and phase of the amplified wave on the amplifier parameters. The maximum gain is estimated.

INTRODUCTION

Optical amplifiers based on laser diodes are promising for use in optical communications systems. The amplifier may be used in the output stage of the transmitter both as a power amplifier and as a modulator [1-3]. The amplifier may also be used as a preamplifier for receiving the optical signal [4]. To date the majority of studies in this field have been for regenerative amplifiers (see, for example, [1-6]). However, from the practical viewpoint a traveling wave amplifier is of primary interest [7, 8]. This type of amplifier has a significantly broader spectral amplification band and better operating stability compared to a regenerative amplifier.

Study [7] has provided a theoretical analysis of a traveling wave amplifier within the scope of an intensity analysis of a propagating wave. This calculation completely ignores the phase modulation of output emission due to the dependence of the refractive index of the active region on the intensity of the amplified wave (the concentration of excess carriers).

This study carries out a calculation of a traveling wave amplifier accounting for the dependence of the refractive index on wave intensity. It is demonstrated that the amplification of an optical signal in a semiconductor amplifier may be accompanied by phase modulation of the signal (such as in amplifying a purely amplitude-modulated signal) or in modulating the pumping level of the amplifier.

MODELING OF THE TRAVELING WAVE AMPLIFIER

We will consider an amplifier model in which the wave is propagating in one direction along the axis z. We may use a laser diode with no reflection off the diode face as such an amplifier. A bleaching coating is applied to the diode faces, or the normals of the faces are inclined at the Brewster angle with respect to the optical axis. The analysis is carried out for a wave with a stable transverse field distribution (in the plane x, y). Moreover we will assume that variations in the wave and pumping intensity are sufficiently slow compared to the lifetime of the nonequilibrium electrons in the active region. In this approximation the problem is single-dimensional and quasistationary. We will represent the wave field in the amplifier as

$$E(z,\ t) = \mathrm{Re}\ \mathcal{E}(z)\ e^{ikz-\omega t+i\Phi(z)}, \tag{1}$$

where $\mathcal{E}$ and Φ represent the slowly varying amplitude and phase of the wave, k is the wave vector in the medium and ω is frequency. The quantities $\mathcal{E}$, k and Φ are real.

It is necessary to solve the following wave equation to find $\mathcal{E}$ and Φ:

$$\Delta E - c^{-2}\partial^2 D/\partial t^2 = 0. \tag{2}$$

Due to the short polarization relaxation time ($T_2 \leqslant 10^{-13}$ sec) the field interaction with the active medium is noncoherent in an injection laser. Hence, we may use the following relation to relate the electrical induction D and the field E

$$D = \varepsilon E, \tag{3}$$

where $\varepsilon = \varepsilon_0' + i\varepsilon_0'' - aRN_0 - a(R + i)(N - N_0)$ and $\varepsilon_0' + i\varepsilon_0''$ is the complex dielectric constant of the active region in the absence of pumping and ignoring resonance losses, N is the concentration of injected electrons, N_0 is the concentration of electrons corresponding to the inversion of threshold, a is the quantity proportional to the cross-section σ of the induced transition and related to it by the relation

$$a = \sigma c\sqrt{\varepsilon_0'}/\omega,$$

where c is the speed of light in a vacuum. The coefficient R characterizes the relation of variations in the refractive index and gain with a change in N, i.e.,

The electron concentration may be found from the balance equation

$$\frac{N}{\tau}+\frac{\sigma(N-N_0)}{\hbar\omega}I-P=0, \tag{4}$$

where τ is the spontaneous recombination time, I is the wave intensity and P is the pumping rate.

Subject to equation (4), the expression for the dielectric constant ε may be written in the form

$$\varepsilon=\varepsilon_0'+i\varepsilon_0''-aRN_0-a(R+i)p\tau(1+I/I_0)^{-1}, \tag{5}$$

where $p = P - N_0/\tau$, I_0 is the saturation intensity, $I_0 = \hbar\omega/\sigma\tau$.

Substituting the expression for the field (1) and the expression for ε (5) into wave equation (2), we have

$$2ik\left(\frac{\partial\mathscr{E}}{\partial z}+i\frac{\partial\Phi}{\partial z}\right)+i\frac{\omega^2}{c^2}\left(\varepsilon_0''-\frac{ap\tau}{1+I/I_0}\right)\mathscr{E}-\frac{\omega^2}{c^2}aR\frac{p\tau}{1+I/I_0}\mathscr{E}=0;$$
$$k^2=(\omega^2/c^2)(\varepsilon_0'-aRN_0). \tag{6}$$

The "slowness" condition of variations in $\mathscr{E}(z)$ and $\Phi(z)$ were used to obtain equation (6), i.e., $\partial^2\mathscr{E}/\partial z^2 \approx 0$; $\partial^2\Phi/\partial z^2 \approx 0$.

Separating the real and imaginary parts in equation (6), and converting the equation for amplitude into an equation for intensity, we derive

$$\partial u/\partial z-gu/(1+u)+\alpha u=0, \tag{7a}$$

$$\partial\Phi/\partial z+gR/[2(1+u)]=0;$$
$$u=I/I_0,\ g=\sigma p\tau,\ \alpha=\omega\varepsilon_0''/c\sqrt{\varepsilon_0'}, \tag{7b}$$

where u is dimensionless intensity, g is nonsaturated gain and α is nonresonance losses. Equation (7a) was solved previously in study [9] in connection with lasing in a laser with a saturable absorber. Equation system (7) is easily integrated, and the expressions for u and Φ take the form

$$\frac{u(z)}{|g/\alpha-1-u(z)|^{g/\alpha}}=\frac{u(0)e^{(g-\alpha)z}}{|g/\alpha-1-u(0)|^{g/\alpha}}, \tag{8a}$$

$$\Phi(z)=\Phi(0)-\frac{R}{2}\alpha z-\frac{R}{2}\ln\frac{u(z)}{u(0)}. \tag{8b}$$

The most interesting case corresponds to output signal levels of $u(0) < g/\alpha-1$, and in this case such a signal is amplified.

Fig. 1a gives the output signal $u(z)$ plotted as a function of the input signal $u(0)$ calculated by formula (8a) for various g/α and a fixed αz. It is clear from this figure that effective amplification occurs at input signal levels of $u(0) \ll 1$. At input signal levels of $u(0) > 1$ ($u = 1$ corresponds to an optical flux equal to saturation power I_0) there is a sharp drop in gain, and with $u(0) = g/\alpha-1$ the

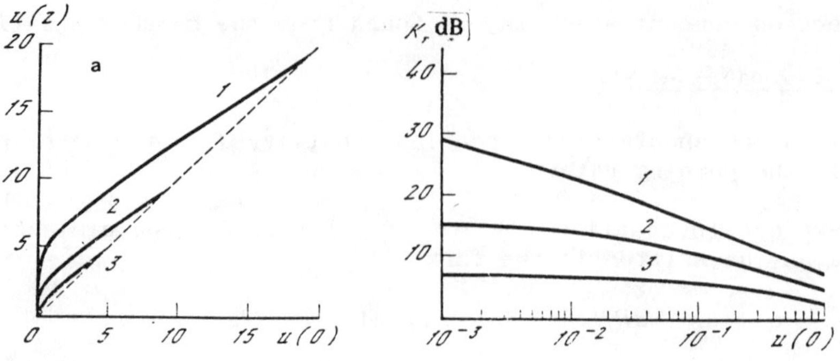

Fig. 1. The output signal $u(z)$ (a) and gain $K = 10 \lg (u(z)/u(0))$ (b) plotted as a function of output signal level $u(0)$.
The following parameters are used for the curves: $\rho = g/\alpha$: 20 (1); 10 (2); 5 (3); $\alpha z = 0.4$.

output signal becomes equal to the input signal, $u(z) \equiv g/\alpha - 1$. This may also be observed directly from equation (8a), as well as equation (7a). In equation (8a) when $u(0) \to g/\alpha - 1$, the denominator $|g/\alpha - 1 - u(0)|$ vanishes and the right half of equation (8a) tends towards infinity. In order for equation (8a) to remain valid, the denominator in the right half of this equation must vanish, i.e., $u(z) \to g/\alpha - 1$. We may also see that when $u(0) \to g/\alpha - 1$ in equation (7a) the derivative $\partial u/\partial z \to 0$, and this means that $u(z) \to \text{const} \equiv u(0)$.

We may see from equation (8a) that with $u(0) > g/\alpha - 1$ the output signal $u(z)$ is always less than the input signal. With a further increase in $u(0)$ the $u(z)$ relation asymptotically approaches the form

$$u(z) \to u(0)\ e^{-\alpha z}.$$

This is because the gain is fully saturated and is equal to zero at the same time that the losses remain the same. Along the same lines it is possible to demonstrate that

$$u(z) \to g/\alpha - 1 \tag{9}$$

when $(g - \alpha)\ z \to +\infty$ and $u(0) = \text{const}$. For example, if $u(0) < g/\alpha - 1$, then $u(z) \to g/\alpha - 1$ from above, and if $u(0) > g/\alpha - 1$, then $u(z) \to g/a - 1$ from below.

Consistent with this discussion, the range of input signals where the output signal exceeds or equals the input signal is limited to the range $0 \leqslant u(0) \leqslant g/\alpha - 1$. The size of this range is determined by the ratio of the nonsaturated gain g (the amplifier pumping level) to linear losses α. The gain relation K is

$$K = 10 \lg (u(z)/u(0))|_{z-1},$$

where l is the length of the amplifier diode; this relation is shown in Fig. 1b.

The curve parameter is the ratio of nonsaturated gain to losses, $\rho = g/\alpha$. We see from the diagram that in this case the range of linear amplification is limited to an input signal level of $u(0) < 10^{-3}$. The higher the value ρ or the nonsaturated gain, the lower the range of linearity of the amplifier at the input signal level. We may see from equation (8) that the independent parameters are the quantities $\rho = g/\alpha$ and the product αz. Hence the amplifier characteristics remain constant for any variations in g, α or z, which maintain the constancy of ρ and αz. The average curve in Fig. 1b is calculated for $\rho = 10$ and $\alpha z = 0.4$. Obviously an amplifier diode with $g = 200$ cm^{-1}, $\alpha = 20$ cm^{-1} and $l = 200$ μm will have such an amplification curve, as will a diode with parameters $g = 100$ $^{-1}$, $\alpha = 10$ cm^{-1} and $l = 400$ μm.

LINEAR AMPLIFICATION

With a low input signal level $u(0) \to 0$, the gain K tends towards a constant value. The smallness conditions of the input signal may be found from equation (8a). This equation may be rewritten as

When $\rho/u(z)$ $(\rho - 1) \ll 1$ expression $|1 - u(z)/(\rho - 1)|^{\rho} \approx 1$. In this case $u(z) \simeq u(0)$ $e^{\rho-1\alpha z}$. Hence the smallness condition of the input signal may be rewritten as:

$$\rho(\rho - 1)^{-1}u(0)\ e^{(\rho-1)\alpha z} \ll 1. \tag{10}$$

It is clear from expression (10) that the range of linearity of the amplifier decreases exponentially with an increase in amplifier pumping.

When condition (10) is satisfied, equations (8) go over to the following simple relations:

$$u\,(z) = u\,(0)\,e^{(g-\alpha)z},\ \Phi\,(z) = \Phi\,(0) - (R/2)\,gz,\ K = 10\,(g - \alpha)\,z\lg e. \tag{11}$$

The first relation (11) is normally used in analyzing threshold lasing conditions. The gain and phase of the amplified wave is independent of its amplitude.

We will perform certain estimates for AlGaAs or InGaPAs. For example, the thermal conditions of the amplifier diode may allow an excitation level such that $g - \alpha$ may reach ~200 cm^{-2}. With a diode length of 500 μm, the gain is ~43 dB. However, in order to implement such gain it is necessary for the residual reflection off of the diode face and the backscattering to be sufficiently low. For example, if it is necessary for the irregularity in the spectral response of the amplifier to be less than 2 dB, it is necessary for the residual reflection coefficient and backscattering to be less than 10^{-5}.

Another factor limiting amplification within the diode itself is the spontaneous emission level. Equations (2) and (3) as well as equation system (7) ignore internal sources of electromagnetic emission; hence, spontaneous emission intensity is not incorporated in the final result (equation system (8)). The correct accounting of the role of spontaneous emission in amplifier operation represents a complex problem, and extends beyond the scope of this study. Nonetheless we may estimate from below the maximum gain level. Indeed, since even in the absence of an input signal there always exists a certain background spontaneous noise emission level, with a sufficiently large gain, this amplified background emission will eliminate inversion in the amplifier and at the same time will saturate amplification. The maximum gain for spontaneous emission may be defined as

$$K_{\text{сп}} \simeq 10 \lg (I_{max}/I_{\text{сп}}), \tag{12}$$

where K_{max} is the maximum output emission intensity from the amplifier. Consistent with (9) its value is determined by the pumping

$$I_{max} = I_0 (g/\alpha - 1). \tag{13}$$

The value $I_{\text{сп}}$ is the effective spontaneous emission intensity at the input signal level of the amplifier. The contribution of spontaneous emission to the mode field was calculated in studies [10, 11]. In accordance with the results of these studies, the effective spontaneous emission intensity may be represented as

$$I_{\text{сп}} \simeq \frac{Nl}{\tau} \hbar\omega \frac{\delta\omega}{\Delta\omega} \eta, \tag{14}$$

where $\delta\omega$ is the spectral bandwidth of the amplifier, $\Delta\omega$ is the spectral linewidth. The geometric factor η may be estimated as

$$\eta = \frac{\Gamma}{4\pi\varepsilon_0''} \frac{\lambda}{d} \frac{\lambda}{w}, \tag{15}$$

where Γ is the optical confinement coefficient, λ is the wavelength in the vacuum. The factor $\delta\omega/\Delta\omega$ in equality (14) results from the fact that the amplified spontaneous emission (superluminescence) has a line whose width is narrower than the spontaneous emission linewidth of the optically thin layer. This factor may be estimated as

$$\delta\omega/\Delta\omega \simeq 1/\sqrt{gl}. \tag{16}$$

Thus, using equalities (12)-(16), for $K_{\text{сп}}$ we have

$$K_{\text{сп}} \simeq 10 \ln \frac{(g/\alpha - 1)\sqrt{gl}}{(\sigma N_0 + g) l\eta}. \tag{17}$$

For example, if nonresonance losses α are 10 cm^{-1}, resonance losses are 100 cm^{-1}, and the pumping level corresponds to double excess over the inversion threshold, $g \simeq 200$ cm-1, the dimensions of the active region are $d \simeq 0.2$ μm, $w = 5$ μm, $l = 500$ μm, the parameter $\Gamma = 0.3$ and $\lambda \simeq 0.8$–1.3 μm, we obtain

$$K_{\text{сп}} \simeq 36\text{–}32 \text{ dB}.$$

This is near the value obtained in study [10].

As noted above, the quantity $K_{\text{сп}}$ is the maximum gain, which is limited by the background spontaneous emission. On the other hand, the gain for a weak signal is 10 lg $e^{(g-a)l}$. Hence, the maximum gain K_{max} is

$$K_{\max} \simeq 10 \lg \min\left\{e^{(g-\alpha)l}, \frac{(g/\alpha - 1)\sqrt{gl}}{(\sigma N_0 + g)\,l\eta}\right\}. \tag{18}$$

However, it may not be advisable to use an amplifier with a gain near K_{max}. The problem is that from the viewpoint of the signal-to-noise ratio, such an operating mode of the amplifier may not be optimum.

AMPLIFICATION SATURATION. PHASE SELF-MODULATION

If the input signal intensity is not an infinitely small quantity, the amplification conditions can no longer be considered linear. Fig. 1b gives results from a calculation of gain K as a function of the input signal level $u(0)$. The curve parameter is the amplifier pumping level (the linear amplification value). It is clear from the figure that the amplification conditions become nonlinear (the value of K changes) with input signals of $u(0) > 10^{-2}$.

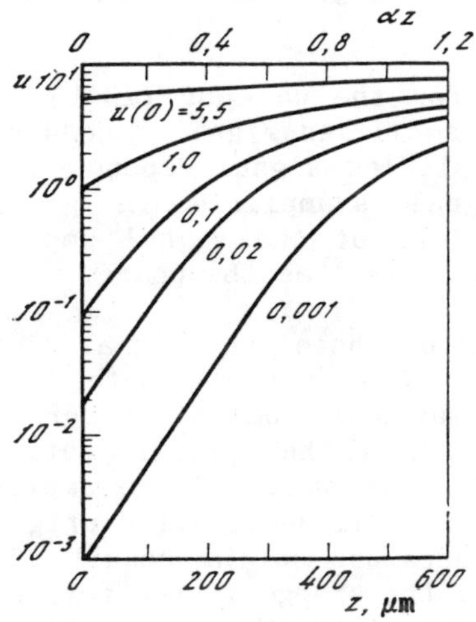

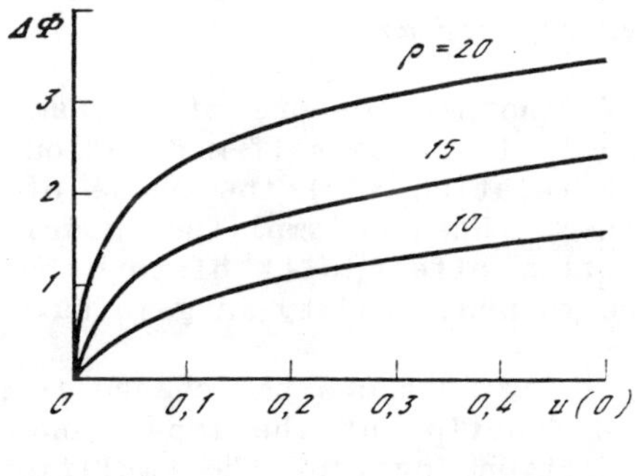

Fig. 3. The change in phase $\Delta\Phi$ of an amplified wave plotted as a function of its intensity level at the amplifier input $u(0)$
The curve parameter is $\rho = g/\alpha$; $\alpha z = 0.3$.

Fig. 2. Output signal $u(z)$ plotted as a function of amplifier length z for various input signal levels $u(0)$ ($\rho = 10$).

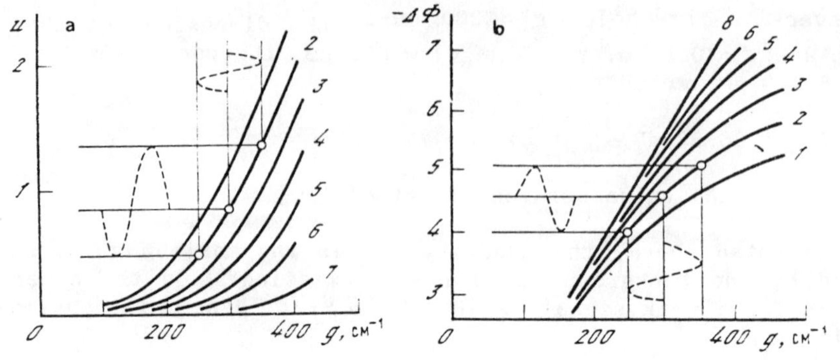

Fig. 4. Output wave intensity $u(z)$ (a) and the phase of the amplified wave (b) plotted as a function of linear gain z (pumping current of the amplifier).
The curve parameter is the input wave intensity $u(0)$; $u(0) \cdot 10^3 = 16$ (1); 8 (2); 4 (3); 2 (4); 1 (5); 0.5 (6); 0.1 (7); 0.01 (8). $\alpha z = 0.4$; $z = 0.02$ cm.

Fig. 2 gives the output signal plotted as a function of amplifier length. This relation may also be interpreted as the distribution of wave intensity along the amplifier length. The curve parameter is the input signal level. It is clear that these curves may be considered fragments of a single curve where the curve for the upper level of the input signal is a continuation of the curve for the lower level of the input signal. As for the case of Fig. 1, the curves in Fig. 2 are calculated for dimensionless quantities ρ and αz, hence their form is preserved for all variations in g, α and z which do not change ρ nor αz.

Another feature of the amplifier for the case of finite input signals is phase self-modulation of the amplified signal. Consistent with relation (8b) the phase of the amplified signal depends on the gain K. When an amplitude-modulated signal is amplified the gain saturation effect will distort both the form of the signal amplitude (due to nonlinearity in gain saturation) as well as the phase.

Fig. 3 shows the change in phase $\Delta\Phi$ of the output signal plotted as a function of the input power level. The curve parameter is the excitation level of the amplifier. We may see from this figure that the most significant change in phase occurs in the initial portion of the curves. The amplitude dependence of the phase of the amplified wave generates parasitic phase modulation. The qualitative origin of parasitic phase modulation is easily understood as the result of the change in the optical path resulting from the change in the refractive index from amplifier saturation. According to Fig. 3 a change of approximately λ in Φ occurs with an input signal $u(0) \simeq 0.3$.

In the discussion above the pumping level of the amplifier was a fixed parameter. It is also interesting to consider the output characteristics of the amplifier as a function of pumping (the linear amplification value). Fig. 4a gives the output signal plotted as a function of linear amplification. It is clear that for a rather high input signal level, the pumping dependence of the output signal is near-linear. Fig. 4b gives the dependence of the output signal phase on linear amplification. This relation approaches linearity when the input signal level is reduced. Using the data from Fig. 4 it is possible to determine the modulation characteristics of the amplifier when it is used in the output stage of an optical transmitter as an amplifier-modulator. It should be noted that the percentage of pumping modulation of the amplifier need not be significant. According to the data from Fig. 4 it is possible to select an input signal level so that it is sufficient to modulate amplifier pumping at 20% in order to obtain an amplitude modulation percentage and phase modulation index near 1.

The analysis carried out above is based on a quasistatic approximation. From the viewpoint of practical applications, modulation frequencies in the order of 1 GHz are of interest. In this case, a quasistatic approximation, generally speaking, is incorrect. Nonetheless it is obvious that qualitatively the behavior of the gain and phase will manifest the same nature. Quantitatively the amplifier characteristics may vary somewhat. The quantity $\Omega\tau$ will be the characteristic parameter of the problem where Ω is the modulation frequency. With very rapid changes in the amplitude of the input signal or the pumping current of the amplifier, the carrier concentration N will not be capable of fully following these changes. However, at moderately high values of the parameter $\Omega\tau$ we may also assume that the error will not be excessive.

Thus, on the basis of a theoretical analysis of the amplifier, we may derive the following conclusions.

1. The maximum gain is determined by the natural spontaneous noise level and based on estimates for standard values of the parameters of active elements, it may be reduced to approximately 40 dB.

2. The range of linear amplification at the input signal level is reduced with an increase in gain and for gain values of 30 dB it is limited to 10^{-3} of the saturation power.

3. It is possible to produce efficient amplitude-phase modulation of the amplified signal by modulating the pumping current of the amplifier.

4. In saturation conditions, the amplifier produces "parasitic" phase modulation in amplifying an amplitude-modulated signal.

In conclusion we note that in the practical implementation of an optical amplifier the problem of optimum amplifier gain in view of the signal-to-noise ratio arises. It is possible that it will not be advantageous to use an amplifier with maximum gain; optimum gain may be significantly less than 30 dB.

The use of the amplifier for optical preamplification in receivers is of primary interest for phase-modulated signals. In this case, the amplifier may operate with minimum distortion. At the same time, preamplification of an amplitude-modulated signal may be accompanied by significant nonlinear distortions and "parasitic" phase modulation. Hence, with amplitude modulation it is preferable to use digital-code modulation. Analog modulation is not promising due to the significant nonlinearity of the amplifier.

The dependence of the amplitude and phase of the amplified wave on the excitation level of the amplifier makes it possible to use the amplifier in the output stage of a transmitter both as a power amplifier and as an efficient modulator.

BIBLIOGRAPHY

1. Yamada, J., Kobayashi, s., Machida, S., Kimura, T. "Improvement in intensity fluctuation noise of a 1 Gbit/s-modulated InGaAsP by light injection technique" JAPAN J. APPL. PHYS., 1980, V. 19, No. 11, p. L689-L692.

2. Iwashita, K., Nakagawa, K. "Suppression of mode partition noise by laser diode light injection" IEEE J. QUANT. ELECTRON., 1982, V. 18, No. 10, p. 1669-1674.

3. Goldberg, L., Taylor, H. F., Weller, J. F., Sideband, F. M. "Injection of diode lasers" ELECTRON., LETT., 1982, V. 18, p. 1019-1020.

4. Kobayashi, S., Kimura, T. "Gain saturation power of resonant AlGaAs laser amplifier" ELECTRON. LETT., 1980, V. 16, p. 230-232.

5. Kobayashi, S., Nishimoto, H., Lang, R. "Experimental observation of asymmetric detuning characteristics in semiconductor injection locking" ELECTRON. LETT., 1982, V. 18, p. 54-55.

6. Nakai, T., Ito, R., Ogasawara, N. "Asymmetric frequency response of semiconductor laser amplifiers" JAPAN. J. APPL. PHYS., 1982, V. 21, No. 11, p. L680-L682.

7. Goldobin, I. S., Luk'yanov, V. N., Solodkov, A. F., et al. "An injection traveling wave laser amplifier based on a double GaAlAs-heterostructure" KVANTOVAYA ELEKTRON., 1984, V. 11, No. 2, p. 375-381.

8. Annenkov, D. M., Bogatov, A. P., Eliseev, P. G., et al. "Spectrally-matched modulation of emission from an injection laser in a

2 GHz traveling wave amplifier" KVANTOVAYA ELEKTRON., 1984, V. 11, No. 2, p. 231-232.

9. Letokhov, V. S. "The generation of ultranarrow light pulses in a laser with a nonlinear absorber" ZHETF, 1968, V. 55, p. 1077-1095.

10. Letokhov, V. S., Suchkov, A. F. "The dynamics of an instantaneous Q switching laser" TR. FIAN, 1968, V. 53, p. 169-186.

11. Petermann, K. "Calculated spontaneous emission factor for double heterostructure injection lasers with gain-induced waveguiding" IEEE J. QUANT. ELECTRON., 1979, V. 15, No. 7, p. 506-570.

AN OPTICAL TRAVELING WAVE AMPLIFIER

A.P. Bogatov, P.G. Eliseev, O.G. Okhotnikov, M.P. Rakhval'skiy, K.A. Khayretdinov

Abstract: An optical travelling wave amplifier consisting of an AlGaAs/GaAs laser diode whose stripe contact is inclined at an angle of approximately 10 degrees to the perpendicular of the diode face is investigated experimentally. The investigation of amplifier characteristics demonstrated that its total "equipment" gain is 14 dB. The spectral width of the amplification band is at least 3-6 10 GHz with a 3dB nonuniformity in the frequency response. The gain is optimized based on the signal-to-noise ratio. Spectrally-matched modulation of injection laser emission is achieved in the travelling wave amplifier where the width of the optical spectrum does not exceed twice the frequency of the modulating signal.

1. INTRODUCTION

Amplifiers based on injection lasers, in spite of the desire to use such devices in communications and optical memory systems as well as in optical logic information processing, etc. have not yet been widely employed, and are the focus of less attention than laser oscillators. As in other laser amplifiers, these devices have a powerful spontaneous emission background which complicates amplification of weak signals if additional spectral filters are not used. Moreover, the active region in laser diodes is quite small; its cross-section normally amounts to a few square micrometers and narrow focusing and careful tuning are required to effectively couple emission to the amplifier-diode. Finally, in semiconductor lasers the natural crystal surfaces easily produce significant optical reflection and eliminating feedback becomes a special problem requiring precision technology. This is required for a traveling wave amplifier whose maximum gain is limited by residual reflection and is equal to $1/R$, where R is the mean geometric reflection coefficient at the incoupling and outcoupling faces of the diode-amplifier. Self-excitation occurs with high gain, i.e., lasing arises.

Early research on diode amplifiers [1-3] examined traveling wave conditions in diodes with bleached faces. However to date the majority of published experimental studies have dealt with resonance (regenerative) amplifiers in which feedback is not minimized and gain is achieved by pumping in the immediate vicinity of the lasing threshold (1-2% above the threshold). Such a design is comparatively easily implemented in laboratory conditions [4-6]. A comparison of the characteristics of resonance amplifiers and traveling wave amplifiers is given in studies [7, 8] with respect to their prospects for application in laser communications systems.

A laser amplifier may be used in a receiver as a preamplifier in front of the photodetector, in a repeater as well as in the output stage of a transmitter as an amplifier-modulator. Due to the rapid progress in optical communications technology, interest in such devices has increased significantly. The development of these applications depends on the level of gain characteristics that are achievable (with respect to the gain, the signal-to-noise ratio, and the coefficient of harmonic distortion) and what operating conditions must be maintained in this case. It should be noted that a number of serious practical difficulties exist with respect to resonance amplifiers, including:

- the narrow spectral amplification band (and consequently the strong wavelength dependence of the gain) requires precise tuning of the amplifier to the signal frequency;

- the strong dependence of the gain on the pumping current (or more precisely on its relative magnitude with respect to the threshold current) requires that the pumping level be carefully maintained;

- due to the variety of resonances for orthogonal polarizations in the waveguiding resonator of the resonance amplifier, the gain is sensitive to the polarization of the signal emission; this means that the application of a resonance amplifier is limited to systems with stabilized polarization (this remains a problem for optical communications technology);

- the resonances shift with a change in temperature; hence, temperature stabilization of the active region of the amplifier (on the order of better than ±0.1 K) is required.

From the practical viewpoint the advantage belongs to traveling wave amplifiers having a broad amplification band and that are less sensitive to temperature and polarization variations and small changes in pumping current. Thus, a traveling wave amplifier is operationally more stable and, moreover, operates at a higher speed and has a broader dynamic range than a resonance amplifier. Study [7] has noted that in the amplification band up to 20 dB, the signal-to-noise ratio is noticeably better in a traveling wave amplifier. As noted in study [8] such an amplifier may be used to improve the sensitivity of a receiver by 13 dB in the 1.55 μm range at signal frequencies of

~1 GHb/s. Regarding dynamic range, study [9] has shown that the conversion to a traveling wave amplifier produces an increase in saturation (with all other levels remaining equal) of ~18 dB. The analysis of traveling wave amplifier conditions was carried out in studies [9, 10]. The following data for a GaAlAs traveling wave amplifier laser are given as sample calculated characteristics in study [9]:

1) maximum gain from amplification of spontaneous emission and gain saturation in superluminescent emission: ~40 dB;

2) maximum gain from residual reflection on the order of 0.1%: up to 20.6 dB;

3) at a current density of 26 ka/cm^2 preferably comprising 96% of the threshold current, gain with a diode length of 1 mm is 15 dB; the gain factor is equal to 34 cm^{-1};

4) the saturation intensity is 3.7 Mw/cm^2 (in resonance amplifiers this value is 0.34-0.1 Mw/cm)

It follows from these data that the gain characteristics achieved in the traveling wave amplifier are quite attractive for practical applications.

Possible configurations of traveling wave diode-amplifiers may be reduced to two primary variants: 1) a Fabry-Perot configuration with bleached faces; 2) a "Brewster" configuration in which the longitudinal axis of the amplifying medium is inclined at the Brewster angle with the incoupling and outcoupling mirrors or at a smaller angle to allow outcoupling of the reflected beam from the amplifying medium. This second design is investigated in this study. In order to quench natural lasing, a 10-15 degree deviation from perpendicularity is introduced between the active stripe and the mirror ends of the diodes (in the active layer plane). This makes it possible to reduce the feedback level to values of less than 10^{-3}-10^{-4} although it is necessary to bear in mind that a polarization sensitivity of the gain arises in the "Brewster" configuration, since the Fresnel transparency of the inclined mirrors depends on polarization.

This study investigates a traveling wave amplifier employing a CW GaAlAs/GaAs stripe laser operating at 0.85 μm at room temperature. Such a device has already been used to build a two-stage transmitter [11] in which the amplifier is used as an amplifier-modulator of a monochromatic signal. This device allows implementation of spectrally-matched emission modulation.

2. EXPERIMENTAL ASSEMBLY

A GaAlAs/GaAs double heterostructure with an active region ~0.2 μm in thickness and a stripe contact ~8 μm in width was used in the fabrication and investigation of amplifier diodes for application in optical communications systems. The diodes operated in CW conditions at room temperature. In order to eliminate the natural resonances of the diode, the stripe contact that determines the position of the active region was inclined at 10° to the normal of the outcoupling faces of the diode.

In order to determine the residual reflection coefficient at the ends of the diode, the superluminescent spectra of the diode were investigated. A sample such spectrum is shown in Fig. 1b. For comparison Fig. 1a also gives the superluminescence spectrum of a test diode with a standard stripe contact (perpendicular to the faces) but with a bleached face. Both spectra are recorded when the lasing threshold in the external cavity is exceeded by 7% (in a linear cavity for a bleached diode and in a ring cavity for a cleaved diode). Assuming equal reflection losses and emission coupling to the active region of the diode in both types of cavities, we may estimate the residual reflection coefficient for a diode with a cleaved stripe contact and for a test diode with a bleached face. The estimates demonstrate that the residual reflection coefficient is less than 10^{-5} for a diode with a cleaved stripe contact and 10^{-3} for the test diode with a bleached face.

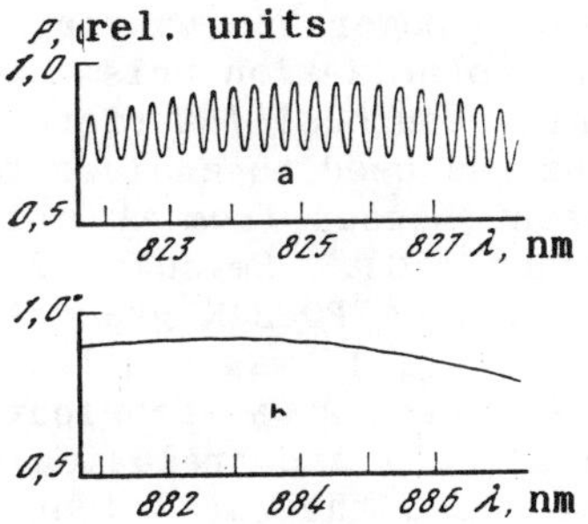

Fig. 1. The superluminescence spectra of a diode with a single bleached face (a) and a diode with a cleaved stripe contact (b).

Fig. 2. Experimental set-up to investigate a traveling wave amplifier.

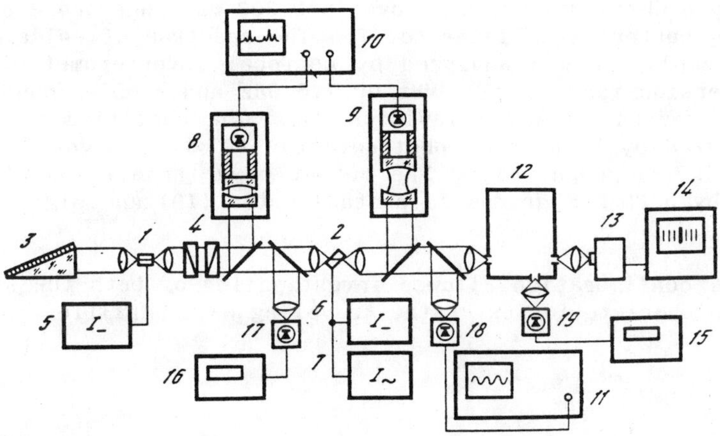

In practice the 10^{-5} level is the upper boundary for the residual reflection coefficient, since in the superluminescent spectrum, no periodic modulation from the natural cavity of the diode was recorded. Hence the level 10^{-5} corresponds to the accuracy level of our measurements (the photodetector noise level) in measuring the spectral density of superluminescence. Such a low value of residual reflection in the diode with a cleaved stripe contact results from the fact that with inclined incidence (~10°) of the waveguide mode emission from the diode on the face, the reflected emission did not strike the waveguide aperture (whose capture angle is ~3°). We note that this also indicates the absence of backscattering of emission incident on the diode from the lens side. In this case emission reflected off of the diode face is not reflected back to the lens aperture.

The length of the amplifier diode was varied over 200-250 μm.

The experimental set-up for investigating the characteristics of the traveling wave amplifier is given in Fig. 2. A single-frequency injection laser with an external dispersive cavity was used as the master oscillator; the dispersive cavity included a laser diode 1, a collimating lens and dispersing element 3 (a holographic total internal reflection selector). The laser diode is a regular CW AlGaAs injection laser operating at room temperature (T = 300 K). D.c. from power supply 5 in a 70-200 ma range was used to power the master oscillator. Attenuator 4 consisting of two Glan polarization prisms was used to vary master oscillator power. Scanning confocal interferometer 8 with a free dispersion range of 3.75 GHz was used to analyze the spectrum of the master oscillator. The output signal from this interferometer was sent to the S8-13 oscillograph (10). Emission from the master oscillator whose level was measured by a FD-24K graduated photodetector (17) and a ShCh31 digital voltmeter (16), was coupled to traveling wave amplifier 2 by a microlens. Amplifier 2 was fabricated from the same wafer as laser diode 1 which allowed a correlation of the lasing wavelength of the master oscillator and the amplification spectrum of the amplifier. The amplifier design was described above. the amplifier was powered from two sources: a d.c. component from power supply 6 which was variable over a 0-300 ma range and a sinusoidal modulating current from generator 7. The spectrum of emission output from the amplifier was analyzed by confocal interferometer 9 with a free dispersion range of 500 MHz or 7.5 GHz and a MDR-4 monochromator (12) with camera 13 and television 14. The amplified emission was also analyzed by high-speed photodetector 18 whose signal was fed to the S7-8 oscillograph (11). The output power from the amplifier was measured by a FD-24K graduated photodetector (19) and digital voltmeter 15.

This configuration allowed investigation of both the static and modulation characteristics of the traveling wave amplifier.

3. GAIN

An important characteristic of the amplifier is the dependence of the output power of the emission $P_{ВЫХ}$ on the input power $P_{ВХ}$. Such experimental relations as indicated by the points are given in Fig. 3 for three different wavelengths of the master oscillator. The amplifier length was l = 200 μm. The curve parameter is the amplifier pumping current. The wavelengths λ = 811.0 nm correspond to the maximum of the luminescent band of the amplifier diode at 200 ma. It is clear from the graphs that the $P_{ВЫХ}/P_{ВХ}$ relation is significantly nonlinear. Fig. 4 gives a more detailed $P_{ВЫХ}/P_{ВХ}$ relation for λ = 811.0 nm and an amplifier pumping current of I_{yc} = 200 ma for low input power levels. The graph may also be used to determine the maximum "equipment" gain (accounting for emission coupling losses to the amplifier), which amount to 14 ±1 dB.

The solid lines in Fig. 3 show the calculated $P_{ВЫХ}/P_{ВХ}$ relations by [22]

$$P_{\text{ВЫХ}}\left(\frac{g}{\alpha}-1-\frac{P_{\text{ВЫХ}}}{P_0}\right)^{-g/\alpha}=\varkappa P_{\text{ВХ}}e^{(g-\alpha)z}\left(\frac{g}{\alpha}-1-\frac{\varkappa P_{\text{ВХ}}}{P_0}\right)^{-g/\alpha}, \quad (1)$$

where g is nonsaturated gain, α is nonresonant losses, P_0 is saturation power, $\varkappa$ is the coupling coefficient. It is clear that the experimental points lie on the theoretical curves.

Saturation power P_0, coupling coefficient $\varkappa$ and $\rho = g/\alpha$ were used as the primary alignment parameters. In our case, correlation of experimental data and calculated data occurred for P_0 = 8 mw and $\varkappa$ = 8%. The difference of nonsaturated gain and nonresonant losses $g - \alpha$ were matched to experimental data for the $P_{ВЫХ}/P_{ВХ}$ relation for low $P_{ВХ}$. It is demonstrated below that the P_0 value derived in this manner from experiment is in good agreement with the direct theoretical estimate of this value. Regarding the value of $\varkappa$ obtained, we also note that it is within reasonable correlation with the theoretical estimate. Indeed, the coupling coefficient $\varkappa$ may be estimated as

$$\varkappa = (1 - r)\ d/D;\ D = 1.2\lambda/A, \quad (2)$$

where d is the thickness of the active region, D is the diameter of the scattering spot produced by an ideal lens, A is numerical aperture, λ is wavelength in a vacuum and r is the reflection coefficient. In our case these quantities are: d = 0.3 μm, λ - 0.8 μm, A = 0.65, $r \simeq 0$. Consequently for $\varkappa$ we have

$$\varkappa \simeq 20\%.$$

This corresponds to experiment if we remember than an actual lens may produce a scattering spot D twice that indicated by Abbe's formula. If we know $\varkappa \simeq 8\%$, we may find the gain achieved within the amplifier itself. For example, in our case, the maximum equipment gain for a fixed pumping current was 14 dB. This indicates that the internal gain is 25 dB.

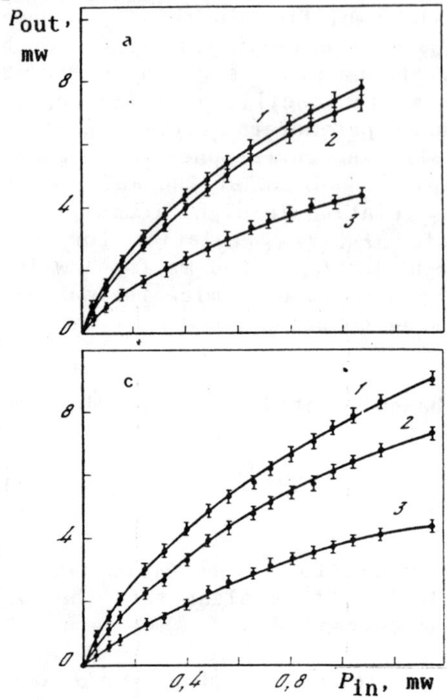

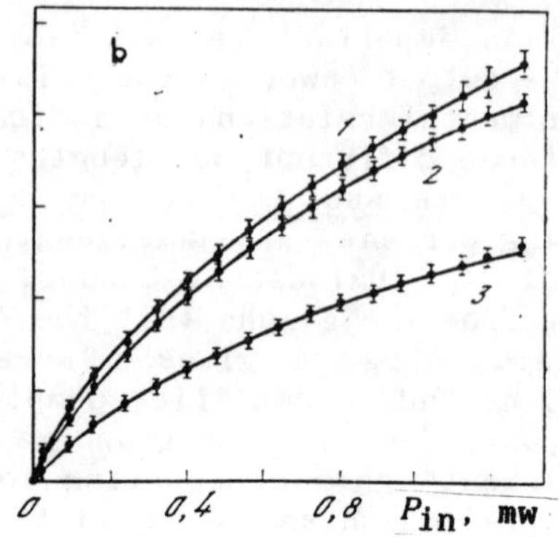

Fig. 3. Experimental relations plotting the output signal power $P_{\text{вых}}$ against the input signal power $P_{\text{вх}}$ of the amplifier for various pumping currents and for three of the master oscillator frequencies (points indicate error range)

a–c: λ = 807.7 nm (a); 811.0 (b); 816.3 (c). Solid lines: calculation. Parameter values used in calculation: α = 20 cm^{-1} $\varkappa$ = 8%, P_0 = 8 mw, l = 200 μm. 1-3: I_{yc} = 200 ma (1); 150 (2); 100 (3). a - g = 293 cm^{-1} (1); 288 (2); 258 (3). b - g = 295 cm^{-1} (1); 290 (2); 260 (3). c - g = 293 cm^{-1} (1); 295 (2); 250 (3).

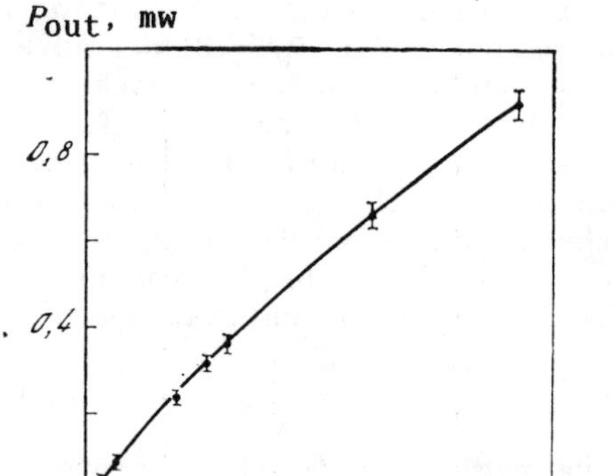

Fig. 4. Experimental relation plotting output signal power $P_{\text{вых}}$ versus input signal power $P_{\text{вх}}$ of the amplifier for low input signal levels.
Amplifier pumping current I_{yc} = 200 ma. Wavelength of master oscillator λ = 811.0 nm.

It is also clear from Fig. 3 that the difference in the gains under pumping at 150 and 200 ma is significantly less than for the case of currents of 100 and 150 ma. This is due to the reduction in local gain resulting from diode heating. Hence in this case the gain reached a maximum at 14 dB for a pumping current of 200 ma. A further increase in pumping reduced gain due to amplifier heating.

In order to eliminate amplifier heating, experiments were performed to investigate gain with pulsed excitation of the amplifier. Fig. 5 gives the gain plotted as a function of the pumping current for

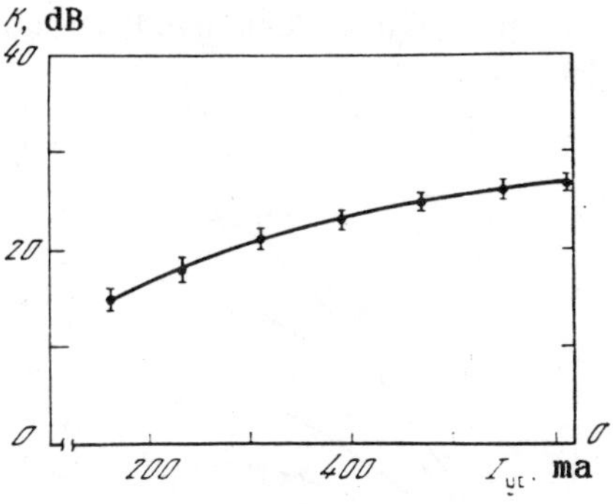

Fig. 5. Experimental relation plotting the total (accounting for coupling losses) gain versus the pulsed pumping current.
P_{BX} = 0.05 mw; λ = 790.0 nm; $t_{ИМП}$ = 0.2 μs, f = 0.5 kHz.

such conditions. We see that the gain increases with current, reaching values in the order of 27 dB at 600 ma, although the increase in gain at this current ceases and saturation begins. Reducing the input signal does not increase the gain. This indicates that gain saturation by natural spontaneous emission occurs.

We may conclude from these measurements that in our case the maximum equipment gain of 14 dB for CW operation of the amplifier is the maximum gain and is limited by thermal conditions, and the gain saturation effect by natural spontaneous emission limits this level to 27 dB for pulsed operation. We find that for such amplifiers, amplification may be increased to ~20 dB and ~33 dB for CW and pulsed excitation emission, if coupling losses are reduced to say a level of ~5 dB.

The $P_{BЫX}/P_{BX}$ relations for various wavelengths (see Fig. 3) also make it possible to find the spectral bandwidth of the amplifier. In this case, the bandwidth is $3.6 \cdot 10^3$ GHz (9 nm) with a gain nonuniformity of 3 dB.

4. SATURATION POWER

Fig. 6 gives the output power $P_{BЫX}$ plotted as a function of the pumping current of the amplifier I_{yc}. Three different wavelengths of the master oscillator correspond to these families of curves. The curve parameter is the output power P_{BX}. The output power increases with an increase in pumping current, reaches a maximum, and then drops off. In the 50-150 ma range these relations are near-linear which indicates gain saturation at these input powers and currents. The drop in gain at high currents results from the heating of the amplifier diode. As the wavelength of the master oscillator increases, the maximum of this relation shifts towards higher currents. This is because heating of the diode results not only in a decrease in the gain but also a shift in the gain maximum towards longer wavelengths. These two facts produce a situation where for longer wavelengths (λ = 816.3 nm) the maximum gain is achieved at a current of ~240 ma, at the same time that at λ = 807.7 nm, the maximum gain is achieved at a current of ~170 ma.

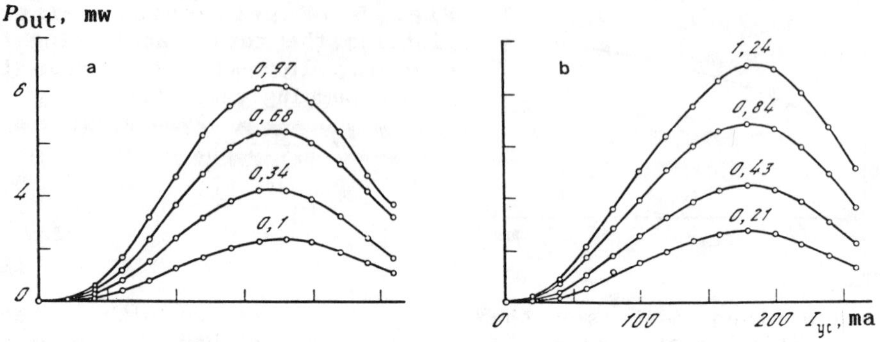

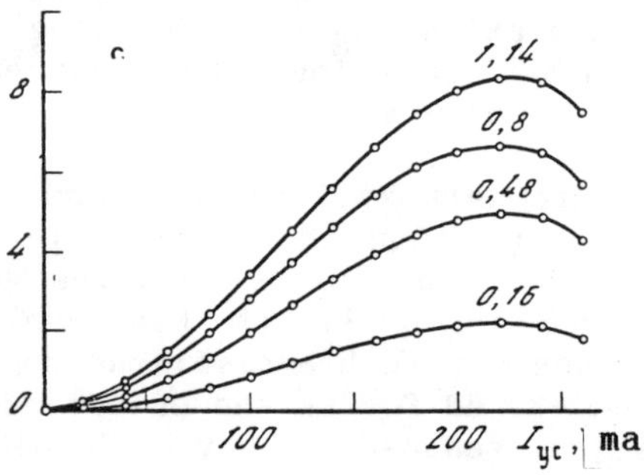

Fig. 6. The experimental relation plotting the output signal of the amplifier $P_{вых}$ as a function of pumping current I_{yc} for different wavelengths of the master oscillator.
a–c: λ – 807.7 nm (a) 811.0 (b); 816.3 (c). The digits above the curves indicate the input signal power in mw.

The dependence of the output power on the pumping current of an amplifier makes it possible to estimate the saturation power which is determined by [12]:

$$P_0 = wd\hbar\omega/\sigma\tau, \tag{3}$$

where w is the width of the active region occupied by the field, d is the thickness of the active region, σ is the cross-section for the induced transition, and τ is the lifetime of the injected electrons. For $\sigma\tau$ we have

$$\sigma\tau = \frac{\partial g}{\partial N}\tau = \frac{\partial g}{\partial I}\frac{\partial I}{\partial N}\tau = \frac{\partial g}{\partial I} lS\,de, \tag{4}$$

N is the concentration of injected electrons, I is pumping current, l is diode length, S is the effective width of the diode excitation region and e is electron charge.

If we assume that the input power and pumping current of the amplifier are sufficiently low so that nonsaturated amplification is achieved,

$$P_{вых} = \varkappa P_{вх} e^{(g-\alpha)l}. \tag{5}$$

From here we find

$$\frac{\partial g}{\partial I} = \frac{1}{P_{\text{вых}}} \frac{\partial P_{\text{вых}}}{\partial I} \frac{1}{l} . \tag{6}$$

Thus, we may derive a formula from (3), (4) and (6) for the saturation power in the form

$$P_0 = \frac{w}{S} \frac{\hbar\omega}{e} \frac{P_{\text{вых}}}{\partial P_{\text{вых}}/\partial I} . \tag{7}$$

The quantity $P_{\text{ВЫХ}}\ (P_{\text{ВЫХ}}/\partial I)^{-1}$ was found from experimental data in Fig. 4b and is approximately $1 \cdot 10^{-2}$ amps. In this case, consistent with (7), saturation power P_0 is 8 mw with $w/S \approx 0.5$. This is a permissible value for w/S, if we remember that significant current spreading under the stripe contact occurs in the test samples. The width of the active region covered by the field w may be two to three times less than the effective width of the p-n-junction region S, through which the primary current flows. Thus, we may state that P_0, established from experimental data (see Fig. 3), is in agreement with the theoretical estimate of this value.

5. AMPLIFIER SPONTANEOUS NOISE

An important issue in the practical use of the amplifier is its internal (intrinsic) noise. In this case this includes the internal spontaneous or superluminescence emission of the amplifier. Fig. 7 gives the spectra of such emission for various pumping levels. It is clear that as the pumping current increases both the gain of the amplifier and its noise increase. We may also see from this diagram that at a pumping current above 150 ma, its thermal conditions change. As was noted above, this is also evident in its gain (see Figs. 3 and 6).

First we will consider the influence of such optical noise on the characteristic of the amplifier when used for optical preamplification of a weak signal in the receiver of a communications link. The existence of such emission in conjunction with the amplified legitimate signal may produce reception errors. Indeed, since the spontaneous radiation process has a random nature, its intensity experiences certain fluctuations about the mean. Hence there is some possibility for a situation where the amplitude of such fluctuations will be sufficient to be comparable to the signal amplitude. In this case there will be errors in the reception of this signal.

For this analysis we will consider the optical signal reception scheme shown in Fig. 8. A monochromatic signal is sent to the input of the optical amplifier. From here the amplified signal $E_s \cos \omega_0 t$ and the background spontaneous emission $\mathcal{E}(t)$ are sent to an optical filter (such as a Fabry-Perot etalon) with a transmission band $\Delta\omega$ tuned to the optical frequency of the signal ω_0. The optical signal is taken from the filter output and is sent to a square-law detector (a photodetector). The electrical signal from the photodetector is the output signal of our receiver.

We will examine how the internal spontaneous emission level of the amplifier influences the amplitude fluctuation of the output signal. We will carry out this analysis for the case of discrete amplitude modulation of the input signal. Moreover, we will assume that the spontaneous emission of the amplifier is a normal stationary process. We note that this assumption is not rigid. In principle even if the amplitude of the spontaneous field does not have a normal distribution law, the amplitude of a wave passing through a filter with a sufficiently narrow transmission band will have a normal distribution. The necessary transmission band of the filter is such that over a correlation time of this filter of $\tau = 1/\Delta\omega$ ($\Delta\omega$ is the spectral width of the filter) several spontaneous emission acts occur. This is virtually always realized.

The signal at the square law detector input may be represented as the sum of the optical signal and filtered quasimonochromatic spontaneous emission:

$$E = E_s(t) \cos \omega_0 t + E_n(t) \cos (\omega_0 t + \varphi),$$

where $E_s(t)$ is a determinant positive function of time adopting a value 0 and $\mathring{E}_s$, while E_n is an arbitrary positive function of time. The function P_ω of the spectral density of spontaneous emission is other than zero only within the transmission band of the filter $\Delta\omega/2 = \omega_0 + \Delta\omega/2$.

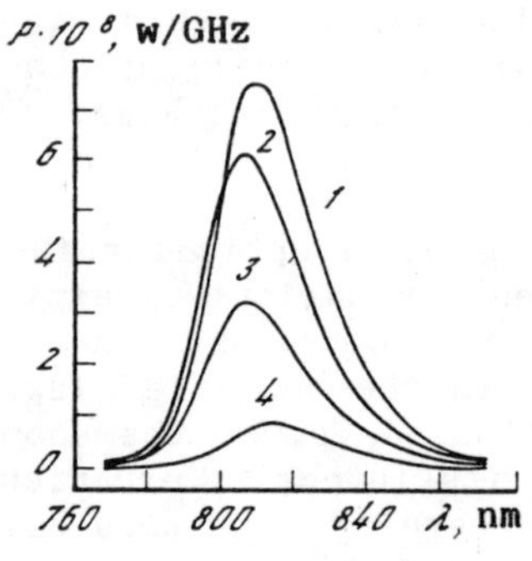

Fig. 7. The spectral noise density of the traveling wave amplifier $P(\omega)$ plotted as a function of emission wavelength.
1 - I_{yc} = 200 ma;
2 - 150;
3 - 100;
4 - 50

Fig. 8. Block diagram of signal receiver with optimum preamplification

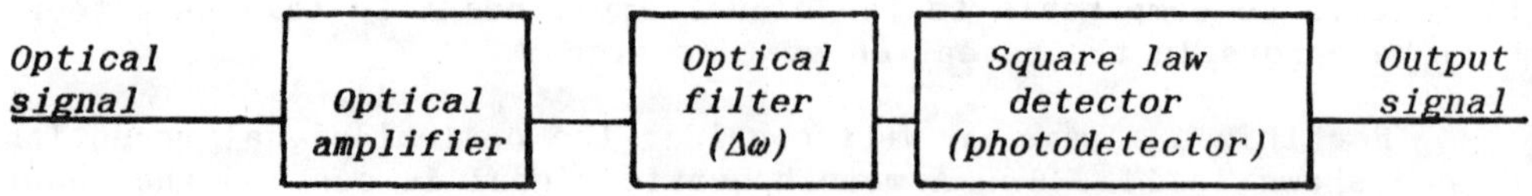

After the square-law detector, the photocurrent signal j is equal to

$$j = \eta\,[E_s^2(t) + 2E_s(t)\,E_n(t)\cos\varphi + E_n^2(t)], \qquad (8)$$

where η is the detector sensitivity.

Relation (8) reflects averaging of the optical frequency ω_0. It is clear from this equality that three terms are incorporated in the photocurrent. The first is proportional to the signal intensity, the second results from signal/spontaneous emission beating, and the third term is the spontaneous emission intensity. We will examine fluctuations δj caused by spontaneous emission, i.e.,

$$\delta j = \eta\,[2\mathring{E}_s E_n(t)\cos\varphi + E_n^2(t)]. \tag{9}$$

Reception error may result form two false readings. The first is that in the absence of the signal $E_s = 0$, a fluctuation in δj is greater than a certain threshold value j_0. The second amounts to the absence of the output signal, i.e., $j < j_0$ with a signal $E_s \neq 0$ at the input. We will now determine the probabilities of these events. The distribution for the amplitude and phase of the noise field takes the form [13]:

$$W(E_n, \varphi) = \frac{E_n}{2\pi\sigma^2} e^{-E_n^2/2\sigma^2}. \tag{10}$$

First we will calculate the probability of recording a false "one", when an output photocurrent pulse is observed in the absence of an input signal. If we use the following value for the threshold level of the photocurrent for recording a one

$$j_0 = \eta k \mathring{E}_s^2, \tag{11}$$

$0 < k < 1$, where k characterizes the threshold dividing a "zero" and a "one." In this case the probability of recording a false "one" may be represented as

$$W(\delta j > j_0) = \int_0^{2\pi} d\varphi \int_{E_0}^{\infty} dE_n W(E_n, \varphi), \tag{12}$$

where the quantity E_n according to (9) and (11) satisfies the inequality

$$\delta j = \eta E_n^2 > j_0 = \eta k \mathring{E}_s^2,$$

or

$$E_n^2 > k\mathring{E}_s^2,\ E_0^2 = k\mathring{E}_s^2. \tag{13}$$

From relations (12), (10) and (13) we obtain

$$W(\delta j > j_0) = e^{-k\mathring{E}_s^2/2\sigma^2}. \tag{14}$$

We will now find the probability of a false "zero", i.e., the probability of an event when $j < j_0$ in the absence of a signal $E_s \neq 0$. In this case E_n and φ satisfy the inequality

$$j = \eta\,[E_s^2(t) + 2E_s(t)E_n(t)\cos\varphi + E_n^2(t)] < j_0 = \eta k \mathring{E}_s^2. \tag{15}$$

In this case, fluctuations in the output signal occur not only due to fluctuations in the intensity of spontaneous emission but also due to beating between the signal and the spontaneous emission. From inequality (15) we may obtain a relation for the spontaneous emission amplitude where an output signal will not occur

$$\mathring{E}_s\left[-\cos\varphi-\sqrt{\cos^2\varphi-(1-k)}\right]<E_n<\mathring{E}_s\left[-\cos\varphi+\sqrt{\cos^2\varphi-(1-k)}\right]. \quad (16)$$

Since we assume E_n and $\mathring{E}_s$ to be positive quantities, inequality (16) may be satisfied if the phase shift φ between the signal field and the spontaneous field alone satisfies the relation

$$\pi/2-\varphi_0<\varphi<3\pi/2-\varphi_0,\ \varphi_0=\arcsin(1-k). \quad (17)$$

From the inequalities (16), (17) and (10) we may derive the probability $W\ (j < j_0)$ of the absence of an output signal when an input signal is present:

$$W(j<j_0)=\frac{1}{\pi}\int_{\varphi_0}^{\pi/2}\left\{\exp\left[-\left(\frac{\mathring{E}_s}{\sigma}\right)^2\frac{(\sin\varphi-\sqrt{\sin^2\varphi-(1-k)})^2}{2}\right]-\right.$$
$$\left.-\exp\left[-\left(\frac{\mathring{E}_s}{\sigma}\right)^2\frac{(\sin\varphi+\sqrt{\sin^2\varphi-(1-k)})^2}{2}\right]\right\}d\varphi. \quad (18)$$

Since the integral in equality (18) is not expressed as analytical functions, it is most convenient to use a majorant function experiment $[-(\mathring{E}_s/\sigma)^2(1 - \sqrt{k})^2/2]$ and thus to estimate the probability $W\ (j < j_0)$ in the form

$$W(j<j_0)\leqslant\left(\frac{1}{2}-\frac{\arcsin(1-k)}{\pi}\right)\exp\left[-\left(\frac{\mathring{E}_s}{\sigma}\right)^2\frac{(1-\sqrt{k})^2}{2}\right]. \quad (19)$$

Thus, the probability of a signal reception error resulting from the internal noise of the amplifier may be represented as

$$W=p(0)\,W(\delta j>j_0)+p(1)\,W(j<j_0), \quad (20)$$

where $p(0)$ and $p(1)$ are the probabilities of transmitting a zero and a one, respectively. If a zero and one are transmitted on the average at an identical frequency, the error probability is

$$W={}^1/_2[W(\delta j>j_0)+W(j<j_0)], \quad (21)$$

where the quantities $W(\delta j > j_0)$ and $W(j < j_0)$ are found from relations (14) and (19).

Further, we note that

$$\overline{E_n^2}=2\sigma^2=\int_{-\infty}^{+\infty}P(\omega)\,|f(\omega)|^2\,d\omega, \quad (22)$$

where $P(\omega)$ is the spectral density of spontaneous emission, $f(\omega)$ is the spectral transmission function of the optical filter. If the transmission band of the filter $\Delta\omega$ is much narrower than the spontaneous emission spectrum (this is the most interesting case), equality (22) may be rewritten as

$$2\sigma^2 = P(\omega_0)\Delta\omega. \tag{23}$$

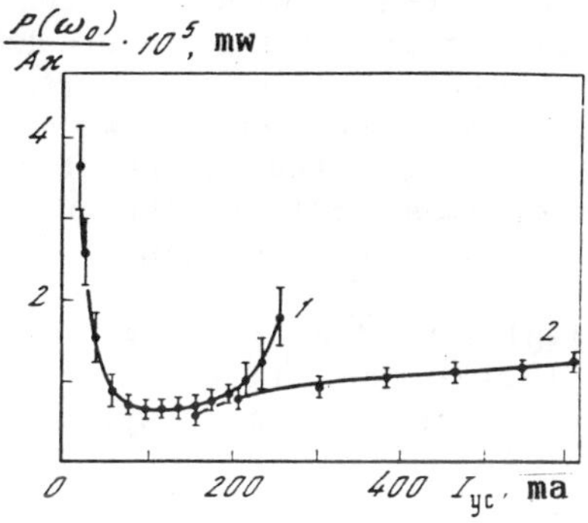

Fig. 9. The parameter $P(\omega_0)/A\varkappa$ plotted as a function of amplifier pumping current.
1 - λ = 811.0 nm, CW operation;
2 - λ = 790.0 nm, pulsed excitation. Pulse duration - 0.2 μs; pulse repetition frequency - 5 kHz

Accounting for equality (23) the error probability (expression (21)) is fully represented by the amplified signal level $\mathring{E}_s$ and the spectral spontaneous emission density $P(\omega)$ as well as the transmission band of the filter $\Delta\omega$ and the threshold k dividing the one and the zero. It is easy to see that with k = 1/4 the signal intensity dependence of $W(\delta j > j_0)$ and $W(j < j_0)$ is identical and for W according to (21) we have

$$W \simeq 0{,}6 \exp\left\{-\frac{P_{\text{сигн}} A}{4P(\omega_0)\Delta\omega}\right\}, \tag{24}$$

where $P_{\text{сигн}}$ is the input signal power, A is the linear gain of the signal, $P(\omega_0)$ is the spectral spontaneous emission density at the signal frequency and $\Delta\omega$ is the spectral transmission band of the optical filter. According to (24) for $P_{\min}$ - the minimum level of an input signal (or sensitivity threshold) for a signal detected with an error probability of less than W - the following relation is valid

$$P_{\min} = \frac{(\lg 0{,}6 - \lg W)\cdot 4\Delta\omega}{\lg e}\,\frac{P(\omega_0)}{A}. \tag{25}$$

Thus, from the viewpoint of the signal-to-noise ratio or the error probabilities, the quantity characterizing the amplifier is the ratio of the spectral density of spontaneous emission to the linear gain A. Consequently optimization of an optical amplifier from the viewpoint of signal-to-noise ratio should be carried out by minimizing the parameter $P(\omega_0)/A$. We will perform certain estimates. According to experimental data on the spectral density of spontaneous emission (see Fig. 6) the quantity $P(\omega_0)$ for an amplifier excitation current of 200 ma is $P(\omega_0) \lessapprox 8\cdot 10^{-8}$ w/GHz. For the same current, the gain within the amplifier (ignoring coupling losses) according to data from Fig. 3 is A = experiment $(g - \alpha)l \simeq 220$ for λ = 811 nm. If the data transmission is, say, 500 Mbits/sec, $\Delta\omega$ may be ~1 GHz. In this case $P_{\min}$ is approximately $3\cdot 10^{-8}$ watts with an error probability of $W = 10^{-9}$.

Fig. 9 gives experimental data for optimization of amplifier pumping current as a function of the parameter $P(\omega_0)/A\varkappa$.

It is clear from the diagram that initially this parameter decreases, since the noise increases slowly with current, while amplification increases exponentially; the parameter then reaches a minimum and begins to rise again. This increase in the parameter is related to gain saturation. With d.c. this occurs primarily due to breakdown of thermal conditions of the amplifier, while with pulsed excitation it results from gain saturation by spontaneous emission. However, we may see from this diagram that although saturation for CW operation is related to amplifier overheating, nonetheless elimination of this overheating does not significantly increase the optimum gain.

In our case the optimum gain is implemented at pumping currents of 100-200 ma and the maximum equipment gain was ~14 dB. By optimizing emission coupling to the amplifier, this level may be increased to values of ⩾20 dB.

Our analysis has assumed that the polarization is identical for the signal wave and the spontaneous emission. This is not always the case, however. It may be demonstrated that accounting for polarization effects does not significantly change the estimate of the sensitivity threshold of the amplifier. Hence in order to simplify this analysis, no difference was assumed in the polarization of the signal wave and that of the spontaneous emission.

6. SPECTRALLY-MATCHED MODULATION OF EMISSION FROM AN INJECTION LASER IN A TRAVELING WAVE AMPLIFIER

When a light pulse propagates along an optical fiber, unavoidable broadening of the pulse occurs due to dispersion. This is particularly significant at information transmission speeds above 400 Mbits/sec when the permissible light guide length is determined not only by losses but also by the dispersion and, consequently, by the optical spectral width of the radiator. The direct pumping current modulation of emission in an injection laser, as a rule, excites a large number of longitudinal modes. As a result, the width of the optical spectrum may exceed that of the spectral band of the modulating signal by a factor of greater than 10^3. The problem of narrowing the optical spectrum of a radiator has been discussed for this reason.

Studies [14-16] have investigated modulation of emission from an injection laser "locked" to the frequency of another single-frequency laser. This has made it possible to reduce the optical spectral width to a band determined by the modulation frequency and thereby to obtain spectrally-matched modulation of emission in an injection laser. The locking arrangement assumes a close correlation of external illumination and resonance frequencies in the laser, whereby the laser becomes highly sensitivity to external frequency drifting.

This study modulates emission from a GaAlAs-laser by modulating the pumping current of a traveling wave amplifier. The experimental set-up is described in Section 2. The experiments were carried out in frequency ranges of 190-500 MHz where the G4-13 generator was used as modulating current generator 7 (see Fig. 2), and 1.75-2 GHz using a generator specially developed for these applications.

Introducing a modulating current into the power supply network of the amplifier serves to modulate its gain and optical length, which in turn modulates the amplitude and phase of the wave passing through the amplifier. Fig. 10 shows typical output emission spectra of the amplifier. The spectrum in static amplification is shown in Fig. 10a; this spectrum traces the spectrum of the master oscillator (accurate within the interferometer resolution). In the absence of a modulating signal (Fig. 10b, c) in addition to the optical frequency of the master oscillator (in this case this frequency is the carrier) two lateral components appear at frequencies $\pm f_M$ from the carrier (the modulating signal frequency). The component 190 MHz from the carrier (Fig. 10b) correspond to the first modulation harmonic. Weak peaks (120 MHz from the carrier) represent the second harmonic of the neighboring transmission orders of the interferometer. The spectral amplitudes of the lateral component (the first harmonic) become comparable to the carrier with an output power from the modulating signal oscillator of: 50 mw for 190 MHz and 300 mw for 2 GHz (Fig. 10b and c, respectively).

The light wave field at the output of the amplifier-modulator may be given as

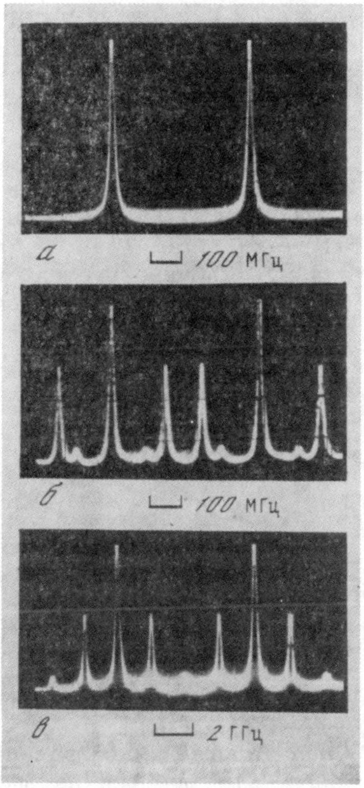

Fig. 10. Output emission spectra from the traveling wave amplifier.
a - zero modulation;
b - f_M = 100 MHz, I_{yc} = 150 ma, P_{BX} = 1.0 mw, P_{Bblx} = 7.3 mw;
c - f_M = 2 GHz, I_{yc} = 150 ma, P_{BX} = 1.2 mw, P_{Bblx} = 8 mw
$P_{BX} = P_{in}$, $P_{Bblx} = P_{out}$

where ω_0 is the angular frequency of the carrier, $\Omega = 2\pi f_M$ is angular frequency of the modulating signal, E_1, E_2, φ_1, φ_2 are the amplitudes and phases of the lateral components. The intensity of this wave at the modulation frequency may have a maximum value with amplitude modulation ($\varphi_1 = \varphi_2$) and a zero value with phase modulation ($\varphi_1 - \varphi_2 = \pi$).

Studies were performed to establish the nature of modulation in a traveling wave amplifier. Fig. 11a gives the intensity of the carrier frequency $P_0 = |E_0|^2$ and the intensity of lateral component $P_1 = |E_1|^2$ plotted as a function of the modulating current. Fig. 11b shows the modulating current dependence of the amplitude modulation percentage $A_\sim/A_-$. Both relations correspond to a modulation frequency of 190 MHz. With a modulating current $I_\sim$ = 20 ma the relation between the intensity of the carrier and the lateral component is $P_0/P_1 = 4$, or $E_0/E_1 = 2$. The amplitude modulation percentage in this case is equal to 15%. If only amplitude modulation was present, with such a relation between the intensity of the carrier and the lateral component, the amplitude modulation percentage would be 100%. We may conclude on this basis that phase modulation predominates in the amplifier.

The dependence of the modulation percentage (the ratio of the intensity of lateral component to the intensity of the carrier) on various parameters was investigated. Fig. 12a gives the P_1/P_0 relation at 190 MHz plotted against the wavelength of the master oscillator. In these measurements the modulating current and input power were fixed ($I_\sim$ = 30 ma, P_{BX} = 0.83 mw). The wavelength of the master oscillator was tuned in the 150 Å range. We see from Fig. 12, a that the modulation percentage has its maximum (60%) at λ = 816.6 nm corresponding to the spectral maximum of the amplifier gain. Tuning served to reduce modulation percentage to 45%. The measurement range was limited by the capacity to obtain single-frequency lasing from the master oscillator at a fixed power level.

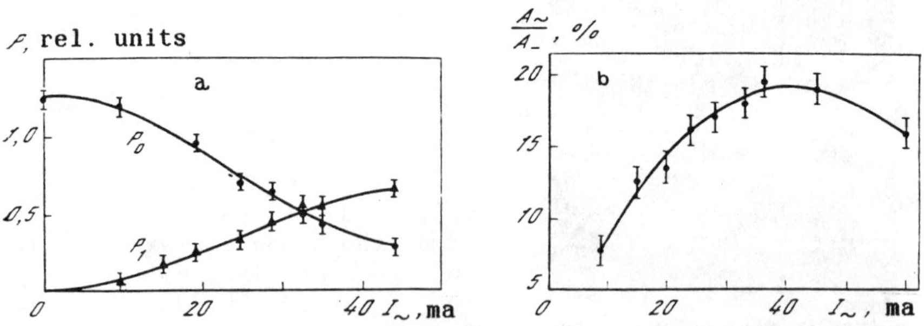

Fig. 11. The intensity of the carrier P_0 and lateral P_1 component (a) and the amplitude modulation percentage $A_\sim/A_-$ (b) plotted as a function of modulation current in the traveling wave amplifier.
f_M = 190 MHz, P_{BX} = 1.0 mw, I_{yc} = 150 ma, λ = 816.6 nm ($P_{BX} = P_{in}$)

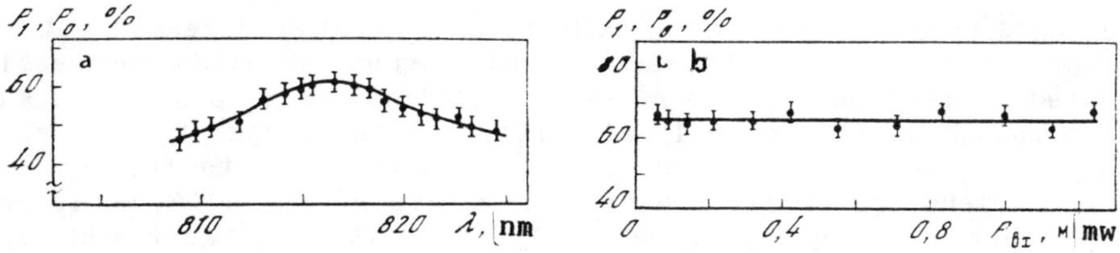

Fig. 12. The ratio of lateral component intensity to carrier intensity P_1/P_0 plotted against master oscillator wavelength at P_{BX} = 0.83 mw (a) and versus input signal power for λ = 816.6 nm (b).
f_M = 190 MHz, I_{yc} = 150 ma, $I_\sim$ = 30 ma. (P_{BX} = P_{in})

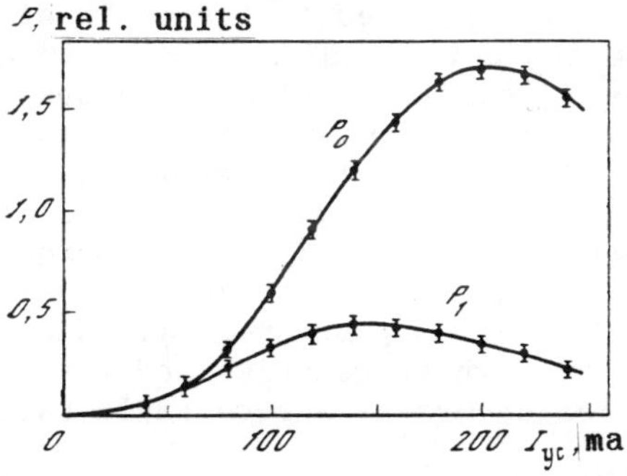

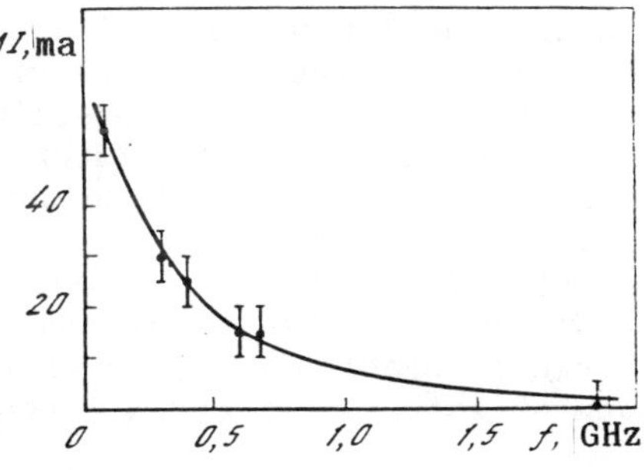

Fig. 13. Intensity of lateral component P_1 and carrier P_0 plotted as function of amplifier pumping current.
P_{BX} = 1.1 mw; λ = 811.0 nm, f_M = 190 MHz (P_{BX} = P_{in})

Fig. 14. The modulation frequency dependence of the difference between pumping currents of the amplifier ΔI corresponding to maximum intensity of the lateral component P_1 and the carrier maximum P_0.
P_{BX} = 1.1 mw, λ = 811.0 nm (P_{BX} = P_{in})

Fig. 12b shows the modulation percentage P_1/P_0 plotted as a function of input power of the master oscillator. The measurements were carried out for a fixed modulation current $I_\sim$ = 30 ma at a modulation frequency f_M = 190 MHz. The wavelength of the master oscillator corresponded to the gain maximum of the amplifier (λ = 816.6 nm). We see from the graph that accurate within measurement results no dependence of the modulation percentage on the input power is observed. This is consistent with the calculation carried out in [12] if we account for the fact that the input signal level in this case is low and phase modulation is used.

This study investigated the dependence of the carrier intensity P_0 and lateral component intensity P_1 on the pumping current level of the amplifier. Fig. 13 also gives the relation for a modulation frequency of 190 MHz and an input power of 1.1 mw. It is clear that this relation has a maximum. The reduction in the P_0 intensity at high currents is related to the reduced gain caused by diode overheating, as noted in section 3. The current dependence of P_1 also has a maximum, although it shifts to lower currents with respect to the maximum of P_0. If only the diode overheating factor were to influence the current dependence of P_1, clearly the maxima of P_1 and P_0 would occur at an identical pumping current. The fact that P_1 has a maximum at lower currents than P_0 indicates that an additional lateral component suppression mechanism exists. These same relations were obtained for other modulation frequencies as well. Fig. 14 gives the modulation frequency dependence of the difference in the pumping currents of the amplifier corresponding to the maximum of P_0 and P_1. The magnitude of this difference decreases with an increase in modulation frequency. This means that the additional lateral component suppression mechanism has a frequency dependence. This may be explained on the basis of the field interaction mechanism in a semiconductor medium as described in study [17]. As the intensity of the P_0 carrier increases, a spectral valley arises in the gain profile, which, in conjunction with thermal overheating, reduces the intensity of the lateral component. This mechanism has a spectral dependence (see [17]). The magnitude of this valley in the gain profile is proportional to $\Gamma/(\Omega^2 + \Gamma^2)$, where Γ is the probability of band-to-band transitions, Ω is the angular modulation frequency. Hence, at low modulation frequencies, the intensity of field interaction is greater than at high frequencies and the saturation of lateral component intensities occurs at a more rapid rate.

Thus, this study has investigated a CW optical amplifier with an equipment gain up to 14 dB operating at room temperature. It is demonstrated that the maximum internal (intrinsic) gain (ignoring coupling losses) is 38 dB and this value is limited by saturation of the amplifier due to intrinsic spontaneous emission.

The characteristic saturation power of the amplifier is determined. For the test samples this value was 8 mw.

An optimum equipment gain of the amplifier of ~14 dB for the signal-to-noise ratio was found experimentally.

We note in conclusion that since emission coupling losses to the amplifier in these experiments were relatively high (~11 dB), obviously it is possible to achieve an optimum equipment gain for this amplifier on the order of 20 dB by improving the emission coupling quality. The experimentally measured maximum gain of 38 dB and saturation power of 8 mw are in good agreement with calculations performed in study [12].

The amplifier was used as an amplifier-modulator to achieve spectrally-matched modulation of emission from an injection laser where optical spectral width did not exceed twice the frequency of the modulating signal. It is demonstrated that both amplitude and phase modulation were present with phase modulation predominating.

This modulation method has several advantages over other methods, particularly those implemented in studies [14-16], specifically:

- high speed (modulation capability at high frequencies);

- elimination of lasing of the modulating diode and, consequently, better stability of the circuit with changes in external parameters, particularly the pumping current and temperature in the modulating diode;

- absence of a Faraday cell (or other elements) for optical isolation of the master oscillator and the amplifier diode.

The use of spectrally-matched modulation makes it possible not only to minimize dispersion spreading of the light pulse, but also to implement spectral multiplexing of optical communications channels to the near-theoretical limit.

Another important advantage of such modulation is the capability to use such high-sensitivity optical signal receiving methods as optical heterodyning and preamplification in the optical amplifier.

BIBLIOGRAPHY

1. Crowe, J. W., Ahearn, W. E. "SEMICONDUCTOR LASER AMPLIFIER" IEEE J. QUANT. ELECTRON., 1966, V. 2, p. 283-289.

2. Kosonocky, W. F., Cornely, R. H. "GaAs laser amplifier" IEEE J. QUANT. ELECTRON., V. 4, p. 125-131.

3. Oraevskii, I. N., Popov, Yu. M. "Semiconductor optical amplifier" PHYS. STATUS. SOLIDI, 1969, V. 21, p. 55-60.

4. Kobayashi, S., Kimura, T. "Gain and saturation power of resonant AlGaAs laser amplifier" ELECTRON. LETT., 1980, V. 16, p. 230-232.

5. Luk'yanov, V. N., Semenov, A. T., Yakubovich, S. D. "The stationary characteristics of a GaAs injection laser with a narrowband input signal" KVANTOVAYA ELEKTRON., 1980, V. 7, p. 2460-2466.

6. Yamamoto, I., Tsuchiya, H. "Optical receiver sensibility improvement by a semiconductor laser amplifier" ELECTRON. LETT., 19870, V. 16, p. 233-235.

7. Simon, J. C., Favennec, J. L., Charil, J. "Comparison of noise characteristic of Fabry-Perot type and traveling-wave semiconductor laser amplifier" ELECTRON. LET., 1983, V. 19, p. 288-290.

8. Fye, D. M. "Practical limitation on optical amplifier performance" J. LIGHTWAVE TECHNOL., 1984, V. 2, No. 4, p. 403-406.

9. Mukai, T., Yamamoto, Y. "Gain, frequency bandwidth and saturation output power of AlGaAs DH laser amplifiers" IEEE J. QUANT. ELECTRON., 1981, V. 7, p. 1028-1034.

10. Goldobin, I. S., Luk'yanov, V. N., Solodkov, A. F., et al. "A double (GaAs)As-heterostructure injection laser traveling wave amplifier" KVANTOVAYA ELEKTRON., 1984, V. 11, p. 375-381.

11. Annenkov, D. M., Bogatov, A. P., Eliseev, P. G., et al. "Spectrally-matched emission modulation for an injection laser in a 2 GHz traveling wave amplifier" KVANTOVAYA ELEKTRON, 1984, V. 11, p. 231-232.

12. Bogatov, A. P. "A laser diode traveling wave optical amplifier" TR. FIAN, 1986, V. 166 (see this volume).

13. Rymov, S. M. "Vvedenie v statisticheskuyu fiziku" [An introduction to statistical physics] Moscow: Nauka, 1976. "Ch. 1. Sluchaynye protsessy" [Part 1. Random processes].

14. Yamada, J., Kobayashi, S. "Improvement in intensity fluctuation noise of a 1 Gbit/s modulated InGaAsP laser by alight injection technique" JAPAN. J. APPL. PHYS., 1980, V. 19, p. L689-L692.

15. Iwashita, K., Nakagawa, K. "Suppression of mode partition noise by laser diode light injection" IEEE J. QUANT. ELECTRON., 1982, V. 18, p. 1669-1674.

16. Goldberg, L., Taylor, H. T., Weller, J. F. "FM sideband injection locking by diode lasers" ELECTRON. LETT., 1982, V. 18, p. 1019-1020.

17. Bogatov, A. P., Eliseev, P. G., Okhotnikov, O. G., et al. "Mode interaction and autostabilization of single-frequency lasing in injection lasers" KVANTOVAYA ELEKTRON., 1983, V. 10, p. 1851-1865.

AN INVESTIGATION OF THE DYNAMICS OF STRIPE-LASERS: SELF-MODULATION, NONLINEAR LOSSES, MODE "CHIRPING"

Yu. M. Popov, I.A. Skopin

Abstract: The lasing dynamics of several types of semiconductor lasers with possible broad practical applications are investigated. Select characteristics of self-modulation conditions are given. The self-modulation pulsation pumping process is investigated. The nonlinear optical losses are studied together with the causes of nonsteady-state lasing and its relation to nonlinearities in the watt-current characteristic. The mechanism of nonlinear losses in planar stripe lasers is discussed.

1. INTRODUCTION. A BRIEF SURVEY ON THE DYNAMICS OF INJECTION LASERS

Experimental research on the dynamics of emission from semiconductor lasers may be divided into three periods. The first period began with studies that established the complex pattern of self-modulation phenomena as well as certain of their regularities [1-4]. In these studies, high-speed photoelectron recorders with a temporal resolution of $\sim 10^{-11}$ seconds played a critical role. These devices made it possible to visualize single events of subnanosecond duration, which was near the upper limit of high-speed equipment.

In spite of the fact that in this early period (the 1960's) the objects of research were broad (nonstripe) homolasers operating primarily at low temperatures (the first generation injection lasers), i.e., from the modern viewpoint, primitive devices, the research in this period succeeded in developing the primary methods of controlling high-speed processes as well as to develop ultranarrow regular pulse generators, and clock frequency generators with pulse repetition periods up to 10^{10} Hz, bistable lasers with speeds of better than 1 ns and to demonstrate the execution of fundamental logic operations using optics [5]. Thus, the component base (on the laboratory level) was developed for discrete optical computer technology. The achievement in this area was the development of existing prototypes of multielement logic circuits (quantum-optical integrated circuits) based on GaAs-lasers [6].

The most important physical idea of the first research period was development of the capability to achieve Q-switching and bistability by including the unpumped (or inhomogeneously pumped) region of the semiconductor in the cavity. It is necessary to use bleachable filters with a heterogeneous optical medium for Q-switched lasers employing other media (ruby, neodymium glass, etc.). Regarding semiconductors, by virtue of the strong nonlinearity in the dependence of the gain (absorption) on the concentration of excess carriers (at a fixed wavelength), the unpumped medium has optical characteristics that differ sufficiently from those of the amplifying medium and is therefore suitable as a bleachable filter medium. We will consider, for example, hard lasing in a two-component medium in the cavity. This is simplified to a design where, beginning at a certain photon density, a further increase is accompanied by a sufficiently strong drop in cavity losses. If the release of S_Φ photons in the active region reduces gain by Δg, the absorption of S_Φ protons (or fewer photons) in the passive region will reduce absorption by $\Delta\alpha$ such that $\Delta\alpha > \Delta g$. This is demonstrated in Fig. 1a which gives a schematic representation of the nonlinear dependence of gain on the excess carrier concentration N_e, qualitatively corresponding to a semiconductor medium. Issues of semiconductor laser dynamics were examined in studies [5, 7-9].

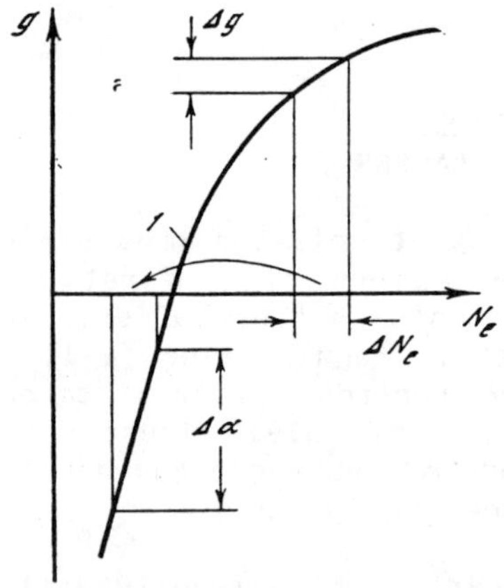

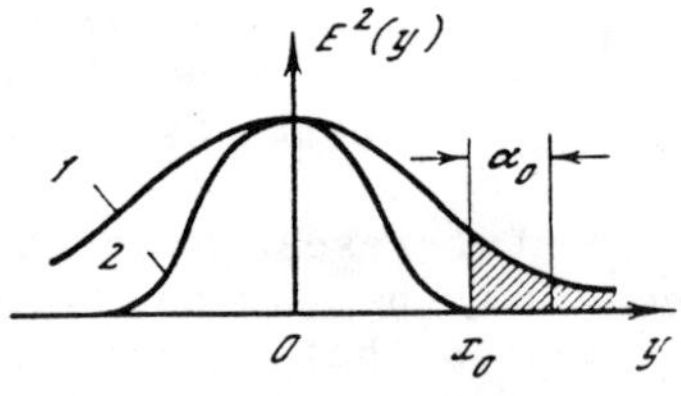

Fig. 1. Diagrams illustrating optical loss nonlinearity producing self-modulation fluctuations.

a - the influence of absorption saturation ($\Delta\alpha$) with a nonlinear concentration dependence of gain (curve 1); b - the influence of self-focusing mode constriction: 1 - initial mode profile; 2 - constricted profile at higher power; x_0 - boundary of the absorbing (α_0) region.

Another fundamental issue in the dynamics of various solid-state lasers, including semiconductor lasers, was establishing the oscillation properties in the laser system (electrons + photons) characterized by significantly different electron and photon relaxation times (10^{-9} sec and 10^{-12} sec, respectively). An analysis of rate equations revealed the existence of limit cycles for the phase trajectories of the oscillators [8, 11, 12] which, with only slight deviations from equilibrium, are characterized by the so-called relaxation resonance frequency $\nu_R \approx (1/\sqrt{\tau_e \tau_\Phi}) \sqrt{Y - 1}$, where τ_Φ is photon lifetime, τ_e is electron lifetime and $Y = I/I_t$ is the relative excess over the lasing threshold. It was noted even earlier that this frequency corresponds to the RF noise peak of the laser [13]. Study [14] demonstrated that the peak of the frequency response of the internal (direct) modulation of laser emission occurs near this frequency. Damped oscillations in transient conditions occur around this same frequency ν_R. A standard relaxation resonance frequency in injection lasers for $Y = 2$ is 1-3 GHz.

The sustained oscillation phenomenon in semiconductor lasers was understood as the manifestation of relaxation pulsations arising from a certain nonlinear mechanism producing instability in the stationary conditions. The instability criterion is reduced to a positive nonlinear loss condition in the cavity (i.e., in the sense of positive feedback). If losses are reduced when emission intensity increases (as in the case of absorption saturation already noted above for a semiconductor), this corresponds to positive nonlinearity and at a sufficient loss reduction rate, produces nonstationary lasing (see, for example, [15]).

The next stage of experimental research marked the transition to more controlled prototypes and laser designs, primarily:

- stripe lasers in which the natural spatial lasing channel is insulated, i.e., the interaction and superimposition of many channels and modes characteristic of broad nonstripe diodes are eliminated (although the spatial heterogeneity factor is not entirely eliminated);

- lasers with selective cavities (compound resonators, diffraction grating cavities, etc.), in which spectral and spatial mode selection is possible, as well as tuning within the amplification band, and controlled variation of losses (feedback).

As we know, the most extensive research on stripe lasers began with the fabrication of CW heterolasers in 1970. Thus, the second stage in research on the dynamics of semiconductor lasers occurred in the 1970's.

New important regularities have been discovered in research on high-speed spectral chronography (using a photorecorder and spectroscope) of laser emission. We will discuss certain such regularities. It has been possible to classify operating conditions from the viewpoint of spectral kinetics in stripe lasers distinct from classifications based on averaged spectra [17, 18, 20, 22].

Spectral kinetics	Averaged spectrum
Piecewise-single-frequency, steady-state spectrum	Multifrequency
Single-frequency, steady-state spectrum	Single-frequency
Multimode, regular fluctuations	Multifrequency
Multimode, irregular fluctuations	Multifrequency

We see that two operating conditions may be differentiated for the averaged spectrum: single-frequency and multifrequency, at the same time that at least four operating conditions may be differentiated in terms of spectral kinetics. Only in one such condition-steady-state single-frequency conditions - do the instantaneous and averaged spectra coincide.

The piecewise-single-frequency operating condition is one variant of the steady-state (total power) condition with predominance of single-frequency lasing, which is nonetheless interrupted by switching of the dominant longitudinal mode. The switching time (i.e., the time when at least two spectral modes have comparable intensity) amounts to a few nanoseconds, which is much less than the duration of single-frequency lasing (when the intensity of any mode is much less than the intensity of the dominant mode). Such an operating condition arises from superluminescent operation as mode competition increases. The fluctuating nature of mode switching not only reflects variations in cavity parameters and temperature, but evidently also is related to the action of spontaneous emission fluctuations. As pumping increases, the role of these fluctuations decreases, resulting in more stable single-frequency operating conditions, since this allows operational fluctuations in external conditions. Mode switching represents an undesirable phenomenon for communications systems, since it generates mode-partition noise.

The transition to pulsation operating conditions, which occurs with a certain increase in pumping, is accompanied by a simultaneous spectral perturbation [20, 22]. In this pulsation operating condition, power dissipation among the many spectral modes increases. It has been established that the instantaneous spectrum measured with a high temporal resolution during the emission peak, is not exactly reproduced from peak to peak, and in the general case differs from the averaged spectrum [18]. RF modulation and self-modulation in transient conditions act in much the same manner [23]. The physical explanation for spectral fluctuations in pulsation operating conditions includes the action of spontaneous emission at the beginning of each spike since there is an intensity increment in several modes and their final intensity depends on the arbitrary number of photons at the initial point in each mode [18, 20]. Simple rate equations are used for multimode operating conditions to calculate the dependence of the average number of excited modes on the pumping current in the presence of pulsations [24]; this was qualitatively confirmed in experiments [25]. The nonreproducibility of the spectrum (with significant emission modulation) indicates a strong constraint on the coherence length, which is more significant than indicated by the number of modes observed. This explains experiments to reduce the coherence length in interferential experiments with injection lasers [26, 27].

Another important experimental observation was the detection of frequency self-modulation of emission accompanying amplitude variations [19]. Studies [16, 18] identified a rapid frequency drift of isolated modes in nonsteady-state operating conditions ("chirping"), although the regularities of this phenomenon could not be established due to the masking of the spectral chronograms from spatially inhomogeneous lasing regions. Studies with stripe lasers [19, 20] have noted that the frequency modulation occurs in sync with amplitude modulation; with regular fluctuations, the frequency shift is $\pi/2$. Moreover, it was discovered that the deviations are synchronous at all observed longitudinal modes. One possible explanation [19] identifies variations in the refractive index of the medium. This explanation was given in greater detail in [28, 29]. This argument accounts for the influence of carriers on the refractive index of the medium and, since the carrier concentration oscillates in nonsteady-state lasing conditions, this influence produces oscillations in the frequencies of the longitudinal modes. The amplitude-frequency interrelation mechanism in semiconductor lasers involves this same process. As this discussion indicates, the mechanism was explored fundamentally in studies [19, 20, 29] and was later reconsidered in a number of studies, primarily [30, 31].

Several important practical circumstances derived from the amplitude-frequency interrelationship, specifically:

- the capability to simultaneously monitor the photon density and the electron concentration in the laser (i.e., both dynamic functions in the rate equations), and to obtain the form of phase trajectories for transient and other nonsteady-state conditions;

- frequency modulation capability employing the pumping current in the 10^9 Hz range for communications systems;

- the need to account for mode "chirping" in dynamic conditions (including single-mode operation) as well as spectral mode broadening [29-31] associated with modulation. Study [32] has demonstrated by calculation that as a result of "chirping" the averaged mode spectra may not only acquire broadening, but may also acquire a complex (such as a split) shape. Today, where long-range fiber-optic communications lines are being developed (greater than 100 km in length without repeaters), mode "chirping" in a 1.55 μm laser operating in a dynamic single-mode condition is among the most significant constraints on communications range (due to the influence of dispersion in the optical fiber in the 1.55 μm range). Thus, in the course of these studies, two primary modulation effects in an injection laser have been identified: mode partitioning (i.e., the transition to multifrequency operating conditions) and "chirping."

A number of other studies may also be attributed to this period, specifically research on the anomalous dynamics of lasers with weak or asymmetric waveguides (see survey in [33], on pulsations in stripe lasers, the influence of operating conditions on emission coherence and on the action of parasitic feedback on noise characteristics and self-modulation).

An important step in understanding the causes of nonsteady-state behavior of injection lasers was the study of self-focusing [34]. It was discovered that as laser emission intensity increases in single-mode operating conditions (when using a selective external cavity to control the mode composition), a speckle structure in the far field appears, corresponding to the formation of spatial self-focusing channels. Estimates have given a rather low self-focusing instability threshold (~5-7% above the lasing threshold). It follows that this type of instability is characteristic of semiconductor lasers and is manifest to a greater or lesser degree in various conditions above threshold. Since in the general case a change in mode configuration is accompanied by a change in optical losses (see Fig. 1b), it is possible that the generation of self-modulation fluctuations is related to self-focusing, as first proposed in study [34]. The contribution of self-focusing was later investigated in greater detail in a number of works [35-38]. We will continue the discussion of this mechanism in Section 10.

The next period, beginning in the 1970's, includes new trends in research and development aimed at eliminating undesirable dynamic processes and improving the characteristics of lasers and laser transmitters. Undesirable phenomena from the viewpoint of practical applications include distortions in conversion to an optical signal, including relaxation transient pulsations and delays, mode partitioning, "chirping", and self-modulation pulsations.

One of the first reports of the possibility for suppressing self-modulation pulsations in an injection laser may be traced to 1970 [17] which established that tuning to single-mode operating conditions in a diode with a compound resonator is accompanied by suppression of sustained oscillations. Later this result was confirmed in other research. This understanding of the influence of emission pulsations on the spectrum was enhanced by understanding the important role of controlling the modal composition in order to avoid self-modulation phenomena.

The current period of research on the dynamics of injection lasers is based on the development of significantly improved designs of single-mode and single-frequency GaAlAs/GaAs and InGaAsP/InP lasers. The most important line of research is to develop lasers operating in a dynamic single-mode condition, i.e., single-mode operation with microwave modulation of emission. Various approaches to this problem have resulted in the formation of the following groups:

1) external cavity resonators;

2) two-stage laser-amplifier/modulator designs;

3) distributed feedback lasers;

4) lasers with compound and C^3-cavities.

Two-stage and "tandem" lasers, as well as C^3-lasers have enhanced capabilities to control emission characteristics by optical interaction and by virtue of their two or more electrical inputs. Various phenomena in compound laser assemblies have been investigated on an extensive scale as early as the 1960's [39]. Today many such designs have been redeveloped [40] and have found practical application (specifically the C^3-laser design).

Results and current research are progressing in the following directions:

1) dynamic single-mode emission;

2) investigation and use of frequency and phase modulation in the microwave range;

3) increasing the maximum modulation frequency to >10 GHz;

4) ultranarrow pulses from short pulse pumping or mode-locking;

5) an investigation of linewidth and "chirping" in single-mode emission;

6) the investigation and suppression of excess noise, including mode partitioning noise, parasitic optical feedback noise, etc.

7) resonance and traveling wave amplifiers employing laser diodes, locking conditions;

8) the use of spontaneous emission, external injection and feedback to suppress self-modulation;

9) the nature of nonlinear mode interaction and its influence on laser dynamics;

10) self-focusing and lateral spatial instabilities as a factor in laser dynamics.

The experimental research in this chapter covers the most widely used stripe heterostructure designs and was conducted over a number of years as new types of injection lasers appeared. The lasers are divided into four primary classes and are investigated on a comparative basis (section 2). Planar stripe lasers (without lateral optical confinement) and stripe lasers with three-dimensional waveguides, including a buried mesa-stripe structure (known as a BH-structure) are compared.

The regularities of self-modulation are investigated (including the case where a large number of identical lasers, ~100 samples, are used), together with the characteristics of sustained fluctuations (Section 4) and the relationship between lasing dynamics and the emission spectrum (Section 5). The dynamics of chirping are examined in Section 6 and an approach is given that makes it possible to develop diagnostics of nonsteady-state lasing.

The relationship between a static nonlinearity and a dynamic nonlinearity, i.e., one producing pulsations is investigated in lasers with nonlinear watt-current characteristics (planar stripe lasers) (Sections 7-9); the nonstationarity regions are established, i.e., regions where the positive feedback criterion is satisfied. The nonlinear mechanisms of this relation are discussed and covered in Sections 9, 10.

2. LASER DIODE SAMPLES FOR RESEARCH

This study utilized various types of double heterostructure CW GaAlAs/GaAs injection lasers operating at room temperature between 830-890 nm, as well as InGaAsP/InP structures operating between 1200 and 1600 nm. For purposes of simplicity, we will differentiate four types of lasers as shown in Table 1.

1. The first type of lasers include GaAlAs/GaAs fine mesa-stripe heterostructure diodes. The double heterostructure was fabricated by liquid phase epitaxy onto silicon-doped n-GaAs substrates. The $Ga_{1-x}Al_xAs$ broadband layer composition was $x = 0.35$ and the composition of the narrowband layer was $x = 0.07$ [41]. The width of the stripe contact for the various samples varied from 10-30 μm. The diodes were soldered using low temperature solder to a heat-radiating copper electrode either on the side with the heterostructure or on the substrate side. Either d.c. heating or CW emission (1-3 watts) from a CO_2-laser ($\lambda = 10.6\mu$m) focused on the holder were used to solder the crystals to the copper holder and to solder the second contact (normally a gold or silver wire ~100 μm in diameter) to the crystal. In both cases, the crystal was attached to the holder using a mounting screen in microscopic increments and monitored through the MBS-2 microscope. Standard dimensions of the active region of these types of lasers are $0.15 \times 20 \times 250$ μm with an operating spectral range of 830-870 nm. The minimum threshold current was 60 ma; the various samples maintained linearity of the watt-current characteristic up to 3 mw in CW operation and up to 10-15 mw in pulsed operation with a single mirror.

Fig. 2 gives the watt-current characteristic for an injection laser of the first type in a pulsed lasing mode at room temperature. This diagram gives the relation of the width of the radiating spot (at half amplitude) on the cavity mirror w_y in the p-n-junction plane. The insert shows the position of the coordinate axes.

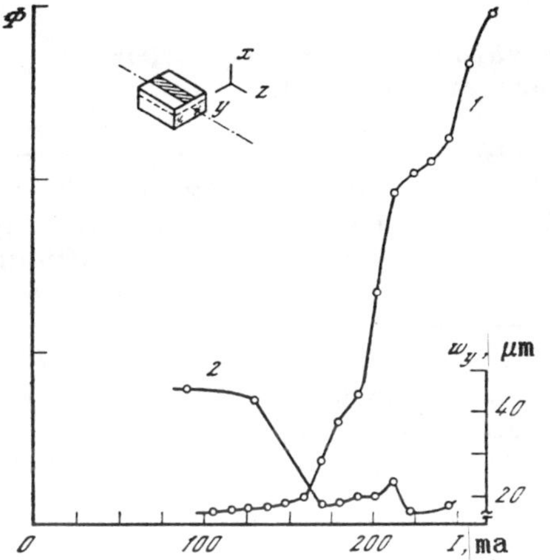

Fig. 2. The watt-current characteristic of a GaAlAs/GaAs planar stripe heterolaser [41] for pulsed lasing (1) and the width of the radiative spot on the cavity mirror (2).
Pumping current pulse duration: $\tau_{И}^{Н}$ = 100 ns; pulse repetition frequency $f_{И}^{Н}$ = 10 kHz; T = 300 K. The width of the stripe contact is ~20 μm. Shown above is the laser and position of coordinate axes.

2. The second type of laser samples are so-called channel lasers [42] or stripe lasers with a "transverse p-n-junction" (TJS lasers). The active region of such a laser is localized at the intersection of the thin narrowband layer of the heterostructure and the p-n-junction given by the diffusion front of the acceptor impurity. With a sufficiently planar heterostructure and diffusion front, the active region is a narrow stripe, virtually the width of a fiber. The thickness of the narrowband layer establishes the width of the active region through the p-n-junction, while its other transverse dimension is determined by the penetration of the injected carriers from the p-n-junction into the narrowband layer. Due to the small transverse dimensions and fixed laser channel, a very low threshold current is achieved in such lasers, and a stable mode composition is preserved (with rather high pumping levels over the lasing threshold), which in turn provides output power linearity.

This study investigated the high-efficiency low-threshold channel heterolasers shown in Fig. 3. The features of such lasers include the use of semiinsulating GaAs in the substrate, the use of a substrate heat sync assembly where the electrical contacts are attached to the p- and n-regions of the diode on one side, and the use of two-stage diffusion to form the transverse p-n-junction of the heterostructure. The substrates had a resistivity $\rho \approx 10^6$ ohm·cm, with a (100) orientation. The heterostructure was fabricated by liquid phase epitaxy; the $Ga_{0.65}Al_{0.35}As$ passive regions were 2-5 μm in thickness, while the active region was 0.4-0.5 μm in thickness and with a composition similar to GaAs. All layers had n-type conduction (as a result of tin and teluride doping). The upper layer of the heterostructure (layer 1, Fig. 3) was used to obtain the ohmic contact. Zn was diffused through the windows in the SiO_2-mask to form the transverse p-n-junction; the diffusion conditions were selected so that the depth of penetration of Zn somewhat exceeded the total thickness of all epitaxial layers. The desired distribution profile was obtained by additional diffusion annealing to obtain the p^+-p-n-junction.

Table 1

Primary characteristics of the emission from various types of stripe injection lasers at T = 300 K

Laser type	Threshold current ma	Lasing wavelength nm	Active layer thickness μm	Width of stripe contact μm	External Differential efficiency w/a
GaAlAs/GaAs system					
Planar stripe laser	60-80	830-870	0.15-0.30	10-30	0.3-0.4
Channel laser	30-60	870-890	0.4-0.5	-	0.3-0.5
Laser with zigzag mesa-stripe structure	30-80	850-870	0.2-0.3	6-15	0.2-0.4
InGaAsP/InP system					
Buriedmesa-stripe structure	30-75	1280-1320	0.15-0.20	5-15	0.5-0.8

Laser type	Dimensions of radiative spot on cavity mirror in the p-n-junction plane with I=1.2 I_t, μm	Threshold for appearance of nonlinearity in the watt-current characteristic in CW lasing		Multimode lasing threshold pumping excess, %	Divergence at half amplitude of emission in the far field in plane of narrowband layer with I=1.2 I_t,°
		Pumping Excess, %	At emission power mw		
GaAlAs/GaAs system					
Planar stripe laser	15	10-30	3	10-30	10
Channel laser	2.0	50	5	50	5
Laser with zigzag mesa-stripe structure	10	100	10	100	8
In GaAsP/InP system					
Buried mesa-stripe structure	7	200	10	200	12

Selective etching of the upper layer is used to separate the p- and n-regions in the diode, as shown in the insert to Fig. 3. This eliminated surface leakage through the contact layer. The samples were soldered to a copper heat sync on the substrate side; wire contacts were attached by thermal compression.

These types of lasers operate in both CW and pulsed operating conditions with a differential of less than 5 ma in the threshold

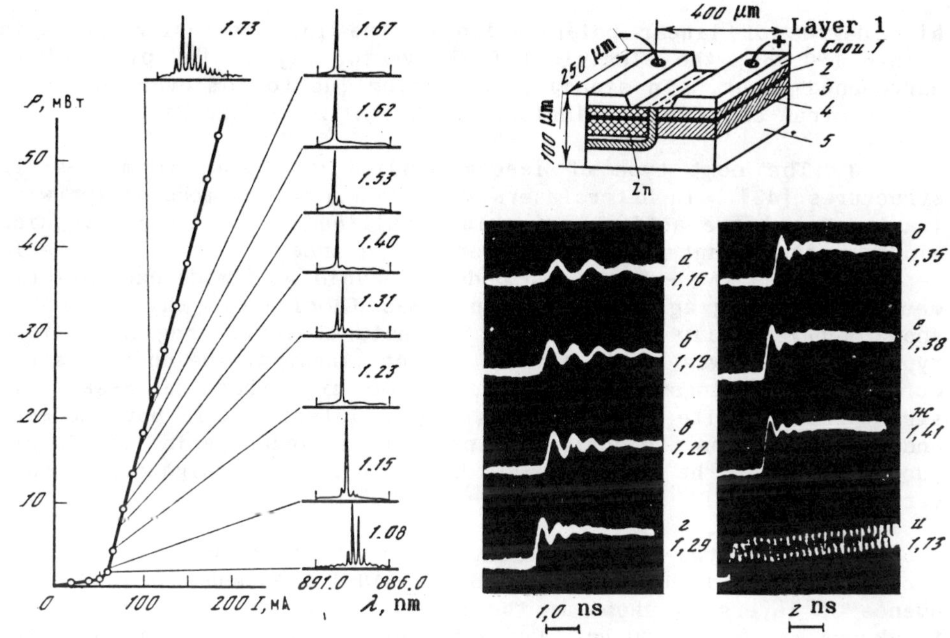

Fig. 3. Watt-current characteristic and spectral behavior of a channel heterolaser in pulsed operating conditions.

$\tau_{И}^{H}$ = 100 ns; $f_{И}^{H}$ = 1 kHz. Right - shape of the emission light pulse; the figures give the relative excess over the lasing threshold. Operating temperature T = 20°C [42]. Above - laser layout: layer 1: GaAs:Sn; $N_e = 4 \cdot 10^{18}$ cm^{-3}; d = 2.0 μm; 2 - GaAlAs:Te; $5 \cdot 10^{17}$; 2.0; 3- GaAs:Sn; $3 \cdot 10^{17}$; 0.45; 4 - GaAlAs:Te; $5 \cdot 10^{17}$; 4.5; 5 - GaAs; $\rho \geqslant 10^6$ ohm·cm

current; the minimum threshold current was 30 ma with standard values of 50-60 ma with a narrowband layer thickness of 0.47 μm. Thus, the minimum threshold current density was approximately 27 ka/cm^2, which is characteristic of homolasers at 300 K.

The watt-current characteristics were linear above the threshold in both conditions (in pulsed conditions linearity was observed up to 50 mw on one side, and with CW it was observed up to 5-6 mw (see Fig. 3)). The maximum differential efficiency was 0.65. The temperature dependence of the threshold current (-60 to +40°C) may be represented by an exponent with a characteristic constant T_0 = 72 K (see [42]).

The angular width of the directional pattern in a plane perpendicular to the heterostructure was similar to a diffraction pattern, and amounted to approximately 30°; in the orthogonal plane it increased from 4° near threshold to 8° at twice threshold. The radiating spot on the cavity mirror had a near-Gaussian shape. Here, the high degree of linear polarization of the laser emission was maintained (~0.96); the electrical field vector lay in the plane of the narrowband layer. In steady-state lasing (up to 70% over threshold) the spectrum contained a single longitudinal mode (see Fig. 3).

3. The next type of lasers studied were special mesa-stripe structures [43]. In heterolasers with even a narrow mesa stripe width (say, 4-6 μm) the active region is a multimode dielectric waveguide, since a significant jump in the refractive index exists along the lateral surfaces (Δn ~ 2.6). In order to avoid multimode excitation, a mesa-stripe zigzag heterostructure was fabricated; this device is shown schematically in Fig. 4. The natural active region of these types of lasers was ih the form of a rectilinear stripe under a stripe contact of analogous shape at the same time that the mesa-stripe structure was quite broad and its zigzag lateral walls both approach and move away from the active stripe (with a zigzag period of ~50 μm). Such a shape of the lateral walls helps to suppress high order transverse modes.

A regular two-sided heterostructure of the normal type was fabricated by liquid phase epitaxy onto a (100) GaAs-substrate. The sequence of layers is shown in the insert to Fig. 4. The active layer thickness was 0.20-0.30 μm. Contact photolithography was used to form the zigzag mesa-stripe on the heterostructure surface; etching was performed to a depth greater than or equal to the depth of the active layer. Then a silicon dioxide layer was applied to the relief surface of the heterostructure; a stripe window for contact to the *p*-side of the heterostructure was then etched into this layer. Two designs were investigated: in one, the width of the mesa structure was 35 μm, while the width of the stripe contact was 15 μm; in the other, more narrow strip, these dimensions were 15 and 6-8 μm, respectively. The ohmic contacts were applied by vacuum deposition and burn in. The laser diodes were produced by cleaving; the cavity length was 200-250 μm.

Standard threshold current values for 300 K in CW and pulsed lasing conditions were 75-90 ma in broad structures and 35-55 ma in narrow structures. The lowest threshold in the second case was 30 ma. These values somewhat exceed the thresholds in rectilinear mesa-stripe heterostructures of the same width, although in this case in lasers with a zigzag mesa-stripe heterostructure with good reproducibility, it is possible to achieve a stable and simple shape of the near field as well as linear watt-current characteristics. In certain broad structures and virtually all narrow structures, zero-peak lasing was observed over a broad dynamic range (up through 2.5 times threshold). Fig. 4 gives the watt-current characteristics in pulsed (1) and CW (2) operation, together with the shape of the emission light pulse for various pumping levels over the threshold value.

4. The fourth type of lasers used in this study were InGaAsP/-InP buried mesa-stripe heterostructure lasers [44]. These lasers are most promising for optical communications systems. By selecting the proper composition of the InGaAsP solid solution used in the active

region, it is possible to achieve lasing in such lasers at 1.3 and 1.55 μm, which are most suitable for the transmission characteristics of existing optical fiber lightguides. The geometry of the active region in the buried mesa-stripe heterolasers nearly ideally (as is the case for lasers with a transverse *p-n*-junction) corresponds to the geometry of the fiber optic light guide, which makes it possible to achieve efficient matching.

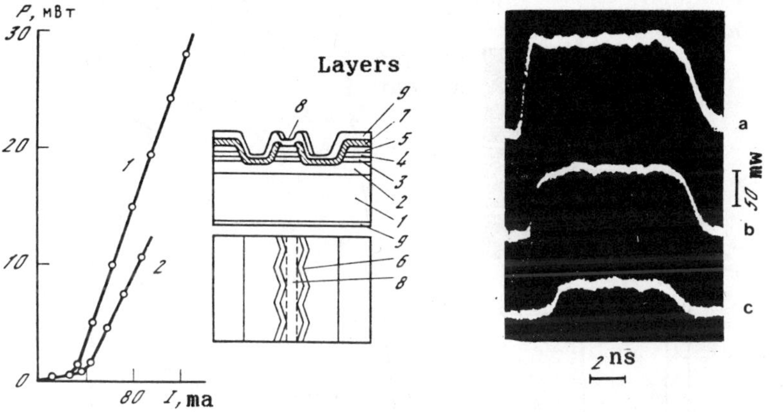

Fig. 4. Watt-current characteristics of a GaAlAs/GaAs zigzag mesa-stripe heterostructure laser [43] in pulsed (1) and CW (2) operation.

$\tau_И^Н$ = 10 ns; $f_И^Н$ = 30 kHz; duration of the leading edge of the current pulse (at 0.1-0.9 amplitude) ~1.5 ns; T = 16°C. Center - layout of the laser diode: layer 1 - *n*-GaAs, substrate; 2 - *n*-GaAlAs, broadband *n*-emitter; 3 - *n*-GaAlAs, undoped layer; 4 - *p*-GaAlAs, broadband *p*-emitter; 5 - P^+-GaAs, contact layer; 6 - mesa-stripe structure; 7- insulating SiO_2-layer; 8 - stripe contact on the *p*- and *n*-sides of the diode; 9 - contact metal-coated layers on the *p*- and *n*-sides of the diode. Right - shape of the light emission pulse for various pumping current levels over threshold: a - 2.25 I_t; b - 1.68 I_t; c - 1.38 I_t; I_t = 35 ma.

The insert to Fig. 5 shows the buried mesa-stripe heterolaser structure based on a four-component InGaAsP solid solution using the two-stage liquid-phase epitaxy process [44]. *P*-type monocrystalline indium phosphide wafers (100) were used as the substrates. After burying the epitaxial planar two-sided heterostructure (the first epitaxy) protective SiO_2 stripes are developed on the heterostructure. A mesa-stripe of given width and height was fabricated by chemical etching of the heterostructure in a CH_3COOH-$Br_2$3 (9:1) solution. The second epitaxy involved burying additional etched nonprotective sites by indium phosphide within the indium melt, combined with zinc diffusion (to a depth of ~1 μm) to establish an opposite *p-n*-junction outside the active region. The threshold current density of the lasers used was 1-5 ka/cm^2. The width of the stripe contact varied from 5-15 μm.

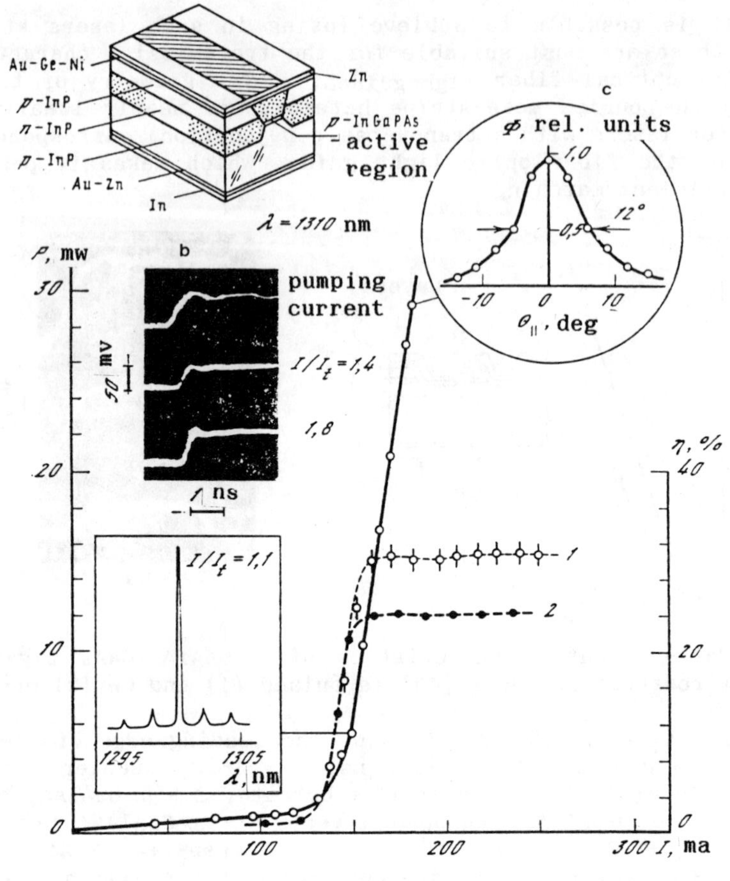

Fig. 5. The watt-current characteristic of the buried mesa-stripe injection InGaAsP/InP heterolaser in pulsed operation.

$\tau_И^Н$ = 100 ns; $f_И^Н$ = 10 kHz, T = 300 K. The coupling efficiency η to a single-mode optical fiber is plotted as a function of pumping current: 1 - with microlens matching ($R_{МЛ}$ ~10 μm); 2 - with direct matching. Inserts: a - laser configuration; b - shape of emission current and light pulses; c - intensity distribution in the plane of the p-n-junction at the cavity mirror; d - lasing spectrum. Width of stripe contact is ~4 μm.

It was noted earlier that the significant reduction in threshold current in buried mesa-stripe heterostructure lasers compared to other types of stripe lasers is achieved primarily due to the development of optical and electron confinement in the lateral direction (in the direction along the p-n-junction plane). The lateral optical confinement effect in these lasers results from the jump in the refractive index on the boundary between the active region and the buried InP layer. As a result, the active region of the laser is in effect a rectangular dielectric waveguide. This serves to reduce diffraction losses and provides spatial stabilization of the lasing channel. The electrical confinement effect produces conditions where all the current flowing through the laser diode goes towards injection of nonequilibrium carriers into the active region. In the buried mesa-stripe structures used in this study, this was achieved by fabricating a *p-n*-junction under the entire contact surface (with the exception of the active region stripe); this junction is reverse biased during laser operation.

Fig. 5 gives the watt-current characteristic of the InGaAsP/InP buried mesa-stripe heterolaser operating in pulsed conditions (pumping current pulse duration of $\tau_{И}^{H}$ = 100 ns; pulse repetition frequency f = 100 kHz) at 1.3 μm at room temperature. We note that a good linear relationship between output power and pumping current is preserved up through twice the lasing threshold. Here, a power of ≥30 mw from one end of the laser was achieved.

We note that the appearance of nonlinear sections in the watt-current characteristic, often observed in planar stripe heterolasers, is related to the deformation of the optical power distribution along the *p-n*-junction. Deformation occurs as a result of self-defocusing of emission [34] when a certain power level is achieved. A "inbuilt" jump in the refractive index exists on the boundaries of the active region along the *p-n*-junction in buried mesa-stripe heterolasers. Such a jump sharply increases the self-focusing threshold, which provides linearity of the lasers.

The inserts to Fig. 5 show the shape of the emission light pulse for various pumping levels over the lasing threshold, together with the intensity distribution of the radiating spot on the cavity mirror.

An investigation of lasers based on mesa-stripe structures in InGaAsP/InP solid solutions grown on *p*-type InP substrates has demonstrated that stable single-mode lasing and linearity of output power with a change in pumping current are characteristic of such lasers. In order to achieve lasing at the fundamental transverse mode (the TE_{00}- or H_{00}-mode) the width of the active stripe should be small, up to 2.5-3 μm, although single-mode lasers operating at the H_{00} mode may be found among laser samples with a width up to 4.5 μm. If InP is not used as the burying material, but rather a quaternary solid solution whose refractive index is closer to that of the active region, it is possible to achieve single-mode lasing with an active stripe width of up to ~9 μm.

Single-mode low-threshold lasers based on an InP-buried mesa-stripe heterostructure have, as a rule, a regular symmetric near field pattern with dimensions of the radiating spot in the order of 3 × 1.5 μm; here, emission divergence in the vertical plane (normal to the active layer plane) is 40-60° and in the horizontal plane is 10-20°. By varying the thickness and width of the active layer of the laser, it is possible to change these values; for example, it is possible reduce the vertical divergence to 24° and the horizontal divergence to 8°. A typical thickness of the active layer, optimized for threshold current, was ~0.3 μm.

3. AN EXPERIMENTAL SET-UP AND MEASUREMENT METHOD FOR MEASURING RADIATIVE CHARACTERISTICS OF INJECTION LASERS

This study investigated lasing dynamics, the spectral composition of emission as well as other radiative characteristics of the injection stripe heterolasers described above. The layout of the experimental assembly is given in Fig. 6. Current pulses 3-100 ns in duration with a pulse rise time of ~1.5 ns and a pulse repetition frequency of 1-10 kHz were used for laser pumping. The G5-58, G5-19, G5-54, and G5-11 series manufactured nanosecond pulse generators were used for pumping. The signal at the generator output was less than 15-20 volts. A special-purpose transistor pulse generator with a pulse rise time of 0.2-0.5 ns together with a harmonic oscillator employing Gunn diodes with a pulse repetition frequency up to 5 GHz were used for a more comprehensive investigation of the dynamic and modulation characteristics of the lasers. The optical emission signal from

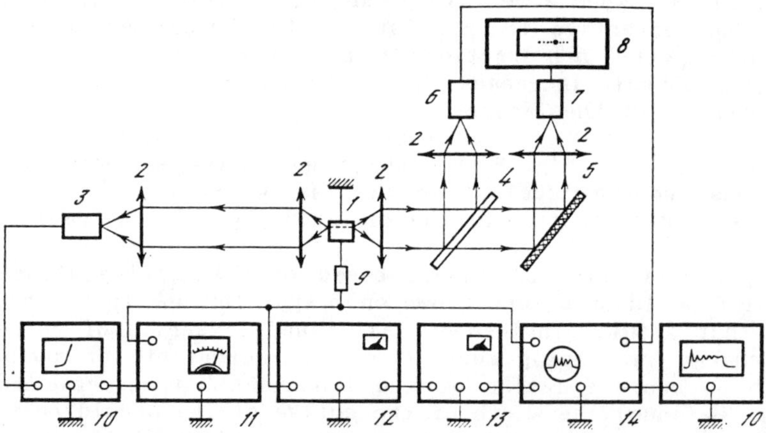

Fig. 6. Block diagram of the experimental assembly for investigating the radiative characteristics of injection lasers.

1 - test laser; 2 - microlenses; 3 - FD-7K (or FD-24K) silicon graduated photodiode; 4 - semitransparent mirror; 5 - diffraction grating (600 lines/mm); 6 - LFD-2 (or LFD-3) avalanche photodiode; 7 - "Elektronika-L-50" television camera; 8 - "Elektronika VL-100" television; 9 - matching load resistor R_H = 51 ohms; 10 - PDS-021 X-Y plotting potentiometer; 11 - V4-14 pulsed millivoltmeter; 12 - G5-48 pumping pulse generator; 13 - G5-15 firing oscillator; 14 - S7-8 stroboscopic oscillograph (S7-9, S7-11, S7-13).

the lasers was recorded by the FER-2M high-speed photorecorder with a temporal resolution of approximately $5 \cdot 10^{-11}$ sec, and the image was recorded on film (or by the LFD-2 avalanche germanium photodiode). Film photometry was then used to determine the rising edge, the period, the frequency and the percentage of modulation of the emission peaks.

In addition to the various existing high-speed emission detectors a Au-nn^+-GaAs surface-barrier structure photodiode may also be used successfully in optical communications links; this structure is designed for a range of 0.25-0.9 μm. The rise and decay time of the photoresponse signal from the injection laser when a reverse bias of -50 v is injected to the photodiode is 0.1 ns with an avalanche multiplication of 5-10. The primary characteristics and parameters of these photodiodes are described in study [45].

A special-purpose transistorized d.c. generator with current amplitude control of ±1 ma was used to investigate the radiative characteristics of lasers in a CW operating condition or under combined pumping.

The laser emission was focused onto the receiving area of the avalanche photodiode (see Fig. 6) by means of microlenses 2, and was fed through a fiber segment to investigate the emission coupling efficiency. The power supply voltage of the avalanche photodetectors was 18.5 v for the LFD-2 and 69-70 v for the Schottkey barrier GaAs-photodetector; the signal was taken from a 50 ohm load. In these conditions, the temporal resolution of the photodetectors was on the order of 0.1-0.2 ns.

Stroboscopic oscillograph 14 (see Fig. 6) (S7-8 or S7-13) was used on the two input channels; a signal taken directly from the photodetector was injected to one channel, while a signal proportional to the amplitude of the pumping current with identical time delays in the feed cables was injected to the other (or to the V4-14 pulsed millivoltmeter). In these conditions it was possible to monitor the amplitude of the pumping current pulses and light pulses as well as to record their shape on film together with their temporal characteristics. The PDS-021 (10) X-Y plotting potentiometer was also used for convenience; this unit may accept a signal from the oscillograph output.

Absolute measurements of the luminous flux radiated by the lasers were carried out using the FD-7K (3) graduated silicon photodiode with a sensitivity of ~0.4 a/w for an average current at λ = 880 nm. In order to support automatic recording onto plotting paper, the signal from the photodiode proportional to the laser radiated power was injected the y-coordinate of the PDS-021, while the output signal from the V4-14, proportional to the amplitude of the current flowing through the laser, was injected to the x-coordinate. The emission spectra were recorded by spectrometer 5 formed by a 600 line/mm diffraction grating with a resolution of ~0.03 nm.

The laser emission reflected off of the diffraction grating was focused through telelens 2 to camera 7 and to television 8. The line sampler was then used to feed the signal from the television to the S1-72 oscillograph. Thus, the spectrum (or the emission intensity distribution in the near field) could be observed visually. The change in the emission directional pattern and emission intensity was monitored on the other end of the laser using the FD-7K calibrated photodiode (2).

4. A COMPARATIVE INVESTIGATION OF THE LASING DYNAMICS OF INJECTION LASERS

Relaxation pulsations in transient conditions in the majority of lasers of the first type are excited when pumping exceeds the threshold level by only a slight amount (≤1%) with a modulation percentage of >10%. Fig. 7 gives the dependencies of the repetition period (curve 1), the duration of the first peak at half intensity ΔT^1 (curve 2) and the rise time of the emission pulse $\tau_{\Phi p}$ (curve 3) in relaxation pulsation conditions on the pumping current level over threshold. Current pulses of duration $\tau_{И}^{H} \sim 200$ ns with a pulse repetition frequency of ~ 1 kHz and a pulse rise time of ~0.2 ns were used in this case for laser pumping. The dependence of the pulsation repetition period on the pumping current, plotted on a dual logarithmic scale, has a slope of 0.5 for small pulsations (modulation percentage of approximately 20%). With developed pulsations, this slope increased to 1.5 and higher (Fig. 8), which was consistent with calculations [46, 47]. The insert to Fig. 8 gives the light pulse shapes in relaxation and self-modulation pulsation conditions as measured by the FER-2M optical recorder for various pumping levels. It should be noted that the lasing conditions may change when the pumping level of the laser changes; for example, weak pulsations may be replaced by developed pulsations. In this case a break occurs in the frequency or pulse repetition relation of the pulsations. Fig. 8 gives the pulsation repetition period plotted as a function of pumping excess over the threshold level for a regular stripe laser (laser of the first type) and a laser with a transverse *p-n*-junction (curves 1 and 2, respectively). Fig. 8 also demonstrates the difference in the excitation threshold of developed pulsations for lasers of the first two types.

At a specific pumping current (for lasers of the first type, 3-5% above threshold; for lasers of the second type, ≥70%) self-modulation pulsations of emission intensity are excited (see the emission light pulse waveforms in Fig. 3). An analysis of the lasing dynamics of 100 samples of lasers of the first type demonstrated that 36 samples in their initial state demonstrated spiking with a variable modulation percentage. An analysis of the characteristic spiking times in pulsing samples of laser diodes made it possible to plot the distribution function of the period and relative pulse duration of the pulsations. This function is given in Fig. 9 and is in good agreement with calculation [11, 12, 48-50].

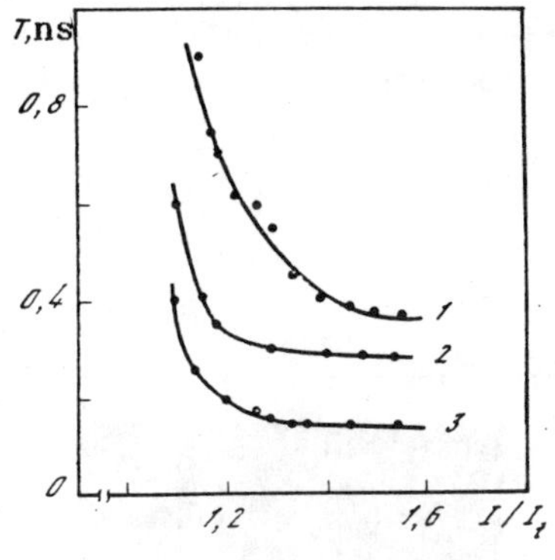

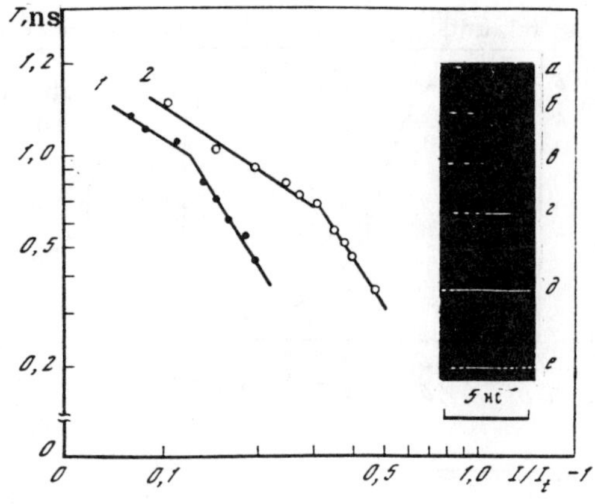

Fig. 7. The pumping current dependencies of the pulse repetition period (1), duration (2) and rise time (3) for the first relaxation pulsation spike in emission intensity from a planar stripe laser

Fig. 8. Relaxation pulsation period plotted as a function of pumping current over threshold for lasers of the first (1) and second (2) type Right - shape of the laser emission light pulse for lasers of the second type for various pumping levels: a - 1.05 I_t; b - 1.11 I_t; c- 1.13 I_t; d - 1.16 I_t; e - 1.20 I_t; f - 1.73 I_t.

Lasers of the third and fourth type have demonstrated significant relaxation (see the inserts to Figs. 4 and 5) pulsations and the absence of sustained pulsations across the entire test range of pumping currents: up to 2.5-3 times over threshold.

The pulsation period, extrapolated for twice threshold, was used to estimate the photon lifetime in the cavity in lasers of the first two types. This time was ~$2 \cdot 10^{-12}$ seconds. The electron lifetime τ_e, estimated by the decay time of the transient process, was 1.3-5 ns with a spontaneous lifetime τ_s found by the delay τ_d of lasing with respect to the current pumping pulse, and was 1.5-8 ns (see Table 2). A more precise measurement of the dynamic characteristics of emission was carried out using the FER-2M photorecorder with a temporal resolution of ~$5 \cdot 10^{-11}$ seconds. Fig. 10 gives τ_d plotted as a function of the normalized injection current for a laser of the second type. This relation is accurately approximated by the logarithmic graph of the function $\tau_d = \tau_s \ln [I/(I - I_t)]$, $\tau_s = 2$ ns. The shape of the emission light pulse for various pumping currents is shown in the diagram.

The quick response of radiative processes in semiconductor heterolasers has attracted and continues to attract attention for applications in the modulation of injection laser emission. This has become the focus of attention since the development of low-threshold single-frequency CW lasers operating at room temperature.

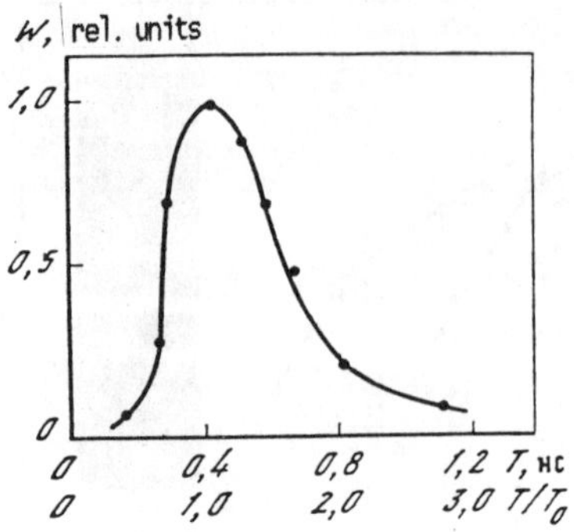

Fig. 9. The distribution function W of the period T and relative pulse duration T/T_0 of pulsations in an injection laser.

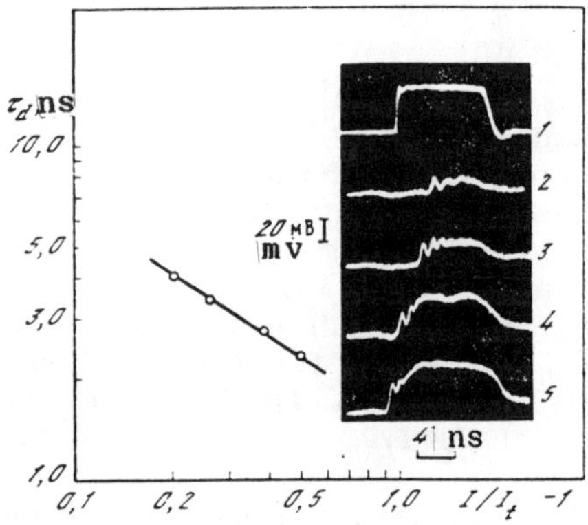

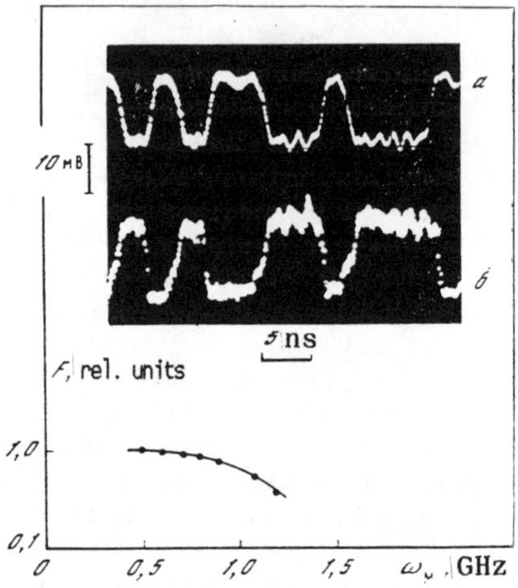

Fig. 10. The delay τ_d of an emission light pulse from a GaAlAs/GaAs channel heterolaser [42] plotted against the current pulse with variations in the pumping level
$\tau_{\text{и}}^{\text{н}}$ = 12 n2; $f_{\text{и}}^{\text{н}}$ = 10 kHz; T = 300 K. Right - Waveform of the current and light emission pulse; 1 - pumping current pulse; 2-5 - I/I_t = 1.06 (2); 1.11 (3); 1.20 (4); 1.39 (5).

Fig. 11. Pulsed modulation of a GaAlAs/GaAs channel laser [42] by a pseudorandom pulse sequence signal at ~200 Mbits/sec.
a - waveform of the current pumping pulses; b - waveform of the emission light pulses; c - the internal modulation efficiency F versus modulation frequency ω_{M} for the case of modulation by a sinusoidal signal.

Our experiments demonstrated the possibility of modulating pumping power to achieve modulation of a channel laser. For example, modulation by a signal with a pseudorandom sequence of pulses at ~200 Mbit/sec (Fig. 11) was achieved as well as microwave modulation using a sinusoidal signal at frequencies up to 3 GHz. Fig. 11 also gives the curve plotting the internal modulation percentage or output emission amplitude F as a function of the frequency of the sinusoidal signal ω_{M}. We note that this curve has no peak near the characteristic pulsation frequency of ~0.5·1010 sec^{-1} characteristic of regular GaAlAs/GaAs planar stripe heterolasers.

It should be noted that the transient process in semiconductor lasers resulting from the quasiresonance interaction of nonequilibrium carriers and photons in the laser cavity, causes the resonance-like behavior of the modulation efficiency $F(\omega_M)$. In a channel laser, the frequency of relaxation pulsations in emission intensity in transient operating conditions was ~1.5-3 GHz, i.e., approximately 3-5 times greater than the relaxation pulsation frequency in regular planar stripe lasers. This may qualitatively explain the absence of a resonance peak on the $F(\omega_M)$ relation up through frequencies of ≤1.5 GHz.

Table 2

The primary characteristics of lasing dynamics for injection lasers of various types at T = 300 K.

Laser type	Duration of transient process, ns	Frequency of relaxation pulsations GHz	Excitation threshold in spiking I/I_t	Frequency of self-modulation pulsations in emission intensity GHz	Pulsation modulation percentage	Delay of emission with respect to pumping pulse (10 ns, 10 kHz), ns
GaAlAs/GaAs System						
Planar stripe	3-5	0.1-2.0	1.05-1.3	0.1-0.6	100	1.5-10
Channel	1.3-2	3-5	1.7	2.0-4.0	100	1.0-2.0
Zigzag mesa-stripe	1	-	2.5	-	3	8
InGaAsP/InP System						
Buried mesa-stripe	1	-	3.0	-	3	3

For standard GaAlAs/GaAs planar stripe heterolasers, when the frequency of the modulating signal coincides with the pulsation frequency or is an even multiple of this frequency, the latter is either divided or multiplied. These relations may also be applied to pulsation locking; the frequencies of these pulsations may also be locked to even order multiples (or even order fractions) of the repetition frequencies of an external modulating signal. Thus, analog reproduction of a pumping signal into emission from a regular planar stripe heterolaser has been limited by nonlinear distortions in the vicinity of frequencies near the frequency of the modulating signal.

The primary characteristics of the lasing dynamics of test laser samples, as well as the possibility for using such devices in optical

Table 3

A comparison table of injection lasers of various types for use in optical communications links

Laser type	Nonlinearity of volt current characteristic (suitability for analog systems)	Simplicity of device and potential low cost	Anticipated service life, h	Maximum radiated power in the linear range of the volt-current characteristic, mw: CW	Pulsed operating conditions
Planar stripe	No	Yes	10^5	3	10
Channel	Yes	No	10^5	5	50
Zigzag mesa-stripe	Yes	Yes	10^4-10^5	5	50
Buried mesa-stripe	Yes	Yes	10^4	10	100
(λ - 1300 nm)		No			

Laser type	Attenuation constant in multimode fiber-optic lightguides dB/km: Minimum value	Standard	Laser emission coupling efficiency in pulsed mode with microlens coupling to optical fiber, %: Single mode ($R_{\text{МЛ}}$=10 μm)	Multimode ($R_{\text{МЛ}}$=10 μm)	Power delivered to multimode fiber with microlens matching ($R_{\text{МЛ}}$=10 μm): CW	Pulsed operation
Planar stripe	2	4-6	10	90	1-2	3-6
Channel	2	4-5	45	90	3-5	5
Zigzag mesa-stripe	2	4-5	40	90	10-12	12
Buried mesa-stripe (λ = 1300 nm)	0.5	1	36	90	16	16

communications links are given in Tables 2 and 3. These results from an investigation of the lasing dynamics of various types of low-threshold single-mode GaAlAs/GaAs and InGaAsP/InP injection lasers have demonstrated that transverse *p*-*n*-junction lasers, zigzag heterostructure lasers as well as buried mesa-stripe heterolasers are the most promising optical oscillator-modulators for long-range fiber-optic communications links. The selection of any of a variety of laser modifications will be based on a comparison of the various radiative and dynamic characteristics.

5. AN INVESTIGATION OF THE SPECTRAL COMPOSITION OF EMISSION FROM GaAlAs/GaAs AND InGaAsP/InP INJECTION LASERS

The lasing spectra of test heterolasers of the second, third and fourth types have certain common regularities:

1) with the working current only slightly exceeding the threshold (up to ~10%) several longitudinal modes are excited;

2) with an increase in current, a relative increase in emission intensity is observed in one of the longitudinal modes, and with Y = 1.1-1.3 lasing becomes single-frequency lasing up to significant levels over the threshold current;

3) a spectral shift towards the lower wavelength mode or a serial (occasionally across several mode intervals) switching to a longer wavelength mode is observed with an increase in pumping current due to heating of the active region of the laser.

The coexistence of several longitudinal modes near the lasing threshold evidently is related to the fact that the spectral profile of the amplification line in these operating conditions is gently sloping, and the gain is similar in the neighboring 3-5 modes near its maximum. This factor in conjunction with fluctuations in the intensity of spontaneous emission in these modes, produces gain conditions nearly identical for several longitudinal modes. The important role of spontaneous emission in the spectral constriction of the envelope was demonstrated theoretically in study [13]. A further increase in pumping current activates the mode competition mechanism characteristic of a homogeneously-broadened emission line from a semiconductor medium. Stabilization of the spatial distribution of optical power in a three-dimensional waveguide as well as spatial pumping uniformity are provided in lasers with good current and lateral optical confinement, i.e., a uniform gain distribution is achieved in the active region. This makes it possible to achieve single-frequency stable lasing conditions without using external selection elements.

Single-frequency lasing in test lasers of the second, third, and fourth types was observed through a two to three fold level over threshold current at 300 K; in this case, an increase in pumping current produced a shift in the lasing wavelength towards longer wave frequencies (both continuous variation and in jumps), which was related to the heating of the active region of the laser. Fig. 3 shows the position of the spectral line of a channel heterolaser in single-frequency pulsed operation at 300 K with variations in the pumping current. A continuous variation in the lasing wavelength resulting from the temperature change in the refractive index of the material in the active region was observed in a current range of 2-4 ma in CW conditions and 10-30 ma in pulsed operation. Analogous results are also characteristic of other types of test single-mode lasers. The shift in the spectral gain maximum of a channel heterolaser (including buried mesa-stripe heterolasers and zigzag stripe heterolasers) operating in single-mode steady-state conditions produces a sequential switching of the lasing line from one longitudinal mode to the neighboring, longer-wavelength mode, as the pumping current increases. The mode switching process continues as the pumping current increases further. As regular self-modulation pulsations are excited in a channel laser (Fig. 3 where $Y \simeq 1.73$) the spectral line shifts towards shorter wavelengths over a single mode interval, lasing becomes multi-mode and the separate modes as well as the entire spectrum are simultaneously broadened. After excitation of chaotic pulsations (with a further increase in pumping) the entire spectrum evens out, and it is difficult to identify separate modes within the spectrum. The following values were obtained from graphs for a channel heterolaser: $d\lambda/dT$ = 0.04 nm/K, $d\lambda/dI$ = 0.04 nm/ma. In this case, a current level three times over threshold in pulsed operation produced an increase in temperature of 17-20 K in the active region of the channel laser.

6. FREQUENCY SELF-MODULATION OF EMISSION IN INJECTION LASERS AND THE LASER LIMIT CYCLE

In investigating the lasing dynamics of semiconductor lasers, it is possible to observe not only the emission intensity but also the concentration of free carriers in the active region of the laser. Such a capability is due to the influence of the current carriers on the refractive index and frequency of each mode. The localized concentration of excess carriers in the active region of the laser depends on the local electromagnetic field intensity, since it is determined by the balance between pumping and the rate of radiative transitions.

Earlier we discussed two experimental facts: first, the frequency of an isolated mode experiences drift [16, 18, 20], reaching a rate of up to 60 GHz/ns; second, a broadening of separate modes [29-31] resulting from frequency self-modulation of emission is observed in the time-averaged spectrum,. An analysis of spectral chronograms [19] with high temporal resolution ($\sim 5\cdot 10^{-10}$ to $5\cdot 10^{-11}$) as well as high spectral resolution (~0.05 nm) of an insulated lasing channel in a stripe laser has demonstrated that frequency self-modulation is a characteristic instability in single-mode lasing. Frequency self-modulation accompanies the excitation of pulsations in single-mode and the breakdown of single-mode lasing in the transition to nonsteady-state (spiking) multimode lasing.

Synchronous frequency deviations in neighboring longitudinal modes have been observed in stripe lasers, as well as a relationship between intensity pulsations and frequency drift, or chirping. The maximum rate of change in wavelength occurs at the intensity spikes. An interrelationship also exists between the concentration of excess current carriers and the refractive index of the medium, where the intensity spikes correspond to a drop in electron concentration, as may be anticipated from solutions of rate equations. Qualitative results may be obtained by analyzing rate equations using phase planes [31].

The simplest system of rate equations, ignoring spatial and multimode mode effects as well as spontaneous emission, takes the form

$$dS_\Phi/dt = GS_\Phi - S_\Phi/\tau_\Phi, \quad dN_e/dt = J - GS_\Phi - N_e/\tau_e, \tag{6.1}$$

where $G = G(N_e)$ is the coefficient describing the rate of the stimulated transitions, τ_e is the lifetime of the electron-hole pairs (recombination time), τ_Φ is the lifetime of the photons in the cavity, N_e is the electron density, F_Φ is the photon density in the cavity and J is the rate of excitation of the current carriers by pumping.

If we incorporate the dimensionless variables

$$t' = t/\tau_e, \quad N = N_e/N_{et}, \quad S = S_\Phi/S_{\Phi t}, \tag{6.2}$$

system (6.1) is written in the form

$$\dot{S} = \frac{\tau_e}{\tau_\Phi} S\,[G(N_e) - 1], \quad \dot{N} = Y - N - \frac{\tau_e}{\tau_\Phi} SG(N_e), \tag{6.3}$$

where $Y = I/I_t$, the point above numbers designates differentiation with respect to t'.

We may reduce rate equation system (6.3) to a single second order nonlinear equation

$$\ddot{x} = \dot{x}^2/x + x\,(x + 1) = -\varepsilon\dot{x}\,[1 + (Y - 1)\,x], \tag{6.4}$$

where

$$x = S\tau_e/(Y - 1), \quad t^* = t\,[(Y - 1)/\tau_e\tau_\Phi]^{1/2}, \quad \varepsilon = [(\tau_e/\tau_\Phi)(Y - 1)]^{-1}, \tag{6.5}$$

the dot over characters designates differentiation with respect to t^*. In view of the fact that the parameter $\varepsilon \ll 1$ with $Y > 1$ (since for injection lasers $\tau_e \simeq 10^{-9}$ sec, $\tau_\Phi \simeq 10^{-12}$ sec), the solutions of equation (6.4) are similar to the solution of an analogous equation with a zero right half, investigated in detail in [12, 49, 50]. Such an equation has the integral of motion

$$(\dot{x}/x)^2 = C + 2\ln x - 2x \tag{6.6}$$

(where C is the integration constant), determining the phase trajectories.

With a significant pulsation amplitude ($C \gg 1$), an approximate solution of equation (6.4) with a zero right half yields

$$x_{\min} = \exp(-C/2), \quad x_{\max} = C/2. \tag{6.7}$$

The time dependence of intensity in the range $x_{\min} \leqslant x \leqslant 1$ may be represented as

$$x(t) = \exp\left\{\frac{C}{2}\left[\frac{C-2}{C^2}(t^*)^2 - 1\right]\right\}, \quad 0 \leqslant t^* \leqslant t_1^* = \sqrt{C}. \tag{6.8}$$

In the vicinity $1 \leqslant x \leqslant j_{\max}$

$$x(t) = (C/2)\,\mathrm{ch}^{-2}\,[\sqrt{C}\,(t^* - t_1^* - t_2^*)/2], \quad t_1^* \leqslant t^* \leqslant t_1^* + 2t_2^*, \tag{6.9}$$

where

$$t_2^* = (2/\sqrt{C})\ \text{arch}\ (\sqrt{C}/2). \tag{6.10}$$

It is clear that intensity pulsations with a high amplitude occur in two stages. The first stage is a slow increase in the field amplitude from x_{min} to 1, then a rapid rise in the field from 1 to x_{max} and a return to the initial state. In the first stage, a short emission pulse is released with a duration

$$T_0 = 3{,}52\ C^{-1/2}\,[\tau_e\tau_\Phi/(Y-1)]^{1/2}. \tag{6.11}$$

For the period through the limit cycle, we derive

$$T = 2\,(C + 2\ln C)C^{-1/2}\,[\tau_e\tau_\Phi/(Y-1)]^{1/2}. \tag{6.12}$$

Substituting solutions (6.8), (6.9) into equation (6.1) produces the temporal dependence of electron concentration as well as the change in the concentration of carriers over a single spike, which are as follows:

$$\frac{N_e(t)}{N_{e0}} = \begin{cases} 1 + \tau_\Phi t\Omega\,[(C-2)/C], & 0 \leqslant t \leqslant t_1^*, \\ 1 - \sqrt{C}\,\tau_\Phi\Omega\,\text{th}\left[\sqrt{C}\,(t - t_1^* - t_2^*)/2\right], & t_1^* \leqslant t \leqslant t_1^* + 2t_2^*, \end{cases} \tag{6.13}$$

$$\delta N_e = 2N_{e0}\tau_\Phi\Omega\sqrt{C-2}, \tag{6.14}$$

where N_{e0} is the steady-state value of inversion equal to threshold,

$$\Omega = [(Y-1)/\tau_e\tau_\Phi]^{1/2}. \tag{6.15}$$

Study [51] has derived an expression for mode broadening with well-developed pulsations:

$$\overline{\delta\lambda} = \sqrt{2}\,N_{e0}\frac{\lambda_0}{n_0}\frac{dn}{dN_e}\tau_\Phi\Omega\sqrt{C}, \tag{6.16}$$

where λ_0 is the wavelength of steady-state lasing, N_0 is the refractive index, dn/dN_e = $3.6\cdot10^{-21}$ cm^3 [51].

Using calculated values (6.13), (6.16) in accordance with study [51], we may obtain a system of equations for the limit cycle in the phase plane plotted on photon density S_Φ and electron density N_e:

$$S_\Phi^2 = \begin{cases} \mathscr{E}_0^2\exp(-C/2)\exp\{[(\lambda-\lambda_0)^2/C]/[2(\lambda_0-\lambda_{max})^2]\} & (\text{область } I), \\ \mathscr{E}_0^2\cdot 0{,}5C\,\{[1-(\lambda-\lambda_0)^2(C-2)]/(\lambda-\lambda_{max})^2 C\} & (\text{область } II), \end{cases} \tag{6.17}$$

where

$$\lambda_0 \leqslant \lambda \leqslant \lambda_{max};\quad \lambda_0 - \lambda_{max} = N_e\,(\lambda_0/n_0)(dn/dN_e)\tau_\Phi\Omega\sqrt{C-2}, \tag{6.18}$$

λ_{max} is the maximum frequency deviations over the spike time.

Calculation of photon drift S_Φ (through one end of the cavity) as a function of emission power P was carried out for an injection laser with the following parameters: width of the stripe contact w = 18.5 μm; cavity length L = 320 μm; volume of active region V = $5.92 \cdot 10^{-9}$ cm^{-3}; stripe contact area $s = 1.85 \cdot 10^{-7}$ cm^2; photon lifetime in cavity: $\tau_\Phi \simeq 2 \cdot 10^{-12}$ sec; electron lifetime: $\tau_e = 2 \cdot 10^{-9}$ sec; pumping current density j = 10.5 ka/cm^2; relative pumping current level over threshold: ~30%; ratio of thickness of active layer to optical confinement factor d/Γ = 1 μm. The following expression was used for emission power P

$$P = \hbar\omega \, \mathfrak{N} \, V S_\Phi / \tau_\Phi s, \tag{6.19}$$

where $\mathfrak{N}$ is the translation coefficient, equal to $2.6 \cdot 10^{11}$ $kw^{-1} \cdot cm^{-3}$.

Studies of the influence of temporal nonstationarity on the spectral composition of emission were carried out using standard planar injection lasers (lasers of the first type) as well as channel lasers (lasers of the second type). The method described in Section 3 was used to record the emission light pulses and lasing spectra.

Fig. 12a gives the time base of the laser emission spectrum in pulsing operation; in Fig. 12b the dotted line traces the experimental phase curve (or "phase portrait") of the excitation process of self-modulation pulsations in the emission intensity of a planar stripe injection GaAlAs/GaAs heterolaser. The phase plane is plotted on the coordinates of the power density of emission through a single end of the resonator P and for emission wavelength λ with the following translation consistent with expression (6.9) to the photon density S_Φ

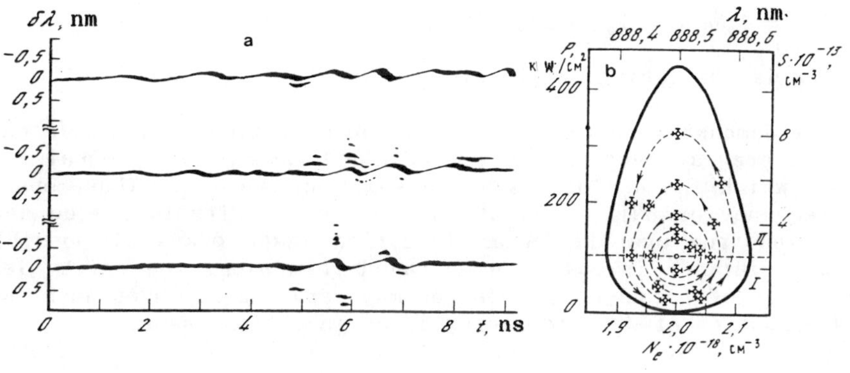

Fig. 12. Time base of the emission spectrum for a laser with resonant rise of pulsations [20] (a) and the phase plane on a (P, λ) or (S_Φ, N_e) plot (b).
Dotted line: experimental phase trajectory for pumping current 30% over threshold; solid line: calculated shape of limit cycle based on formulae (6.16) - (6.19).

and the electron density N_e with translation coefficients $\Re_1 = 2.6 \cdot 10^{11}$ kw$^{-1}\cdot$cm^{-3} (S_Φ (cm^{-3}) = $2.67\cdot10^{11}$ P(kw)) and $\Re_2 = 2.6\cdot10^{11}$ kw$^{-1}\cdot$cm^{-3} ($\delta\lambda$ (nm) = $1.5\cdot10^{-18}$ δN_e (cm^{-3})). The experimental points are plotted every 0.24-0.26 seconds based on an analysis of the spectral chronograms from study [20] (see Fig. 12a). The spectral chronograms were obtained from the FER-2M image converter tube. The solid line in Fig. 12b gives the calculated shape of the limit cycle in the rise of self-modulation intensity pulsations obtained using formula (6.17) for a relative pulse duration of four, a maximum spectral broadening of 0.3 nm and 100% modulation, corresponding to the quantities observed in experiment [31]. It is clear from Fig. 12b that the phase trajectory of the pulsation rise process has two stages (state I and II). The representative point rapidly passes through the upper half-loop of the cycle (region II, Fig. 12b) on the phase plane, corresponding to the release of a short pulse, with slower motion through the lower half loop (region I). In the latter case the emission intensity remains less than steady-state, and since the excess carriers recombine at a lower rate, they are accumulated. The behavior of the representative point on the phase plane is analogous to its motion through the limit cycle. The representative point moves in a counterclockwise direction. The slope of the phase trajectories is significant along the entire plane, with the exception of regions adjacent to the lines $S_\Phi = S_{\Phi 0}$, $N_e/N_{e0} - 1 = 0$. Thus, the phase trajectories take the form of a spiral around the singularity $S_\Phi = S_{\Phi 0}$ and $N_e = N_{e0}$ (N_{e0} is the steady-state value of inversion equal to threshold, $N_{e0} = 1.9\cdot10^{18}$ cm^{-3}). With developed emission intensity pulsations, the theoretical estimate of the maximum relative inversion oscillation is $\delta Nie_{max}/N_{e0} \approx (4\text{-}10)\cdot10^{-2}(C-2)^{1/2}$. In this regard with a relative pulsation duration of 3-6, the relative inversion oscillations may reach 8-30%. Wavelength variation during the emission spike stage is

$$\delta\lambda = \lambda_0 \delta n/n_0 = \lambda_0 \, (dn/dN_e) \, dN_e/n_0, \tag{6.20}$$

where λ_0 is the lasing wavelength.

As demonstrated in study [51], the deviation of an isolated mode in the averaged spectrum may exceed the mode interval, equal to 0.2-0.3 nm, which in turn diffuses the spectrum overall. Thus, observing the frequency or half-width of an isolated longitudinal mode and the entire spectrum overall in an injection laser makes it possible to follow variations in $N_e(t)$ in averaged form along the cavity length. Observation of frequency modulation may serve as an independent method for diagnostics of dynamic lasing in an injection laser.

7. A PHENOMENOLOGICAL MODEL OF THE NONLINEAR WATT-CURRENT CHARACTERISTIC OF AN INJECTION LASER

Consistent with steady-state solutions of rate equations (6.1), the photon density S_Φ in a laser mode increases linearly with an increase in pumping current density j and in proportion to its magnitude, minus the threshold level. The rate of photon generation in the active layer is balanced by photon losses to emission to the external medium and to reabsorption within the active layer. In a linear approximation, the internal losses comprise a fixed portion of total losses. As a result, photon loss to the external medium increases linearly with pumping. The expression for the output (watt-current) characteristic of a semiconductor laser takes the form

$$P(I) = \eta_{\text{д}} (\hbar\omega/e)(I - I_t), \tag{7.1}$$

where $P(I)$ is coherent emission power, $\eta_{\text{д}}$ is differential efficiency, I_t is threshold current, I is pumping current, $\hbar\omega$ is quantum energy, e is electron charge. The threshold current density is normally represented as

$$j_t = j^0 + \alpha_1/\beta + \alpha_2, \tag{7.2}$$

where j^0 is the inversion threshold, β is specific (differential) gain, α_1 is the internal optical loss coefficient, α_2 is the internal loss coefficient converted to the dimensions of the coefficient of absorption. In a planar resonator, the external losses caused by emission release from the ends of the laser diode are equal to

$$\alpha_2 = -(\beta L)^{-1} \ln R, \tag{7.3}$$

where L is resonator length, R is end reflection coefficient ($R_1 = R_2 = R = 0.3$). Substituting formulae (7.2) and (7.3) into (7.1) we may derive an expression for the output power of emission in the form of

$$P(j) = \eta_0 s (j - j^0)/(1 - \alpha_1 L \ln R) - P^*, \tag{7.4}$$

where η_0 is the pumping efficiency, s is the area of the active layer, j is the variable pumping current density. We may write the expression for the parameter P^* in this case in the form

$$P^* = -\eta_0 s \ln R/\beta L. \tag{7.5}$$

Consistent with expression (7.4) the nonlinear watt-current characteristic will have a variable slope in the nonlinearity region with a certain center of rotation of the tangential line at the point $j = j^0$, $P = -P^*$, as shown in Fig. 13.

Linearity of the output characteristic $P(j)$ in semiconductor lasers is also normally observed when steady-state solutions of rate equations are unsuitable, such as in spiking. Deviations from a simple linear form of relation $P(j)$ may occur due to variations in the modal composition of emission, as well as changes in the volume of the portion of the active region producing the output, overheating of the active region as well as various degradation phenomena. As a result, lasers with nonlinear characteristics have a variable threshold current value.

Since the average pulsed emission power in steady-state and spiking operation is maintained, we may derive the form of the function $\alpha_1(S_\Phi)$ from the watt-current characteristic assuming that gain saturation occurs at threshold. In the nonlinearity region there is a change in total losses; consequently, the gain g changes and since the intensity of spontaneous emission is related to the gain, the losses also change. Such a nonlinear variation in losses in the active region of the laser may be described as a change in the lasing threshold, i.e., it is possible to incorporate the dependence of the threshold I_t on radiated power P. An analogous technique is used in studies [52, 53]. The mathematical expression for the altered lasing threshold in accordance with Fig. 16 may be represented as

$$j_t' = (P^*j' + P'j_t^0)/(P' + P^*), \qquad (7.6)$$

where P' is the radiated power at a pumping current density j'.

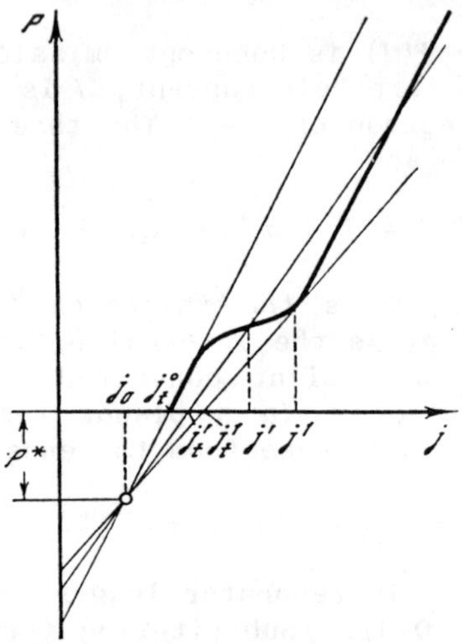

Fig. 13. The calculated watt-current characteristic of an injection laser with the center of rotation of the tangential line located at the point (j_0, P^*) corresponding to formula (7.4). The straight lines plot the slope of the characteristics for various pumping current densities.

According to formula (7.2) the change in the internal optical loss coefficient α_1, accounting for the change in the threshold current density, may be expressed as

$$\Delta\alpha_1 = (\alpha_0 - L^{-1}\ln R)(j_t' - j_0)(j_t^0 - j_0)^{-1} + L^{-1}\ln R, \qquad (7.7)$$

where a_0 is the loss coefficient at the lasing threshold, j_1^0 is the initial threshold current density, j_t' is the changed threshold current density.

The value P' calculated by formula (7.5) for the parameters outlined above was $P^* \approx 24$ mv for conditions where pumping efficiency is $\eta_0 \approx 1$ for stripe injection heterolasers. The internal optical loss coefficient α_1 may be represented through the internal parameters of the laser in the form

$$\alpha_1 = -L^{-1}\ln R\,(\eta_{\text{п}}^*\eta_{\text{п}}^{-1} - 1), \qquad (7.8)$$

where $\eta_{\text{п}}^*$ is the internal differential activity equal to the internal conversion efficiency or pumping efficiency.

With $\eta_{\text{П}}^{*} \to \eta g =_{\text{П}}$ from both higher and lower values, $\alpha_1 \to 0$. Hence, it is necessary to know the value η_0 with significant accuracy. If $\eta_0 = 0.5$, the measured losses vary by a factor of 4.

8. THE STABILITY CRITERION FOR STEADY-STATE LASING IN AN INJECTION LASER

In stripe contact semiconductor lasers, developed regular self-modulation pulsations may be excited even when pumping only slightly exceeds the threshold value. In lasers with well-developed waveguide properties, the spiking excitation threshold is ≳2.5-3.0 times the threshold current.

The nonstationarity of emission from semiconductor lasers noted in previous experiments [1-5] manifests a threshold nature, i.e., it arises only at a certain critical pumping level over the lasing threshold, or at a specific emission power.

It has been noted experimentally that sustained emission intensity pulsations often appear in the vicinity of a nonlinearity. Hence it is assumed that within the active region of the laser, in addition to an amplification region, an absorption region also exists. In the amplification region, the photon density S_Φ increases, while the pumping discontinuity reduces the electron density N_e, which in turn serves to reduce gain g. In the absorption region, an increase in N_e changes absorption. If the reduction in absorption compensates for the reduction in gain, a temporal instability in the form of self-modulation pulsations will arise in the laser emission [1-3, 54]. Rate equation system (6.3) in this case may be represented as [15]

$$\dot{S}_\Phi = (\tau_e/\tau_\Phi) S_\Phi [N_e - \alpha(S_\Phi)], \qquad \dot{N}_e = Y - N_e[(\tau_e/\tau_\Phi) S_\Phi + 1]. \tag{8.1}$$

The total number of equilibrium positions of rate equation system (8.1) depends on the form of function $\alpha(S_\Phi)$ functions as the isoclinal with horizontal tangential lines. The isoclinal $dS_\Phi/dN_e = 0$ on the phase plane will have a shape that depends on the form of the function $\alpha(S_\Phi)$. Ignoring spontaneous emission, it will take the form

$$N_e = \alpha(S_\Phi). \tag{8.2}$$

For the case of positive feedback $dN_e/dS_\Phi < 0$, or $d\alpha/dS_\Phi < 0$. In lasers, an increase in the cavity Q with an increase in emission power P reduces the stability of equilibrium position. For the case of a single intersection of $\alpha(S_\Phi)$ with the boundary stability curve of steady-state lasing on the phase space plotted on coordinates S_Φ and N_e, the nontrivial equilibrium position is singular. If it is stable, this particular point is a focus. In order to make the focus unstable, it is necessary to increase feedback. Linearization of rate equations (8.1) near the singularity and a stability test in accordance with [15] requires that the following equality be satisfied

$$d\alpha/dS_{\Phi} \sim d\alpha/dP \leqslant -(S_{\Phi 0}+1)/GS_{\Phi 0} \leqslant -Q^{-1}+Q^{-1}S_{\Phi 0}^{-1}, \quad (8.3)$$

where $S_{\Phi 0}$ is the steady-state value of S_{Φ}. In this case, the unstable singularity is encompassed by a stable limit cycle and emission takes the form of self-modulation intensity spikes [31].

During the generation of a light pulse, there is a noticeable change in the electromagnetic field intensity distribution along the *p-n*-junction; here, the order of the transverse mode does not change, but its width does change. This result demonstrates that the waveguide properties of a planar stripe injection laser without an inbuilt waveguide in the *p-n*-junction plane are relatively weak and may change with a change in the electromagnetic field intensity in the cavity. Here, the field spreads to passive regions, and additional nonlinear losses arise which may produce a discontinuity in steady-state lasing and may also excite spiking.

Fig. 2 shows a sample nonlinear watt-current characteristic of a planar stripe GaAlAs/GaAs heterolaser (with a stripe contact width of ~20 μm) as well as the intensity distribution at the mirror (near field) and the pumping current dependence of the width of the luminescent spot in the plane of the *p-n*-junction. Fig. 2 shows the change in the position of the spatial lasing channel in the vicinity of the nonlinearity; the width of the luminescent spot on the mirror also changes. The position of the spatial lasing channel changes due to self-focusing [34] which may excite spiking. For this spiking mechanism, only rapid (compared to the period of the spikes of ~0.1-1 ns) nonlinear processes are significant, unlike the slow thermal heating processes. Changes in the refractive index due to saturation are also irrelevant for this mechanism.

For the case of injection lasers and clearly-expressed waveguide properties of the active region (such as channel heterolasers, lasers with a substrate channel) the transverse field distribution is determined primarily by a discontinuity in the refractive index, and not by a discontinuity in the gain. Changes in the gain have only a slight influence on the field distribution in the transverse direction, and there is no self-modulation. However, a change in waveguide properties or a discontinuity in the refractive index during lasing have a noticeable influence on emission losses in the active region. The existence of a nonlinear addition to the refractive index in this self-modulation mechanism indicates that additional Q switching of the active layer occurs due to a change in the refractive index over time: a "breathing" waveguide model [55, 56] or, what amounts to the same thing, a decrease in diffraction losses with an increase in emission intensity due to self-focusing. The associated loss nonlinearity may excite self-modulation pulsations.

9. THE ACTION OF NONLINEAR LOSSES IN THE ACTIVE REGION OF AN INJECTION LASER ON SELF-MODULATION EMISSION PULSATION ZONES

In semiconductor lasers, the active medium is characterized by a significant nonlinearity in the refractive index (or the dielectric constant). In nonsteady-state lasing, the existence of a nonlinear addition to the refractive index changes the level of diffraction losses from the active region with changes in emission intensity over time. Emission losses become nonlinear and may cause emission spiking.

Analysis of the nonlinearity of optical losses makes it possible to demonstrate a wide variety of dynamic operating conditions in injection lasers and to provide related classification rules. Moreover, knowledge of regularities in the various operating conditions of a laser with nonlinear losses may be useful for understanding the properties of actual injection lasers that often have nonlinear, discontinuous or hysteresis-like watt-current characteristics and to solve a number of practical problems (such as suppressing pulsations or, conversely, obtaining a regular sequence of narrow emission pulses).

This study investigates the interrelationship between the nonlinear behavior of internal losses in the active region of lasers of the first type and spiking [57, 58].

Spiking in the test planar stripe lasers of the first type was excited by pumping over threshold by 5-30% (see Table 2). The threshold current density was j_t = 3220 Å/cm^2. The watt-current efficiency was ρ = 0.42 watts/amp per side with a differential quantum efficiency $\eta_{\text{Д}}$ = 0.6. The internal loss coefficient at threshold was α_0 = 40 cm^{-1}, the differential gain was β = 0.05 cm/a, with an inversion threshold j^0 = 1622 a/cm^2 and an end reflection coefficient $R_1 = R_2 = R = 0.3$.

Fig. 14 gives a standard nonlinear characteristic of a planar stripe GaAlAs/GaAs injection heterolaser of the first type in pulsed operation (with a pumping pulse duration $\tau_{\text{И}}^{\text{H}}$ = 20 ns, a pulse repetition $f_{\text{И}}^{\text{H}}$ = 10 kHz and T = 300 K). Also shown here is the change in the waveform of the emission pulse with a change in the pumping current through the laser diode, and the interrelationship between the pulse wavefront and the nonlinearities. It is clear from the diagram that in the range of variation in the pumping currents under examination (up to Y = 2) two zones (I, II) of regular self-modulation emission intensity pulsations are observed with a variable modulation percentage of M (the variation in M is shown in Fig. 15).

The insert to Fig. 14 shows the spike frequency f plotted as a function of the pumping level over the lasing threshold $Y = I/I_t$ in the pulsation zones: the circles correspond to excess pumping levels observed in the experiment, while the dots correspond to calculated values accounting for the change in the threshold current density (see (7.6)). The f $(Y - 1)$ relation shown is on a logarithmic scale in the

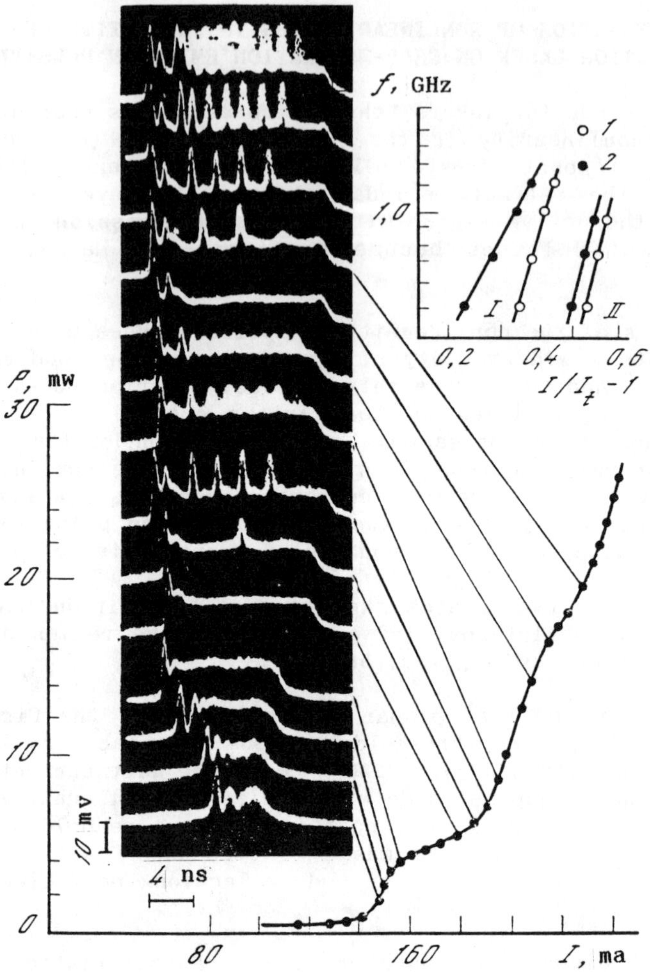

Fig. 14. The watt-current characteristic of a planar mesa-stripe GaAlAs/GaAs injection heterolaser in pulsed operation at room temperature.

$\tau_И^H$ = 20 ns, $f_И^H$ = 10 kHz. Left - waveform of emission pulse for various pumping currents; right - the frequency of spikes in self-modulation pulsation zones I, II plotted as a function of pumping level over threshold: 1 - experiment; 2 - calculation.

spike zones and may be approximately described by a linear function with a slope of ~1.7. With a change in lasing threshold $I_t(P)$ (see (7.6)) this relation has a slope of ~1.5, which is characteristic for heavily modulated spikes. The slope of the $f(Y$ experimental relation derived here is greater than the anticipated calculation slope of 0.5. This slope of relation $f(Y)$ is determined by the limit cycle parame-

Fig. 15. Nonlinear behavior of internal optical losses α, corresponding to the watt-current characteristic shown in Fig. 14 (1) and the loss derivative $d\alpha/dP$ (2) of a planar mesa-stripe GaAlAs/GaAs heterolaser plotted against radiated power P.
Width of the stripe contact: 15 μm; 3- boundary stable lasing curve calculated by formula (8.3) (shaded regions correspond self-modulation pulsations) 4- modulation percentage M in pulsation zones.

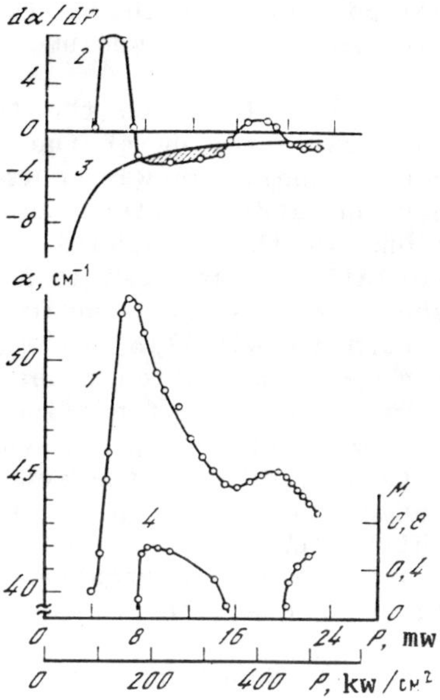

ter. The modulation level of the spikes M or the relative pulse duration T/T_0 may serve as this parameter. Experimental measurements of the temporal characteristics (the period, frequency and relative pulse duration of the spikes) has demonstrated that with an increase in the spike frequency, the modulation percentage and relative pulse duration change (see, for example, Figs. 14, 15). Hence the change in slope or a break on the $T(Y)$ or $f(Y)$ relations with developed or regular pulsations may be explained by a change in the limit cycle parameter (modulation percentage or relative pulse duration) on the phase plane plotted on S_Φ and N_e coordinates. The maximum change in threshold current ΔI in the vicinity of the nonlinearity in our experiments was ~10%. Here, the total loss coefficient changed by ~7-8 cm^{-1}.

It should be noted that the nonlinear behavior of the internal optical loss coefficient α_1, in addition to generating separate spiking zones, may also excite a second, a third, etc. transient process. Fig.14 demonstrates rather significant distortions in the regular transient process with an explicit nonlinearity in the watt-current characteristic (excitation of a second transient process), as well as the interrelationship of nonlinearities in the output watt-current characteristic with various dynamic lasing conditions in a stripe injection laser of the first type. Damping of pulsations in transient conditions is highly dependent on nonlinear losses. Within a specific range of variation in $d\alpha/dS_\Phi$ (see (8.3)) the attenuation time constant goes from infinity to zero. Hence, the uncontrolled sources of optical nonlinearity may be the cause of the variance in the observed transient conditions.

An output characteristic nonlinearity in the form of inflections may clearly be seen in Figs. 2 and 14. On both sides of the inflections, the differential efficiency is significantly greater than in the deflection region. This typical nonlinearity for stripe contact GaAlAs/GaAs lasers may be caused by self-focusing in the inverted semiconductor medium, i.e., the interrelationship between the refractive index of the medium N and the electron concentration N_e.

Fig. 15 gives the form of the internal loss function $\alpha_1(P)$ in the active region of the laser (curve 1) calculated from the experimental nonlinear watt-current characteristic (see Fig. 14) of an injection stripe heterolaser of the first type, accounting for variations in the threshold current density as well as the rate of loss variations, as described by the derivative $d\alpha_1/dP$ (curve 2); also shown here is the boundary stable lasing curve (curve 3) calculated in accordance with [15] (see (8.2)), and the change in modulation percentage M in pulsation zones (curve 4). All curves are accurate within ~30%. The role of nonlinear losses in sustaining the pulsations is as follows. At the spike, due to a reduction in losses, the reduction in intensity occurs at a somewhat higher level than on analogous trajectory with $N_e = \alpha\ (S_\Phi) = 1$. In this case, the top of the limit cycle shifts towards a lower electron density. As a result, the emission pulse produces a greater change in inversion, than with damped pulsations; hence, the process is entirely repeated in each cycle.

It should be noted that in lasers with linear characteristics, such zones cannot be differentiated, and spiking occurs within a single broad zone. In order to excite nonsteady-state lasing in injection lasers, an insignificant change in internal losses (occasionally a few percent), $\leqslant 2\ cm^{-1}$ [59] is all that is necessary; such changes may not produce significant power nonlinearity. However, experience has demonstrated that in regular planar stripe heterolasers where sufficient lateral and optical confinement is not provided, a linear characteristic is virtually never observed in practice.

10. AN ANALYSIS OF POSSIBLE EXCITATION MECHANISMS OF SPIKING IN INJECTION LASERS

We may conclude based on experimental data obtained from a study of numerous planar stripe injection lasers that the mechanism of discontinuities in the watt-current characteristic, based on the nonlinear behavior of optical losses and which generates spiking, is a quick response mechanism. As a result, the spiking regions are closely linked to the characteristic nonlinearity regions. We know that such nonlinearities are caused by the transverse instability in the spatial lasing channel and, consequently, the same nature of the nonlinearity may be used to explain the self-modulation emission mechanism. The theory of lasing in planar stripe heterolasers that explains the nonlinearities incorporates the influence of the injected electron concentration N_e on the dielectric constant and the self-focusing dynamics with an increase in emission power [34]. Nonlinear deformation and switching of the dominant transverse mode in lasers without an inbuilt waveguide significantly increase the amplitude of pulsations [60] and may be responsible for their excitation. The change in the width and position of the spatial lasing channel due to self-focusing serves to reduce field penetration into the passive regions along the

p-n-junction plane and, consequently, may serve to reduce diffraction losses. When self-focusing reduces internal optical losses as well as their rate of change with an increase in emission power, and when these changes are sufficiently large (for example, in the sense of criterion (8.3)), the self-focusing effect may excite self-modulation. We know that the activation time of self-focusing is in the order of a few transit times of the cavity; therefore, this is a rather rapid process.

It should be noted that the saturation absorption effect in passive regions to which the field penetrates is unavoidable in structures without lateral optical confinement, and the contribution of this mechanism is added to the contribution from self-focusing. With the laser locked into lasing conditions and with power hysteresis at threshold, the saturation absorption effect may predominate. This is particularly obvious with pumping discontinuities along the cavity length. In addition to such threshold effects, an analogous phenomenon is observed in planar stripe heterolasers long after lasing and sufficiently removed from the lasing threshold. In such cases a predominant action of the self-focusing mechanism may be anticipated. It was noted earlier that in narrow stripe structures with lateral optical confinement, in addition to transverse stabilization of the spatial laser channel, emission is stabilized over time. Such lasers have, as a rule, linear watt-current characteristic over a broad dynamic pumping range together with stable lasing. As a result, a single mode dominates within the emission spectrum.

Thus, the primary factors responsible for nonlinear losses in a semiconductor laser may be reduced to the following.

1. Saturation absorption in a certain portion of the laser mode. This mechanism is most strongly manifest in special inhomogeneously-pumped diodes. It is postulated that random absorption regions exist in homogeneous diodes; their dimensions may increase with degradation. The contribution of microscopic absorption regions (such as in the surface layer due to surface recombination) is insufficient, although it increases significantly if we account for the local overheating of such regions.

2. The optical nonlinearity that changes the configuration of the mode (deformation) with an increase in intensity, i.e., self-focusing.

The first factor, as we see from the survey in Section 1, is covered most comprehensively in the literature [3, 5, 8-12]. The second factor attracted attention later [34] and its theory is less well-developed. To some extent this is related to the cumbersome nature of the required calculations. It is noted in study [34] that self-modulation may be caused by loss nonlinearity and study [35] gives the instability criterion for steady-state operation, when the losses depend on intensity. Study [36] uses numerical calculations to show that accounting for self-focusing reduces the attenuation of relaxation pulsations.

Study [37] discusses the self-focusing instability factor in a proton bombardment insulation heterolaser. It has been discovered in practice that the width of the beam depends on the change in the carrier concentration during the emission pulse. It increases in the valley between pulses and drops during emission (the variations in width in the near field are between 9 and 7.2 μm in lasers with a nominal contact width of 12 μm). Narrowing of the beam due to self-focusing produces beam concentration in the high local gain region. This reduces the damping of relaxation pulsations and helps to enhance other nonlinear effects such as saturation absorption etc., which play a role in the generation of self-modulation. We note that the thermal waveguide plays an important role in the gain instability produced by self-focusing.

Experiments with planar stripe lasers in [38] have demonstrated that the thermal waveguide (with a lag of 1-5 μm) makes it possible to develop as significant a lateral confinement as desired in the stripe structure, which in a number of test samples made it possible to eliminate the nonlinearity in the watt-current characteristic. This is achieved in pulsed operating conditions by applying preliminary pulses (of both positive and negative polarity). Naturally both the shape and stability of the spot in the near field change. Thus, at least two components of the self-focusing effect exist: the component with greater lag: the thermal component, and the quick response component related to the spatial inversion burning.

A numerical calculation of the change in mode gain due to self-focusing beam narrowing by 20% with total optical losses of 30 cm^{-1} yielded ~10 cm^{-1} [37]; the contributions of free carriers and of the thermal effect were greater in absolute value, although they had a different sign. The thermal waveguide has a positive contribution with the carriers producing a negative contribution. Thus, in study [37] positive feedback from the nonlinearity in gain losses was generated by a slow-response thermal waveguide, and not by carriers. Here, the portion of change in gain corresponding to the unchanged beam width compensated for the self-focusing contribution. In this regard it is clear that a simple homogeneous model [37] does not yield a sufficiently strong self-focusing effect in amplification-losses which would be sufficient for generation of self-modulation.

The source of the significant optical loss nonlinearity is predicted in the self-focusing model for an inhomogeneous medium (for example, with spreading of the absorption regions, see Fig. 22 in study [61]).

As an illustrative example we will consider a simplified model of nonlinear optical losses for the case of single-dimensional self-focusing. We will assume that the profile of the self-focusing channel is determined only by the induced addition to the refractive index. Then the electrodynamic problem has a comparatively simple analytical solution (see, for example, [61, p. 103]), and the transverse profile is described by the function

$$E(y) = E_0/\mathrm{ch}\, ay, \tag{10.1}$$

where E_0 is the amplitude of the electrical field strength in the center of the channel, a is the transverse profile parameter that is interrelated to the amplitude in self-focusing conditions. This relation may be expressed as

$$E_0 w = C(n_2), \tag{10.2}$$

where w is the width of the channel at half amplitude, $C(n_2)$ is a constant whose value is determined by the coefficient of nonlinearity n_2 (the expression for C is given in [61]). We will now assume that an absorption region is located at a certain distance y_0 from the center of the spatial channel such that the addition α to the distributed internal losses is equal to (see Fig. 1b)

$$\alpha(y) = \begin{cases} 0, & y \leqslant y_0, \\ \alpha_0 & y > y_0, \end{cases} \tag{10.3}$$

where α_0 is a constant (linear, nonsaturation absorption). Then the contribution $\delta\alpha$ of this addition to mode losses is

$$\delta\alpha = \int_{y_0}^{\infty} \alpha_0 \frac{dy}{\mathrm{ch}^2(ay)} \left(\int_{-\infty}^{\infty} \frac{dy}{\mathrm{ch}^2(ay)} \right)^{-1} = \alpha_0 \left[1 - \mathrm{th}\left(1{,}76 \frac{y_0}{w}\right) \right]. \tag{10.4}$$

It follows that the losses $\delta\alpha$ may be highly dependent on the width of the self-focusing channel, specifically,

$$\frac{d\alpha}{dw} = \frac{1{,}76\alpha_0 y_0}{w^2\, \mathrm{ch}^2(1{,}76 y_0/w)}. \tag{10.5}$$

The power of the mode P is proportional to the integral of $E^2(y)$ and the quantity E_0. Consequently, the width w is inversely proportional to the power of the mode or the number of photons in the mode S_Φ:

$$w = w_0 (S_\Phi^0/S_\Phi) = w_0 (P^0/P), \tag{10.6}$$

where w_0 and S_Φ^0 are constants convenient for simplified notation (w_0 is the width of the channel with a power P_0 or a photon number S_Φ). We may write

$$\frac{d\alpha}{dP} = \frac{d\alpha}{dw}\frac{dw}{dP} = -\frac{1{,}76\alpha_0 y_0}{P_0 w_0\, \mathrm{ch}^2(1{,}76 y_0 P/w_0 P^0)}, \tag{10.7}$$

which makes it possible to estimate the instability threshold of the rate equations by means of criterion (8.3). We may easily estimate the appearance of the instability on the watt-current characteristic (vertical section). For example, in a stripe laser with an area of $2 \cdot 10^{-5}$ cm^2 and a differential efficiency 0.5 watts/amp, we use the following parameters: w_0 = 10 μm, P_0 = 1 mw, $x_0 \simeq 2$ μm, α_0 = 20 cm^{-1} with a differential gain of $5 \cdot 10^{-2}$ cm/amp (GaAs, 300 K). Then the instability threshold will be observed at 3 mw. This is a rather arbitrary estimate, although it demonstrates that a comparatively low level of absorption along the periphery of the laser channel is capable of producing a significant change in operating conditions.

11. CONCLUSION

An investigation of the spectral characteristics and emission dynamics of injection lasers of various types has demonstrated that self-modulation pulsations are standard in planar lasers (see Fig. 2.14) without spatial stabilization of laser modes (at 1.5 I_t approximately 25% of the lasers operate in self-modulation conditions). For lasers employing lateral optical confinement (so-called channel, mesa-stripe and buried lasers) steady-state lasing is characteristic up to 1.7-3 times threshold, which is manifest in the greater probability of achieving single-frequency emission. It is quite clear that the appearance of sustained pulsations corresponds to a breakdown in single-frequency emission and a conversion to a multimode lasing spectrum. These observations confirm existing concepts of the relationship between the spectral and temporal characteristics of semiconductor lasers. This study provides a comparison of lasers of various types with prospects for broad applications in practice. Select statistical characteristics of self-modulation conditions are given. An interrelationship is established to exist between the self-modulation percentage and frequency.

Phase trajectories are examined at the self-modulation threshold. A new spectral chronographic method is employed that makes it possible to obtain simultaneously both dynamic variables of the rate equations: the emission intensity and the carrier concentration. The trajectory of self-modulation excitation that leads to the limit cycle is demonstrated.

The nonlinear optical losses in semiconductor lasers are investigated and discussed, particularly with regard to the causes of nonstationarity regions and their relation to nonlinearities in the watt-current characteristic of the laser. It is demonstrated that the form of this characteristic may be used to determine the dependence of internal losses on emission pumping or emission power, which in turn makes it possible to determine the self-modulation zones by employing the instability criterion of a solution of the rate equations.

The nonlinear loss mechanism caused by self-focusing in an inhomogeneous medium is discussed and calculated on the basis of a simplified model. Evidently this mechanism makes a contribution to the development of self-modulation processes in planar stripe lasers. Thus, the primary factors behind the generation of sustained self-modulation pulsations are nonlinear effects resulting from:

1) the saturation absorption of a noninverted medium,

2) the self-focusing instability in an inhomogeneous medium (containing irregular regions of optical losses - absorption or scattering).

In conclusion the author wishes to express his gratitude to Professor P. G. Eliseev for helpful discussion of results, and to G. T. Pak, N. D. Zhukov, B. N. Sverdlov for their assistance in this study.

BIBLIOGRAPHY

1. Kurnosov, V. D., Magalyas, V. I., Pleshkov, A. A., et al. "Self-modulation of emission from an injection semiconductor laser" PIS'MA V ZHETF, 1966, V. 4, No. 11, p. 449-453.

2. Basov, N. G., Drozhbin, Yu. A., Zakharov, Yu. P., et al. "The influence of injection current on the temporal characteristic of emission from a GaAs laser" FTT, 1966, V. 8, No. 9, p. 2816-2818.

3. Basov, N. G., Morozov, V. N., Nikitin, V. V., Semenov, A.S. "An investigation of emission pulsation conditions of a GaAs laser" FTP, 1967, V. 1, No. 10, p. 1570-1574.

4. Magalyas, V. I., Pleshkov, A. A., Rivlin, L. A., et al. "Emission modulation and synchronization in a semiconductor laser" PIS'MA V ZHETF, 1967, V. 6, No. 3, p.550-552.

5. Basov, N. G., Nikitin, V. V., Semenov, A.S. "The emission dynamics of semiconductor injection lasers" UFN, 1969, V. 7, No. 4, p. 561-600.

6. Goldobin, I. S., Dobkin, A. S., Kurnosov, V. D., et al. "Quantum optical gallium arsenide integrated circuits" FTP, 1971, V. 5, No. 1, p. 170-171.

7. Zakharov, Yu. P., Nikitin, V. V., Semenov, A. S. et al. "A theory of optically-coupled *p-n*-junction GaAs lasers" FTT, 1966, V. 8, No. 7, p. 2087-2091.

8. Basov, N. G., Morogov, V. N., Oraevskiy, A. N. "A theory of sustained intensity pulsations in lasers" DAN SSSR, 1966, V. 168, No. 3, p. 550-553.

9. Basov, N. G., "Dynamics of injection lasers." IEEE J. QUANT. ELECTRON., 1968, V. 4, No. 11, p. 855-864.

10. Kurnosov, V. D., Pleshkov, A. A., Rivlin, L. A., et al. "The emission dynamics of semiconductor lasers" In: Tr. IX Mezhdunar. konf. po fizike poluprobodnikov [Papers of the Ninth International Conference on Semiconductor Physics] Leningrad: Nauka, 1969, V. 1, p. 582-585.

11. Basov, N. G., Morozov, V. N. "A theory of the dynamics of injection lasers" ZHETF, 1969, V. 57, No. 2(8), p. 617-627.

12. Belenov, E. M., Morozov, V. N., Oraevskiy, A. N. "Issues in the dynamics of lasers" TR. FIAN, 1970, V. 52, p. 237-336.

13. McCumber, D. E. "Intensity fluctuations in the output of CW laser oscillators" PHYS. REV., 1966, V. 141, p. 306-322.

14. Ikegami, T., Suematsu, Y. "Resonance-line characteristics of the direct modulation of a junction lasers" PROC. IEEE, 1967, V. 55, No. 1, p. 122-123.

15. Khanin, Ya. I. "The dynamics of lasers" Moscow: SOV. RADIO, 1975, V. 2, 282 p.

16. Vagrashkin, A. I., Pleshkov, A. A., Trukhan, V. G., Tsvetkov, V. V. "Spontaneous transient processes in the emission spectrum of a semiconductor laser" PIS'MA V ZHETF, 1970, V. 12, No. 1, p. 3-6.

17. Bogatov, A. P., Eliseev, P. G., Man'ko, M. A., Chan Min' Tkhay "The action of a compound resonator on emission coherence of injection lasers" KVAT. SOOBSHCH. PO FIZIKE FIAN, 1971, No. 9, p. 60-65.

18. Bogatov, A. P., Eliseev, P. G., Panteleev, V. I., Shevchenko, E. G. "A comparison of the instantaneous and averaged emission spectra of an injection laser in spontaneous pulsation operation" KVANTOVAYA ELEKTRON, 1971, No. 5, p. 93-95.

19. Eliseev, P. G., Ivanov, L. P., Logginov, A. S., Senatorov, K. Ya. "The frequency self-modulation of emission in an injection laser" KVAT. SOOBSHCH PO FIZIKE FIAN, 1972, No. 6, p. 53-65.

20. Bogatov, A. P., Eliseev, P. G., Manko, M. A., et al. "Study of single mode injection laser" IEEE J. QUANT. ELECTRON., 1973, V. 9, No. 2, p. 392-394.

21. Eliseev, P. G., Ivanov, L. P., Logginov, A. S., Nikitin, E. N., "Internal Q-switching in injection lasers with a single-sided heterojunction" KVANTOVAYA ELEKTRON., 1973, No. 5(17), p. 116-117.

22. Bogatov, A. P., Eliseev, P. G., Ivanov, L. P. et al. "The kinetics of the emission spectrum of an injection laser and the breakdown phenomenon of single-mode lasing" KVANTOVAYA ELEKTRON., 1973, No. 5(17), p. 14-20.

23. Eliseev, P. G., Ivanov, L. P., Logginov, A. S., Senatorov, K. Ya. "The influence of microwave modulation on the emission spectrum of an injection laser" KVANTOVAYA ELEKTRON., 1974, V. 1, No. 1, p. 151-154.

24. Morozov, V. N. "A theory of multimode lasing in semiconductor lasers" KVANTOVAYA ELEKTRON., 1974, V. 1, No. 3, p. 634-644.

25. Eliseev, P. G., Chan Minimum' Tkhay "An investigation of the regularities of multimode excitation in injection lasers" KVANTOVAYA ELEKTRON., 1974, V. 1, No. 5, p. 1138-1144.

26. Smith, A. W., Edmonds, H. D., "Picosecond structure in selfpulsing GaAs injection lasers. IEEE J. QUANT. ELECTRON., 1971, V. 7, No. 2, p. 94-95.

27. Bakhert, Kh. Yu., Eliseev, P. G., Man'ko, M. A., et al. "An investigation of the picosecond structure and ultranarrow pulses of emission from an injection laser using an interferometer method" KVANTOVAYA ELEKTRON., 1974, V. 1, No. 9, p. 1988-1993.

28. Eliseev, P. G. "Semiconductor lasers and converters" In: Itogi nauki i tekhniki. Radiotekhnika [Achievements in science and technology. Radio engineering] Moscow: VINITI, 1978, V. 14, Ch. 2, p. 196-198.

29. Eliseev, P. G. "An interpretation of the frequency self-modulation effect in a semiconductor laser" KVANTOVAYA ELEKTRON., 1979, V. 6, No. 10, p. 2243-2245.

30. Van der Ziel, J. P. "Spectral broadening of pulsing AlGaAs DH lasers" IEEE J. QUANT. ELECTRON., 1979, V. 11, p. 1277-1281.

31. Eliseev, P. G., Morozov, V. N., Pashko, S. A., et al. "Spectral mode broadening in a semiconducting laser with emission intensity pulsations" KVANTOVAYA ELEKTRON., 1980, V. 7, No. 10, p. 2197-2201.

32. Osinski, M., Adams, M. J. "Influence of frequency chirping on the transient time averaged spectra of gain switched stripe lasers" In: Ninth IEEE Intern. Semiconductor Laser Conf. Rio de Janeiro (Brazil.), 1984, Abstr. of Pap., p. 122-123.

33. Eliseev, P. G., Bogatov, A. P. "Phenomena in semiconductor lasers associated with nonlinear refraction and carrier action on the refractive index" TR FIAN, 1986, V. 166. (See this volume).

34. Bakhert, Kh.-Yu., Bogatov, A. P., Eliseev, P. G. "Mode deformation in an injection laser from self-focusing and its relation to nonlinearity of the output characteristic" KVANTOVAYA ELEKTRON., 1978, V. 5, No. 3, p. 603-608.

35. Bogatov, A. P., Gurov, Yu. V., Eliseev, P. G., et al. "Low frequency pulsation in the output intensity and multimode operation of CW diode laser coupled to an external cavity" IEE PROC. PT. I. SOLID STATE ELECTRON. DEVICES, 1979, V. 3, No. 3, p.72-74.

36. Lang, R. "Intensity pulsation enhancement by self-focusing in semiconductor injection lasers" JAPAN. J. APPL. PHYS., 1980, V. 19, No. 2, p.L93-L96.

37. Van der Ziel, J. P. "Self-focusing effects in pulsating AlGaAs DH lasers" IEEE J. APPL. PHYS., 1981, V. 17, No.1, p. 60-68.

38. Vy Van Lyk, Kalshabekov, A. S., Man'ko, M. A., et al. "Preliminary pumping pulse action on the output characteristics of stripe heterolasers" KVANTOVAYA ELEKTRON., 1981, V. 8, No. 12, p. 2697-2699.

39. Eliseev, P. G., Novikov, A. A., Fedorov, Yu. F. "The interaction effects of two diode lasers" PIS'MA V ZHETF, 1965, V. 2, No. 2, p. 58-63.

40. Eliseev, P. G., Lavrov, V.N., Morozov, V. N. "The attenuation of relaxation oscillations in an injection laser by external pumping" PIS'MA V ZHTF, 1978, V. 4, No. 19, p. 1160-1163.

41. Eliseev, P. G., Kobyakova, M. Sh., Pak, G. T., et al. "A mesa-stripe injection heterolaser with thermal flow through the substrate" 1980, V. 7, No. 11, p. 2504-2506.

42. Borsdulin, V. I., Eliseev, P. G., Konyaev, V. P., et al. "The characteristics of a channel injection heterolaser" KVANTOVAYA ELECTRON., 1981, V. 8, No. 1, p. 193-196.

43. Alaverdyan, S. A., Eliseev, P.G., Zhukov, N. D., et al. "The radiation characteristics of a zigzag mesa-stripe heterostructure AlGaAs/GaAs injection laser" KVANTOVAYA ELECTRON., 1981, V. 8, No. 7, p. 1565-1566.,

44. Bezotosnyy, V.V., Drakin, A. E., Eliseev, P. G., et al. "CW buried mesa-stripe InGaPAs-InP heterolasers at 1.24-1.28 μm operating at room temperature" KVANTOVAYA ELECTRON., 1980, V. 7, No. 9, p. 1990-1992.

45. Basil'ev, Yu. A., Dmitriev, Yu. V., Eliseev, P. G., et al. "A high-speed Au-nn$^+$-GaAs surface barrier structure photodiode" KVANTOVAYA ELECTRON., 1980, V. 7, No. 10, p. 2218-2221.

46. Adams, M. J. "The rate equations for semiconductor lasers" OPTO-ELECTRON., 1973, V. 5, No. 2, p. 101-215.

47. Thomas, V., Adams, M. J., Crundorfer, S. "A comprehensive model for Q-switching in semiconductor lasers" IEEE J. QUANT. ELECTRON., 1975, V. 11, No. 7, p. 528-532.

48. Basov, N. G., Morozov, V. N. "Multimode lasing theory in semiconductor" KVANTOVAYA ELEKTRON., 1974, V. 1, No. 3, p. 634-644.

49. Morozov, V. N., Oraevskiy, A. N. "The operating conditions of lasers with the cavity Q dependent on emission field intensity" ZHTF, 1967, V. 57, No. 5, p.947-952.

50. Basov, N. G., Morozov, V. N.,Oraevskiy, A. N. "A theory of the dynamics of a single-mode laser" KVANTOVAYA ELEKTRON., 1974, V. 1, No. 10, p.2264-2274.

51. Bogatov, A. P., Eliseev, P. G., Okhotnikov, O. G., Pak, G. T. "The influence of excess current carriers on the dielectric constant of GaAs at the radiative transition frequency in injection lasers" KVANTOVAYA ELEKTRON., 1976, V.3, No. 7, p. 1609-1610.

52. Bogatov, A. P., Eliseev, P. G., Okhotnikov, O. G., Pak, G. T. "The properties of planar stripe heterolasers. I. Nonlinear and irregular watt-current characteristic. II. Analysis of electrical characteristics" KVANTOVAYA ELEKTRON., 1980, V. 7, No. 8, p. 1664-1676.

53. Bogatov, A. P., Gurov, Yu. V., Eliseev, P. G., Khayretdinov, K. A. "Anomalous lasing dynamics in semiconductor lasers with an asymmetric waveguide structure. I. Experimental investigation with external cavity. II. Theory." KVANTOVAYA ELEKTRON., 1978, V. 5, No. 11, p. 2402-2415.

54. Basov, N. G., Eliseev, P. G., Zakharov, S. D., et al. "The properties of diode lasers fabricated from gallium arsenide" FTT, 1966, V. 8, No. 9, p. 2616-2622.

55. Suchkov, A. F. "The influence of discontinuities on operating conditions of solid state lasers" ZHETF, 1965, V. 49, No. 5, p. 1495-1503.

56. Allakhverdyan, R. G., Morozov, V. N., Suchkov, A. F., "Nonsteady-state excitation of multimode lasing in semiconductor lasers" KVAT. SOOBSHCH PO FIZIKI FIAN, 1972, No.6, p. 3-7.

57. Skopin, I. A. "The influence of nonlinear losses in the active region of an injection laser on the self-modulation intensity pulsation zones" PREPR. FIAN, No. 191, Moscow, 1981, 18 p.

58. Skopin, I. A. "The influence of nonlinear losses in the active region of an injection laser on self-modulation intensity pulsation zones" KVAT. SOOBSHCH PO FIZIKE FIAN, 1982, No. 3. p. 57-64.

59. Hartman, R. L., Logan, R. A., Koszi, L. A., Tsang. W. T. "Pulsation and absorbing defects in (Al, Ga) As injection lasers" J. APPL. PHYS., 1979, V. 50, No. 7, p. 4616-4619.

60. Seki, K., Yanai, H., Kamiya, T., "Structure dependent axial mode stability in GaAlAs/GaAs injection lasers" ELECTRON. LETT., 1981, V. 17, No. 8, p. 288-290.

61. Eliseev, P. G., "An introduction to the physics of injection lasers" Moscow: Nauka, 1983, 294 p.

MODES AND MODE CONVERSION IN ACTIVE SEMICONDUCTOR WAVEGUIDES

M. A. Man'ko, G. T. Mikaelyan

Abstract: A concept based on an analog of the quantum mechanical Schrodinger equation is developed together with an equation describing the propagation of light in inhomogeneous media. A new parameter for an active semiconductor waveguide of complex reduced width is introduced for the relation of the complex dielectric constant in the plane of the p-n junction in the form of an Epstein layer; this parameter is used to tabulate, in the form of three-dimensional graphs, the parameters of the emission directional pattern, and the mode coupling factors are expressed in a model describing a lengthwise-inhomogeneous laser in the form of two (or more) end-joined homogeneous active regions. For this model, analogs of the Franck-Condon principle and the Ramsauer effect are formulated and are used as the basis for proposing a method of active control over the output characteristics of the heterolaser.

INTRODUCTION

Dielectric waveguide theory [1-3] developed within the scope of an optical injection laser model; this theory accounts for the specific nature of the semiconductor active medium, particularly the gain and optical nonlinearity of this medium [4, 5]. This model essentially redeveloped the concept of gain guided modes, i.e., modes whose profile is determined by the imaginary part of the dielectric constant to a greater degree than the real part. Injection lasers with the most significant gain guiding includes planar stripe heterolasers without lateral optical confinement. Optical confinement is an important concept in this model, and refers to the property of a laser structure where the laser emission intensity is confined within the amplifying medium. As a result, the active region of a stripe heterolaser is considered an active waveguide capable of amplifying and emitting an electromagnetic field, of confining this field in the cross-section (x, y), and allowing waveguide propagation in the longitudinal direction (z).

We know that an electromagnetic field in a stripe heterolaser may be considered, from the phenomenological viewpoint, a field in an inhomogeneous medium with a complex dielectric constant dependent on the three spatial coordinates ε (x, y, z) in the most general case;

however, for actual lasers, this relation is homogeneous along the resonator length (on the z axis) and is stratified in the direction of current flow (on the x axis) and is modeled by a certain function (specifically: a rectangular well [4, 6, 7], a parabola [8, 9], an Epstein layer [10-15] or other continuous functions producing relations equivalent to an Epstein layer; see, for example, [16]) across the resonator width (on the y axis). In the direction perpendicular to the *p-n*-junction (on the x axis), the distribution of the dielectric constant $\varepsilon(x)$ is largely determined by the difference in the refractive indices of the active waveguide and the broadband layers of the heterostructure, and assumes a stepwise shape. The model of a rectangular potential well is fully applicable to this direction. In the direction parallel to the *p-n*-junction (on the y axis), the situation is much more complex. Here we must find the distribution $\varepsilon(y)$ such that, on the one hand, it reflects the actual situation to a sufficient degree and likewise makes it possible to analytically find solutions of the equation for the electromagnetic field in an inhomogeneous medium. The model of a rectangular potential well (i.e., the $\varepsilon(y)$ distribution in step form) is not suitable, since this model, although it allows estimation of the parameters of the near field, is too approximate to determine the parameters of the far field (this model does not account for the contribution of the wavefront slope to emission divergence in the cross-section, along the *p-n*-junction). Approximation of the relation $\varepsilon(y)$ by the parabola $\varepsilon(y) = ay^2 + b$ gives solutions of the equation for an electromagnetic field in Hermite-Gauss functions [8, 17] and is satisfactory for a symmetrical emission distribution in the near and far fields. However, this approximation is not suitable for describing an asymmetrical distribution, which may exist in waveguides without lateral optical confinement, when light is emitted in the form of a non-Gaussian beam.

Epstein layer relations, having extrema and constants (generally speaking these differ) at negative and positive infinities are more suitable, since they adequately describe the actual situation with a finite number of modes in a planar heterolaser, and make it possible to analytically solve the wave equation for an electromagnetic field in the cavity. In Epstein's study [18] this model was used to calculate the reflection coefficient off of ionospheric layers. In study [19] the Epstein layer model was applied to a description of the waveguide properties of *p-n*-junction injection lasers accounting for absorption, while studies [20, 10] used the model to describe heterolasers (we note that the authors of studies [19, 20] used the Epstein layer model for the direction of current flow, i.e, on the x axis). For the direction parallel to the *p-n*-junction (on the x axis), the Epstein layer model was used in studies [16, 10] in which the $\varepsilon(y)$ relation was represented as

$$\varepsilon(y) = (\varepsilon_3 - \varepsilon_1)\frac{\exp(y/l)}{1+\exp(y/l)} + 2(2\varepsilon_2 - \varepsilon_1 - \varepsilon_3)\frac{\exp(y/l)}{[1+\exp(y/l)]^2}, \quad (1)$$

where $\varepsilon_1 = \varepsilon(-\infty)$, $\varepsilon_2 = \varepsilon(0)$, $\varepsilon_3 = \varepsilon(\infty)$, l is the layer width parameter. For a symmetrical distribution ($\varepsilon_1 = \varepsilon_3 = \varepsilon_0$), formula (1) is reduced to the traditional "hyperbolic cosine":

$$\varepsilon(y) = 4(\varepsilon_2 - \varepsilon_0)\frac{\exp(y/l)}{[1+\exp(y/l)]^2} = \frac{\varepsilon_2 - \varepsilon_0}{\mathrm{ch}^2(y/2l)}. \qquad (2)$$

The symmetrical Epstein layer model was also employed by a number of authors (a detailed list of these references is given in study [15].

This study describes the results from an investigation of the spatial characteristics of gain guided waveguide modes. Underlying this analysis is an Epstein layer model of the distribution of the dielectric constant. A characteristic "reduced width" parameter of the active waveguide *s* is introduced; the directional pattern of emission (extrema positions and their widths) is calculated depending on the real and imaginary values of this parameter. A plan is presented for selecting specific waveguide parameters for a given emission wavelength with the necessary far field characteristics, and a method of fabricating waveguides with a variable directional pattern such as angular scanning, variable width, etc. is also discussed (Sections 2, 3). The possibility of a reverse analysis is noted; i.e., determining the waveguide parameters based on the spatial emission distribution (Section 4). The nonlinearities and anomalies in planar stripe heterolasers without lateral optical confinement are discussed within the scope of this model (Section 5).

The differential in the power and spatial distribution of emission exiting the opposite ends of a heterolaser means that such a laser may be considered to be two (or several) end joined lasers [21, 11, 22]. This means the heterolaser has a discontinuity in the dielectric constant along the cavity. This discontinuity may be described both by a stepwise relation (sharp jumps in the dielectric constant), which designates several end joined lasers, and by a continuous relation of the dielectric constant in the longitudinal direction (analogous to the Epstein layer model in the transverse direction, for example). Hence in order to investigate such heterolaser output characteristics as the emission power and spectral composition, emission directional pattern and wavefront at the outcoupling mirror etc., it is important to understand the influence of transverse and longitudinal discontinuities in the dielectric constant. Section 6 of this study formulates, for the stepwise case (end joining of two lasers) the principle (see also [22]) that makes it possible to easily identify which mode of the second laser obtains the power concentrated in a given mode of the first laser. This principle is formulated on the basis of an unexpected analogy to the well-known (see, for example, [23]) Franck-Condon principle describing the intensity distribution in the absorption and emission spectra of multiatomic molecules with electron transitions. The possibility for formulating this principle is based on the theoretical concepts presented in Section 1 formally proving the mathematical analogy of the equation describing an electromagnetic field in a heterolaser, and the quantum mechanical Schrodinger equation.

A unique effect (an analog of the Ramsauer effect in quantum mechanics) may exist for the case of end-joined lasers, which is discussed in Section 7. According to this effect the layer interconnecting the two end-joined lasers may be fabricated to be totally transparent or totally nontransparent to electromagnetic emission. The properties of this layer may be controlled by varying the localized temperature, by special doping, etc.

We note that the possibility for employing existing quantum mechanical results for problems described by the scalar wave equation with discontinuities, specifically the use of integrals of motion of nonsteady-state quantum systems, coherent states, the Franck-Condon principle, group methods, etc. was discussed in general terms in study [24]. This approach was employed and developed in study [25] in conjunction with select results from nonstationary quantum mechanics for application to a number of problems in the optics of lightguides.

1. AN EQUATION FOR EMISSION PROPAGATION IN A LENGTHWISE-INHOMOGENEOUS HETEROLASER AND SCHRODINGER'S EQUATION

In accordance with standard mode propagation theory for a heterolaser (see, for example, [17]), we write the scalar wave equation for an electromagnetic field

$$\nabla^2 E(x, y, z) + \varepsilon k_0^2 \dot{E}(x, y, z) = 0, \tag{3}$$

where E is the corresponding vector field component, $k_0 = 2\pi/\lambda$, λ is the wavelength for the given frequency in the vacuum, $\varepsilon \equiv \varepsilon(x, y, z)$ is dielectric constant dependent in the most general case on all coordinates (the axis x is perpendicular to the heteroboundaries, the axis y is parallel to the heteroboundaries, Fig. 1). We consider $\tilde{\tilde{E}}(x, y, z, t)$ fields with oscillations over time:

$$\tilde{\tilde{E}}(x, y, z, t) = E(x, y; z) \exp(-i\omega t). \tag{4}$$

When the dielectric constant is homogeneous on the laser axis (the axis z, Fig. 1) ($\varepsilon(x, y, z) = \varepsilon(x, y)$), the field $E(x, y, z)$ takes the form

$$E(x, y, z) = E(x, y) \exp i\beta z, \tag{5}$$

where $E(x, y)$ is the amplitude of oscillations independent of z, while β is the complex propagation constant on the axis z, $\beta = \beta' + i\beta''$. The imaginary part of this constant describes either an increase in the field amplitude on the laser axis with $\beta'' < 0$, or a diminishing of field amplitude with $\beta'' > 0$. After substitution of (5) into the wave equation, we derive the equation for the field component $E(x, y)$

$$\frac{\partial^2 E(x, y)}{\partial x^2} + \frac{\partial^2 E(x, y)}{\partial y^2} + [k_0^2 \varepsilon(x, y) - \beta^2] E(x, y) = 0. \tag{6}$$

Equation (6) for the laser field is formally analogous to Schrodinger's equation for a particle in two measurements in a potential (in the general case a complex potential) determined by the dielectric constant $\varepsilon(x, y)$. Indeed, Schrodinger's equation for the wave function $\Psi(x, y)$ of the steady-state of a particle of mass m takes the form:

$$-\frac{\hbar^2}{2m}\frac{\partial^2\Psi(x,y)}{\partial x^2}-\frac{\hbar^2}{2m}\frac{\partial^2\Psi(x,y)}{\partial y^2}+V(x,y)\Psi(x,y)=\mathscr{E}\Psi(x,y), \tag{7}$$

where $\hbar$ is Planck's constant; $\mathscr{E}$ is the steady-state energy. By comparing (6) and (7), we see that the equation for the laser field for

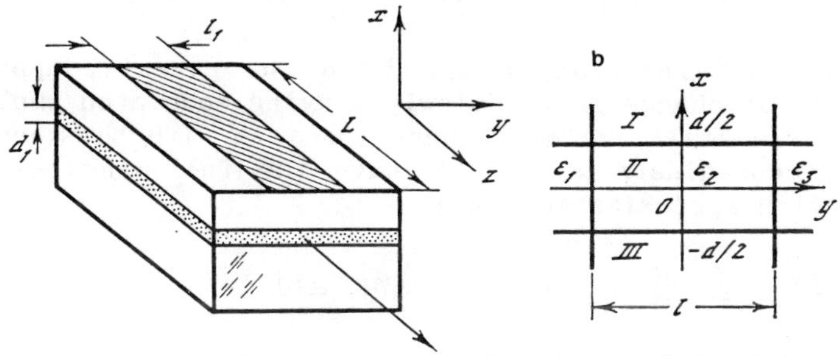

Fig. 1. A planar stripe heterolaser (a) and a model of its dielectric structure (cross-section) (b)
The active region is limited to $|x| \leqslant d/2$, $|y| < l/2$

mally coincides with Schrodinger's equation with the following redesignations, which are obvious if we multiply equation (6) by $(-1/2k_0^2)$, and transpose the term from β^2 to the right half, and multiply equation (7) by m, i.e.,

$$-\frac{1}{2k_0^2}\frac{\partial^2E(x,y)}{\partial x^2}-\frac{1}{2k_0^2}\frac{\partial^2E(x,y)}{\partial y^2}-\frac{\varepsilon(x,y)}{2}E(x,y)=-\frac{\beta^2}{2k_0^2}E(x,y), \tag{8}$$

$$-\frac{\hbar^2}{2}\frac{\partial^2\Psi(x,y)}{\partial x^2}-\frac{\hbar^2}{2}\frac{\partial^2\Psi(x,y)}{\partial y^2}+mV(x,y)\Psi(x,y)=m\mathscr{E}\Psi(x,y). \tag{9}$$

Thus, with the redesignations

$$E\to\Psi,\qquad \frac{1}{k_0}\to\hbar,\qquad \frac{\varepsilon(x,y)}{2}\to-mV(x,y),\qquad \frac{\beta^2}{2k_0^2}\to-m\mathscr{E} \tag{10}$$

The equation for the field in the laser goes over to the two-dimensional Schrodinger equation.

We now account for the possibility of a discontinuity in the dielectric constant on the cavity axis of the heterolaser in a paraxial approximation, when the emission is propagating axially in the cavity, and emission divergence is slight. Here we will modulate the amplitude of the oscillations on the cavity axis, assuming that in place of equation (6) (in which the amplitude of the field on the axis z is independent of z) we have a slight dependence of the amplitude axially on z, i.e.,

$$E(x, y, z) = \tilde{E}(x, y, z) \exp i\beta z. \qquad (11)$$

Formula (11) may be considered a simple transition from the variable E to the variable $\tilde{E}$, although the amplitude of $\tilde{E}$ varies only slightly on the cavity axis against a background of fast oscillations determined by the term experiment $i\beta z$. Substituting (11) into (3) we derive

$$\left\{\frac{\partial^2 \tilde{E}}{\partial x^2} + \frac{\partial^2 \tilde{E}}{\partial y^2} + \frac{\partial^2 \tilde{E}}{\partial z^2} - 2i\beta \frac{\partial \tilde{E}}{\partial z} + [\varepsilon(x, y, z) k_0^2 - \beta^2] \tilde{E}\right\} \exp(-i\beta z) = 0. \qquad (12)$$

This is an exact relation, although in the paraxial approximation, when the rate of change in the slowly varying modulated amplitude with a variable z is small, we may ignore the term with the second derivative of the coordinate z, and we have the following equation in a Fock-Leontovich approximation [26]:

$$\frac{\partial^2 \tilde{E}}{\partial x^2} + \frac{\partial^2 \tilde{E}}{\partial y^2} - 2i\beta \frac{\partial \tilde{E}}{\partial z} + [\varepsilon(x, y, z) k_0^2 - \beta^2] \tilde{E} = 0. \qquad (13)$$

We will compare equation (13) for the field in the heterolaser with discontinuities on the axis of the cavity to the two-dimensional Schrodinger equation, although for the wave functions $\Psi(x, y, t)$ of the nonsteady-states of a particle of mass m moving in the nonsteady-state potential $V(x, y, t)$, where t is time we have:

$$i\hbar \frac{\partial \Psi(x, y, t)}{\partial t} = -\frac{\hbar^2}{2m} \frac{\partial^2 \Psi(x, y, t)}{\partial x^2} - \frac{\hbar^2}{2m} \frac{\partial^2 \Psi(x, y, t)}{\partial y^2} + V(x, y, t) \Psi(x, y, t). \qquad (14)$$

We will carry out a routine with equations (13) and (14) analogous to that performed with equations (6) and (7), rewriting the latter in the form of (8) and (9). Then equation (13) is rewritten as

$$-\frac{i\beta}{k_0^2} \frac{\partial \tilde{E}(x, y, z)}{\partial z} = -\frac{1}{2k_0^2} \left[\frac{\partial^2 \tilde{E}(x, y, z)}{\partial x^2} + \frac{\partial^2 \tilde{E}(x, y, z)}{\partial y^2}\right] + \left[\frac{\beta^2}{2k_0^2} - \frac{\varepsilon(x, y, z)}{2}\right] \tilde{E}(x, y, z), \qquad (15)$$

while equation (14) is written in the form

$$-im\hbar \frac{\partial \Psi(x, y, t)}{\partial t} = -\frac{\hbar}{2} \left[\frac{\partial^2 \Psi(x, y, t)}{\partial x^2} + \frac{\partial^2 \Psi(x, y, t)}{\partial y^2}\right] + mV(x, y, t) \Psi(x, y, t). \qquad (16)$$

We see that the equation for the field in a lengthwise inhomogeneous heterolaser formally coincides with the two-dimensional Schrodinger equation for a particle of mass m moving in a time-independent complex potential, if we carry out the redesignations:

$$\tilde{E} \to \Psi, \quad \frac{1}{k_0} \to \hbar, \quad -\frac{k_0^2 z}{\beta} \to \frac{t}{\hbar m}, \quad \frac{\beta^2}{2k_0^2} - \frac{\varepsilon(x, y, z)}{2} \to mV(x, y, t). \qquad (17)$$

Formally the role of time in the corresponding Schrodinger equation is played by the coordinate proportional to z: the length traversed by light along the cavity of the laser, multiplied by the constant. The role of the time-independent potential is played by the dielectric constant with a distribution discontinuity on the laser axis, offset by a magnitude equal to the constant.

We will now discuss the situation where the equation for the electromagnetic field in the medium may be solved analytically. For example, equation (6) in a lengthwise-homogeneous resonator (the z axis) may be solved by separation of variables [17]. Let $E(x, y) = X(x)Y(y)$, then from (6) we obtain

$$d^2X/dx^2 + [k_0^2\varepsilon(x) - p^2 - \beta^2]X = 0, \tag{18}$$

$$d^2Y/dy^2 + [k_0^2\varepsilon(y) - p^2]Y = 0, \tag{19}$$

where p is the complex constant of the separation of variables. Multiplying (19) by $|X(x)|^2$ and integrating for x going from $-\infty$ to ∞, we obtain

$$d^2Y/dy^2 + [k_0^2\Gamma_0\varepsilon(y) - p^2]Y = 0, \tag{20}$$

where Γ_0 is the optical confinement parameter (on the x axis),

$$\Gamma_0 = \int_{-d/2}^{d/2} |X(x)|^2 dx \Big/ \int_{-\infty}^{\infty} |X(x)|^2 dx. \tag{21}$$

The solutions of $X(x)$ exist [6] and take the form

$$X(x) = \frac{1}{2}\begin{cases} A\exp(-\gamma x) + A^*\exp(-\gamma^* x) & \text{when } x > d/2, \\ \cos(qx + \varphi) + \cos(q^* x + \varphi) & \text{when } |x| \leqslant d/2, \\ B\exp(rx) + B^*\exp(r^* x) & \text{when } x < -d/2, \end{cases} \tag{22}$$

where d is the thickness of the active region; the asterisk designates complex conjugate quantities. The numerical values of the parameters γ, q and r are given in study [6], while the coefficients A, B and φ are determined through γ, q and r (for the symmetric case $\gamma = r$).

We rewrite equation (20) in the form

$$\frac{1}{k_0^2\Gamma_0}\frac{d^2Y}{dy^2} + \varepsilon(y)Y = \frac{p^2}{k_0^2\Gamma_0}Y. \tag{23}$$

It is easy to see that equation (23) has the same form as the Schrodinger equation in quantum mechanics:

$$\frac{\hbar^2}{2m}\frac{\partial^2\Psi}{\partial x^2} - V(x)\Psi = -E\Psi, \tag{24}$$

where m is the mass of a particle in a potential well $V(x)$. If we assume that $-p^2/k_0^2\Gamma_0$ corresponds to an energy E, while $-\varepsilon(y)$ corresponds to a potential $V(x)$, then we may use the extensive literature of quantum mechanics to solve equation (23).

2. THE EPSTEIN MODEL AND THE REDUCED WIDTH PARAMETER OF AN ACTIVE WAVEGUIDE

Our task is to estimate the actual values of the emission near and far field parameters within the scope of the Epstein model for emission exiting an active stripe waveguide. The term parameter is understood to mean the field amplitude and phase distribution in the near field, while for the far field we refer to the angular positions of points characterizing the directional pattern; specifically, the extrema, the half amplitude level, etc.

In accordance with study [27], we will write the solution of equation (23) for the $\varepsilon(y)$ distribution as an Epstein layer (1) in the form

$$Y_n(y) = A_N(n)\frac{\exp(a_n y/2l)}{[\mathrm{ch}(y/2l)]^{s-n}} P_n^{(s-n+a_n,\ s-n-a_n)}(-\mathrm{th}(y/2l)), \tag{25}$$

where $P_n^{(\alpha,\ \beta)}(t)$ are n^{th} order Jacobi polynomials (we are observing the determination of these polynomials given in study [28]) $a_n = k_0^2\Gamma_0 l^2\ (\varepsilon_3 - \varepsilon_1)/(s - n)$ is the asymmetry parameter, $A_N(n)$ is the normalizing factor, while the complex parameter s is determined by the formula

$$s = {}^1/_2\{-1 \pm [1 + 8(2\varepsilon_2 - \varepsilon_1 - \varepsilon_3)k_0^2\Gamma_0 l^2]^{1/2}\}. \tag{26}$$

In formula (26) the "plus" sign corresponds to the case where Re $\varepsilon_2 \geqslant$ Re ε_1, Re ε_3, etc., while the waveguide properties result from the real part of the dielectric constant; the "minus" sign, when the real part corresponds to the antiwaveguide effect, the waveguide properties are sustained by virtue of amplification, i.e., the contribution of the imaginary part of ε_2 (Im $\varepsilon_2 < 0$).

The emission intensity distribution at the cavity mirror of the heterolaser (near field) is determined as $I_n(y) = |Y_n(y)|^2$, while in the far field it is determined as $I_n(\theta) = \cos\theta |F_n(\theta)|^2$, where

$$F_n(\theta) = \int_{-\infty}^{\infty} Y_n(y)\exp(-ik_0 y\sin\theta)\,dy. \tag{27}$$

Substituting (25) into (27), we obtain

$$F_n(\theta) = \mathrm{const}\cdot\Gamma\left(\frac{s-n-a_n}{2} + ik_0 l\sin\theta\right)\Gamma\left(\frac{s-n+a_n}{2} - ik_0 l\sin\theta\right)\times$$
$$\times {}_3F_2\left(-n, s-a_n, \frac{s-n+a_n}{2} - ik_0 l\sin\theta; s-n-a_n+1, s-n; 1\right), \tag{28}$$

where $_3F_2$ is the hypergeometric function, $\Gamma(\approx + ib)$ is the gamma-function (for their definitions see [28]). For the fundamental mode (n = 0) we have

$$F_0(\theta) = \text{const} \cdot \Gamma\left(\frac{s-a_0}{2} + ik_0 l \sin\theta\right)\Gamma\left(\frac{s+a_0}{2} - ik_0 l \sin\theta\right), \tag{29}$$

for n = 1

$$F_1(\theta) = \text{const} \cdot \Gamma\left(\frac{s-1-a_1}{2} + ik_0 l \sin\theta\right)\Gamma\left(\frac{s-1+a_1}{2} - ik_0 l \sin\theta\right) \times \\ \times \left\{1 + \frac{(s-a_1)\,[(s-1+a_1)/2 - ik_0 l \sin\theta]}{(s+a_1)(s-1)}\right\}. \tag{30}$$

The parameter s is included in formulae (28)-(30); this parameter, as we see from (26), is a function of the dielectric constant, and as a result is a complex quantity, i.e., it may be represented in the form $s = s' + is''$. If we know the values s' and s'', it is easy to determine the emission parameters. The magnitude of the imaginary part s' (when $s > 0$) determines the decay in the field amplitude with an increase in y; we may determine the phase based on the imaginary part s''. For example, for the fundamental mode (n = 0) in a symmetric waveguide (a_0 = 0) from (25) we have

$$Y_0(y) = A_0/[\text{ch}\,(y/2l)]^{s'+is''} = E_0(y) \exp[i\Phi(y)], \tag{31}$$

where $E_0(y)$ and $\Phi(y)$ are the amplitude and phase of an electromagnetic field in the far field,

$$E_0(y) = A_0 \,\text{ch}^{-s'}(y/2l), \tag{32}$$

$$\Phi(y) = s'' \ln[\text{ch}\,(y/2l)]. \tag{33}$$

We will examine expression (26) for the complex parameter s; this expression includes the width parameter of the Epstein layer l and the differential of the complex values of the dielectric constant at the extrema (when $y = 0, \pm\infty$). By analogy to Anderson's theory [6] for analyzing the properties of an active waveguide perpendicular to the p-n-junction, and proposing a solution of equation (18) in the form (22), we will assume for our case of analysis parallel to the p-n-junction (the y axis) a reduced width parameter on the order $s = s' + is''$, determined by formula (26).

Comparing the quantity $p^2/k_0^2\Gamma_0$ in equation (23) to the related values of the electron energy in the potential well $V(x) = B_0 \,\text{th}\,\alpha x - u_0 \,\text{ch}^{-2}\alpha x$ [27] makes it possible to determine the discrete values p_n corresponding to the various modes (n = 0, 1, 2,...),

$$p_n = \left[\frac{(s-n)^2}{4l^2} + \frac{k_0^2\Gamma_0(\varepsilon_3-\varepsilon_1)}{2} + \frac{k_0^4 l^2\Gamma_0^2(\varepsilon_3-\varepsilon_1)^2}{4(s-n)^2}\right]^{1/2}. \tag{34}$$

The propagation constant β_n on the z axis is also determined by the complex reduced width parameter of the active waveguide s [11, 12]:

$$\beta_n = (p_n^2 + k_0^2\varepsilon_4 + r^2)^{1/2}, \tag{35}$$

where ε_4 is the dielectric constant of the narrowband layer of the heterostructure next to the active region (on the x axis, see Fig. 1), while r is the exponential decay parameter of the electromagnetic field in this layer (see formula 22)).

3. THE INFLUENCE OF THE ACTIVE WAVEGUIDE PARAMETERS ON EMISSION DISTRIBUTION IN THE FAR FIELD

We will rewrite formula (29) for the fundamental mode ($n = 0$) incorporating the new variable $\chi = 2\pi l\lambda^{-1} \sin\theta$,

$$F_0(\chi) = \text{const}\cdot\Gamma\left(\frac{s-a_0}{2} + i\chi\right)\Gamma\left(\frac{s+a_0}{2} - i\chi\right). \tag{36}$$

It is clear from formula (36) that with a given asymmetry parameter a_0, the singular variable parameter is $s = s' + s''$; hence, it is convenient to represent the calculation results as three-dimensional graphs (plotting the dependence of the quantity under consideration against s' and s''). We used formula (36) to carry out a numerical calculation of the far field parameters, particularly the position of the emission intensity distribution extrema χ_m^1, χ_m^2 and the points corresponding to the half intensity levels of the central and lateral extrema $W_{1/2}^{1,2}$, $V_{1/2}^{1,2}$ as well as the intensity relation between the extrema $A_m^{1,2}/A_m^0$, A_m^1/A_m^2. For convenience the inserts to Figs. 2-10 give a qualitative view of the directional pattern together with specific points characterizing this pattern. The purpose of this study is to determine these points.

The symmetric case ($a_0 = 0$). Figs. 2-5 show the variation in the far field parameters for the fundamental mode of a symmetric waveguide with the following reduced width parameter values: $0 \leqslant s' \leqslant 1$, $-5 \leqslant s'' \leqslant 0$. Fig. 2 shows the surface over the plane (s', s'') that determines the position of the off-axis maxima of the directional pattern χ_m^1 and χ_m^2. The intersecting curve between this surface and the plane (s', s'') divides the plane into two sections: a lower section (the emission distribution maximum is on the waveguide axis) and an upper section (two off-axis lobes exist). The lower section includes waveguides with $s = 0.9 - i0.5$; $0.8 - i0.65$, etc., and, naturally, all waveguides with $s'' = 0$. The angular position of the off-axis lobes $\theta_m^{1,2}$ always existing in waveguides with an induced width parameter s corresponding to the upper section, is determined by the formula

(37)

For example, for a waveguide with s = 0.5 - i3 we find from Fig. 2 χ^1_m = χ^2_m - 1.3 and we use formula (37) to calculate $\theta^2_m = -\theta^1_m$ = arcsin (1.3/$k_0 l$). For λ = 0.88 μm and l = 3 μm $\theta^2_m = -\theta^1_m$ - 3.5°. The existence of off-axis lobes in the directional pattern of a symmetric active waveguide results from the curvature of the wavefront due to the difference of the imaginary part of the dielectric constant in the active region (where gain predominates) and the regions adjoining the p-n-junction (where losses predominate) [29].

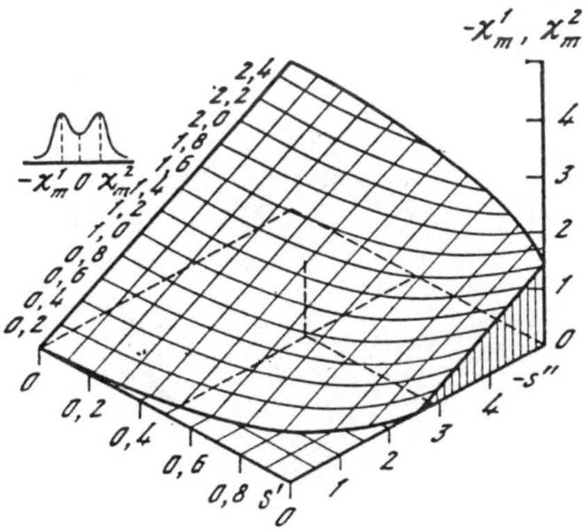

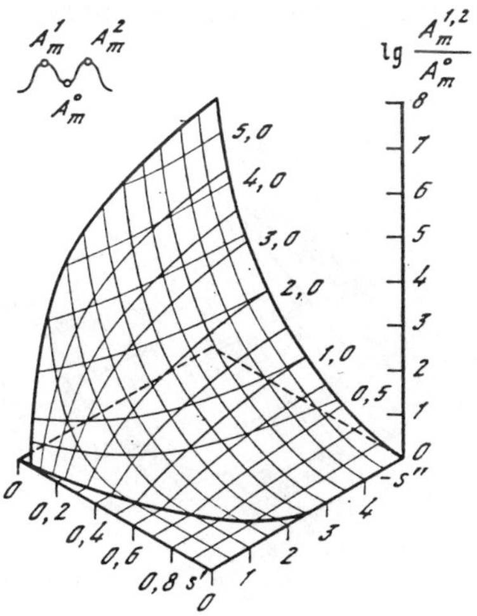

Fig. 2. The position of off-axis directional pattern maxima plotted against the complex parameter $s = s' + is''$ for a symmetric waveguide. $\theta_m = \arcsin(\chi^{1,2}_m/k_0 l)$. The lines of intersection between the surface and the planes parallel to the coordinate planes (s', s'') and (s'', χ_m) are shown.

Fig. 3. The ratio of the intensity at the directional pattern maximum to the intensity at the minimum plotted as a function of the complex parameter $s = s' + is''$ for a symmetric waveguide.
The lines of intersection between the surface and planes parallel to the coordinate planes (s', s''), (s', lg($A^{1,2}_m/A^0_m$)) and (s'', lg($A^{1,2}_m/A^0_m$)) are shown.

Fig. 3 gives the ratio of the emission intensity at the directional pattern maxima to the emission intensity at the center. The intersecting curve between this surface and the plane (s', s'') divides the latter into two sections. For s', s'' lying below this curve, $A_m^0 \geqslant A_m^1, A_m^2$ (i.e., the directional pattern maximum is on the axis). From a comparison of Figs. 2 and 3 it is clear that the intersecting curves between the three-dimensional surfaces and the plane (s', s'' do not coincide. For s' and s'' lying between these curves, $A_m^0 = A_m^1 = A_m^2$, i.e., the apex of the directional pattern has a plateau, rather than a sharp maximum. With low values of s' ($s' \leqslant 0.3$) the $A_m^{1,2}/A_m^0$ relation is a rather large quantity: all the emission is concentrated in the off-axis lobes; their position is heavily dependent on the quantity s''. The quantity s'' may be varied by changing the concentration of free carriers in the active region, i.e., by simply changing the pumping current. Thus, it is possible to scan the directional pattern by modulating the pumping current in a planar stripe heterolaser

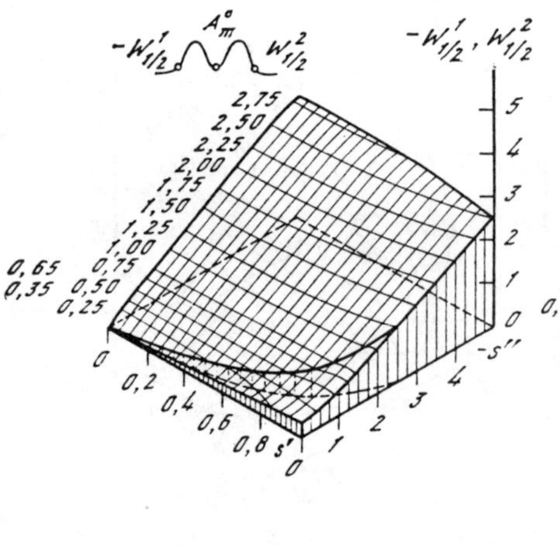

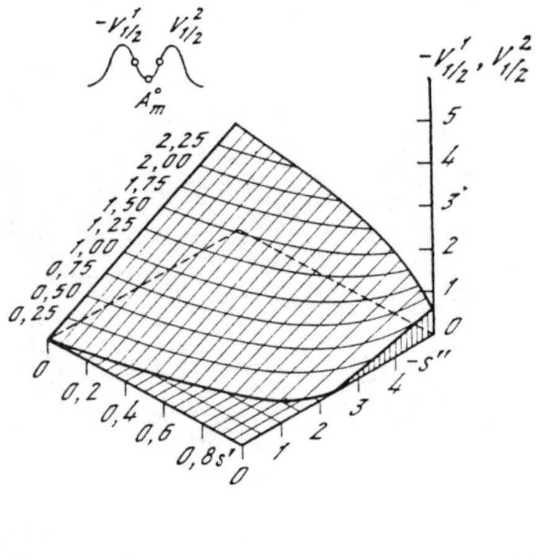

Fig. 4. The distance from the central extremum ($\theta = 0$) to the point characterizing the half amplitude level of the central extremum of the directional pattern plotted against the complex parameter $s = s' + is''$ for a symmetric waveguide.
$\theta = \arcsin(W_{1/2}^{1,2}/k_0 l)$. The lines of intersection between the surface and planes parallel to the coordinate planes (s', s'') and (s'', $W_{1/2}$) are shown.

Fig. 5. The distance from the central extremum ($\theta = 0$) to the point corresponding to half the difference in intensities at the off-axis maxima and the central extremum of the directional pattern plotted as a function of the complex parameter $s = s' + is''$ for a symmetric waveguide.
$\theta = \arcsin(V_{1/2}^{1,2}/k_0 l)$. The lines of intersection of the surface and the planes parallel to the coordinate planes (s', s'') and (s'', $V_{1/2}$) are shown.

(without lateral optical confinement) for a heterolaser-based amplifier.

Figs. 4 and 5 show the points in the directional pattern determining the half-intensity level to provide a complete picture of the emission distribution pattern in the far field.

The asymmetric case ($a_0 \equiv 0$). Calculation results for the fundamental mode (n = 0) of an active stripe waveguide with an asymmetric distribution of the complex dielectric constant on the *p-n*-junction are shown in Figs. 6–10 for asymmetry parameter values a_0 = 0.51 i/s and a reduced width parameter $0 \leqslant s \leqslant 1$, $-5 \leqslant s'' \leqslant 0$. The existence condition of a solution takes the form $s' \geqslant |a_0'|$ [11, 12]. No solution exists in the region between the curve s' = 0 and the curve s' = a_0 (this region is shaded in the diagrams). In waveguides with a reduced width parameter s lying in the shaded region, no electromagnetic

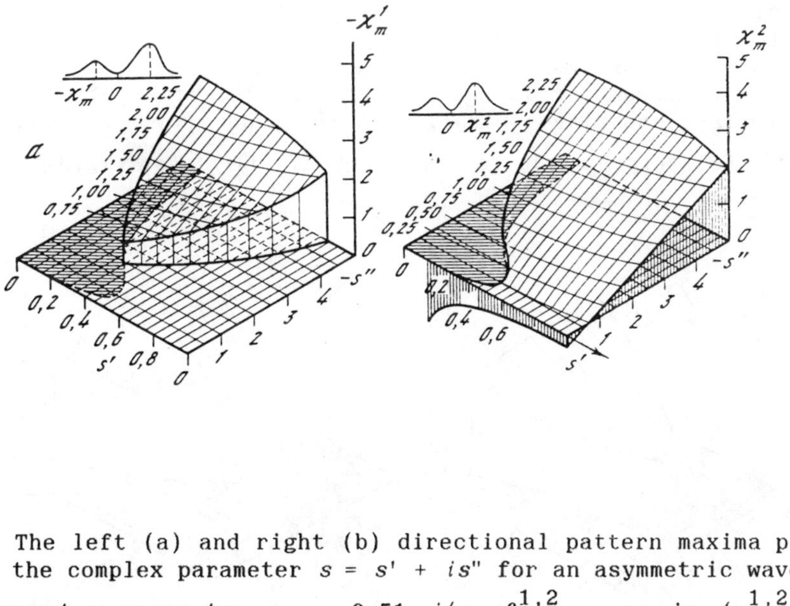

Fig. 6. The left (a) and right (b) directional pattern maxima plotted against the complex parameter $s = s' + is''$ for an asymmetric waveguide The asymmetry parameter a_0 = 0.51 i/s; $\theta_m^{1,2}$ = arcsin $(\chi_m^{1,2}/k_0 l)$. Lines of intersection of the surfaces and planes parallel to the coordinate planes (s', s'') and (s'', χ_m^1) (a), (χ_m^2, s'') (b) are shown.

field propagation is possible. If, with changes in carrier concentration in the active layer, the quantities s' and s'' change so that the point at s' and s'' falls within the shaded region, loss of lasing occurs in the laser (in the amplifier loss of amplification occurs). A curve dividing the existence regions of two off-axis directional pattern lobes and a single maximum, also offset from the axis, is mapped on the coordinate plane (s', s'').

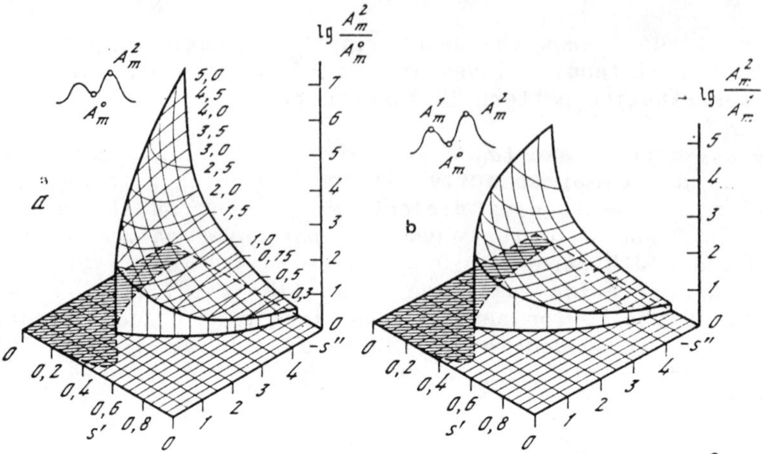

Fig. 7. The ratio of the intensity at the right maximum A_m^2 of the directional pattern to the intensities at the central extremum A_m^0 (a) and at the left maximum A_m^1 (b) plotted against the complex parameter $s = s' + is''$ for an asymmetric waveguide.
The asymmetry parameter is a_0 = 0.51 i/s. The lines of intersection of the surface and planes parallel to the planes (s', s'') and (s', lg (A_m^2/A_m^0)), (s'', (lg(A_m^2/A_m^0)) (a), (s', lg (A_m^2/A_m^1)), (s:, lg(A_m^2/A_m^1)) (b) are shown.

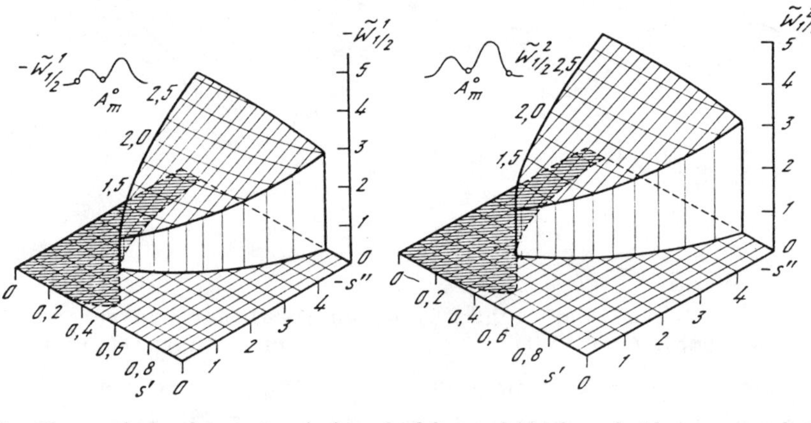

Fig. 8. The point characterizing half amplitude of the central extremum to the left of the left maximum (a) and to the right of the right maximum (b) of the directional pattern plotted against the complex parameter $s = s' + is''$ for an asymmetric waveguide.
The asymmetry parameter is a_0 = 0.51 i/s; θ = arcsin ($\tilde{W}^{1,2}_{1/2}/k_0 l$). The lines of intersection of the three-dimensional surface and planes parallel to the coordinate planes (s', s'') and (s'', $\tilde{W}^1_{1/2}$) (a) (s'', $\tilde{W}^2_{1/2}$) are shown.

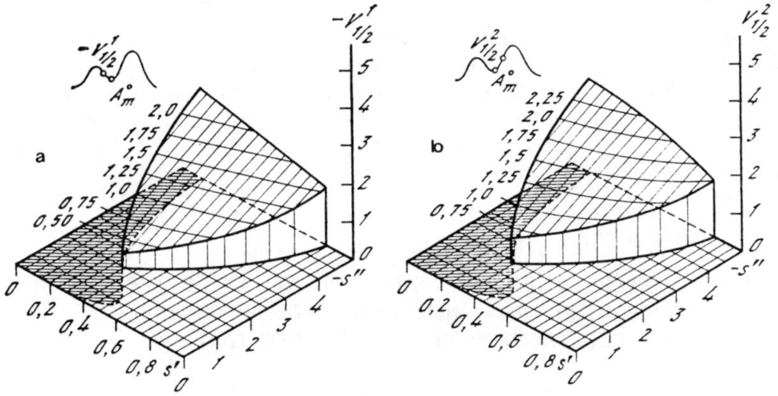

Fig. 9. The point corresponding to half the difference in the intensities at the left maximum (a) and the right maximum (b) as well as the central maximum of the emission distribution in the far field plotted against the complex parameter $s = s' + is''$ for an asymmetric waveguide The asymmetry parameter is $a_0 = 0.51\ i/s$; $\theta = \arcsin\,(V^{1,2}_{1/2}/k_0 l)$. The lines of intersection of the three-dimensional surface and planes parallel to the coordinate planes (s', s'') and (s'', $V^1_{1/2}$) (a) (s'', $V^2_{1/2}$) are shown.

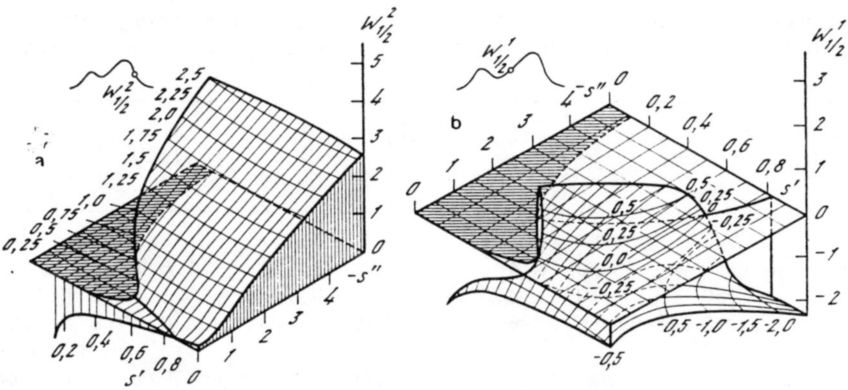

Fig. 10. The point characterizing the half intensity level of the right maximum of the directional pattern to the left (a) and to the right (b) plotted against the complex parameter $s = s' + is''$ for an asymmetric waveguide.
The asymmetry parameter is $a_0 = 0.51\ i/s$; $\theta = \arcsin\,(\tilde{W}^{1,2}_{1/2}/k_0 l)$. The lines of intersection of the three-dimensional surface and planes parallel to the coordinate planes (s', s'') and (s'', $W^2_{1/2}$) (a) (s'', $W^1_{1/2}$) are shown.

The range of s' and s'' values in which $A_m^2/A_m^1 \gg 1$, is an interesting problem. In this region virtually all emission is concentrated in a single lobe of the directional pattern so that, as we see from Fig. 6b, the position of this lobe changes noticeably with a change in s''. As for the case of a symmetric active waveguide, here we may use this effect for scanning the directional pattern by changing the pumping current.

4. THE INVERSE PROBLEM. DETERMINING THE PARAMETERS OF AN ACTIVE WAVEGUIDE FROM AN EXPERIMENTALLY OBSERVED ELECTROMAGNETIC FIELD DISTRIBUTION

In this section we will attempt to solve the inverse problem; specifically, estimating the parameters of an active waveguide from experimental data on emission distribution at the mirror of the laser diode (near field). Fig. 11a gives curves of the phase distribution of the emission field of a planar stripe heterolaser without lateral optical confinement for select pumping current levels. Results from numerical calculations using formula (33) for the curves in Fig. 11a are given in the table which indicates that in a stripe heterolaser without lateral optical confinement, the parameters of an optical waveguide under pumping current change significantly. As noted previously [11] these changes produce nonlinear watt-current characteristics of such lasers (see Fig. 11b).

Calculated values of the parameters of an active waveguide in a planar stripe heterolaser without lateral optical confinement

Curve No. Fig. 11a	*Pumping current ma*	y_1 *μm*	y_2 *μm*	Φ_1	Φ_2	s'' *(calc.)*	l, *μm (calc.)*
1	220	10	4.5	-3.0π	-0.75π	-10.3 ± 0.5	3.2 ± 0.1
2*	240	10	4	-3.75π	-1.25π	-4.5 ± 2.0	1.5 ± 0.5
3	260	10	4.5	-3.00π	-0.75π	-10.3 ± 0.5	3.2 ± 0.1
4	280	10	6	-2.00π	-0.80π	-11.0 ± 0.2	4.3 ± 0.1

* *The discrepancy in determining s'' and l for curve 2 results from the significant deviation of the experimental points (see Fig. 11a).*

We will use our scheme to analyze the experimental data of study [30]; Fig. 3 from this study gives the amplitude and phase distribution of the field at the mirror of an injection laser with a narrow stripe contact. Substitution of experimentally-measured quantities into formula (32) produces a system of equations, and solving these equations we derive $l = 2.5$ μm, $s' = 0.73$. Analogously, using formula

(33), we find $s'' = 2.5$. Thus, the parameters of an active waveguide in the test laser are: $l = 2.5$ μm, $s = 0.73 - i2.5$. If we wish to cal-

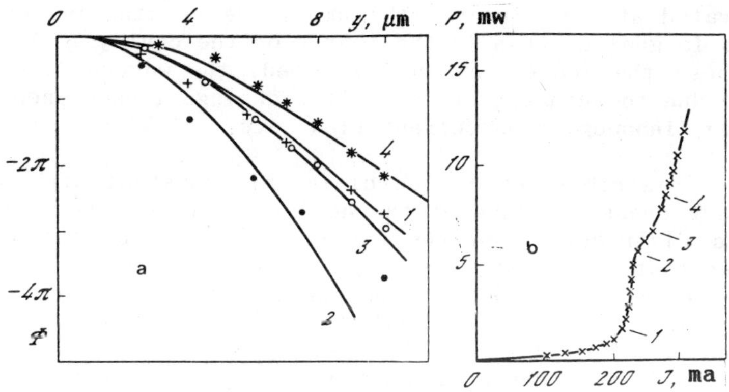

Fig. 11. The wave front distribution at the end mirror (a) of a planar stripe heterolaser without lateral optical confinement for various pumping currents marked on the watt-current characteristic (b).
1 = J = 220 ma; 2 - 240; 3 - 260; 4 - 280; CW operation.

culate the far field parameters, we will use the graphs given in Figs. 2-10. Naturally the proposed scheme includes the possibility for determining parameters of an active waveguide from an experimentally-measured directional pattern of emission (far field) as well.

A comprehensive comparison of the theoretical model examined in this article to experimentally-observed wavefronts and directional patterns for a number of planar stripe heterolaser samples is given in study [14] and in one article from this volume [31].

5. A DISCUSSION OF THE NONLINEARITIES AND ANOMALIES IN STRIPE HETEROLASERS WITHOUT LATERAL OPTICAL CONFINEMENT

A number of nonlinear phenomena occur in stripe heterolasers without additional lateral optical confinement, including the nonlinear dependence of emission power on pumping current, the asymmetric emission distribution in the near and far fields, the change in these distributions with changes in the pumping current, the difference in the distribution of the emission exiting the opposite ends of a laser diode, etc. It is clear that changes in the parameters of the active waveguide of the laser are responsible for all these phenomena.

Study [21] has identified the relationship between the inhomogeneous distribution of active layer parameters along the laser cavity and the difference in the amplitudes and spatial distributions of emission exiting the two opposite ends of the laser diode. In gain saturation conditions, the emission distribution along the length of the cavity has a maximum where losses are concentrated. Even if the active region is homogeneous along the cavity length, the losses will be concentrated at the mirrors (the amplitude of standing waves near the mirrors is greater than in the center of the cavity). In critical regions (where the losses are concentrated) the waveguide parameters also change due to such effects as self-focusing, inhomogeneous inversion burning, inhomogeneous current flow, etc.

If the distributions of the dielectric constant $\varepsilon(y)$ differ in two different cross-sections of the active waveguide, the electromagnetic field distribution on the cavity axis will be inhomogeneous. Such a laser that is inhomogeneous along the cavity length may be considered to be two (or several) homogeneous joined lasers (i.e., in each laser the dielectric constant is independent of z). Emission power losses occur at the joint between these lasers and these losses are determined in accordance with study [32] by the formula

$$\Delta P/P = 1 - | C_{\bar{n}n} |^2 - | a_r |^2, \tag{38}$$

where   are the coupling coefficients between the

modes of the two active waveguides (of the lengthwise-homogeneous laser cavity), the index $\bar{n}$ applies to the first waveguide, and n applies to the second waveguide; the asterisk designates the complex-conjugate quantities, the function $Y_n(y)$ is given by formula (25); $a_r = (\beta_{\bar{n}} - \beta_n)/(\beta_{\bar{n}} + \beta_n)$ is the reflection coefficient at the joint between the two waveguides, and β_n is the complex propagation constant (see (35)). For the case of a symmetric active waveguide ($a = 0$) the coupling coefficient for the fundamental mode ($n = 0$) is transformed in the following manner:

$$C_{00}^{s} = \frac{[\Gamma(2\bar{s}')\,\Gamma(2s')]^{1/2}\,[\Gamma((\bar{s}+s^*)/2)]^2}{\Gamma(\bar{s}')\,\Gamma(s')\,\Gamma(\bar{s}+s^*)}. \tag{39}$$

The coupling coefficient takes the following form for the fundamental mode ($n = 0$) of an asymmetric waveguide ($a_0 = a_0' + i a_0''$):

$$C_{00}^{ns} = \bar{A}_N(0)\,A_N(0)\,2^{\bar{s}+s^*} l\Gamma\left(\frac{\bar{s}+s^*}{2} - \frac{\bar{a}_0 + a_0^*}{2}\right) \times \\ \times \Gamma\left(\frac{\bar{s}+s^*}{2} + \frac{\bar{a}_0 + a_0^*}{2}\right)[\Gamma(\bar{s}+s^*)]^{-1}, \tag{40}$$

where the normalizing factor $A_N(0)$ is defined as

$$A_N(0) = 2^{-s'}\left[\frac{\Gamma(2s')}{l\Gamma(s' - a_0')\,\Gamma(s' + a_0')}\right]^{1/2}. \tag{41}$$

As we see from formulae (39)-(41), the coupling coefficient between the modes of the two joined active waveguides C_{00} in the model of the lengthwise-inhomogeneous laser is unambiguously determined through the parameters of the reduced width of these waveguides $|s|$ and s. This fact, in conjunction with the analysis given in Section 3 for the emission distribution in the near and far fields, again favorably supports the convenience of incorporating the reduced width parameter s.

If increasing the pumping current in a certain portion of the active region of a stripe heterolaser without lateral optical confinement produces a change in the parameters of the waveguide, the additional power losses that occur here serve to change the threshold current. This in turn is accompanied by a change in the watt-current characteristic and the operating point on the watt-current characteristic converts from the initial line to another line corresponding to a new value of threshold current. As a result, the watt-current characteristic is nonlinear; rapid changes in the parameters of the active waveguide due to such effects as self-focusing produce sharp changes in the watt-current characteristic (step nonlinearities produce kinks); while slow changes produce a monotonic deviation away from linearity of the watt-current characteristic.

In order to estimate the scale of power losses at the joint, we will carry out a numerical calculation using formulae (38), (39). For the example we will assume that the reduced width parameter changed from $\bar{s} = 3.9 - i6.6$ to $s = 2.7 - i6.0$, then $\Delta P/P \approx 6\%$. Here if using the formulae from study [32] we calculate the corresponding change in the threshold current, it equals 3%. Naturally if the parameters of the active waveguide change at several points in the active region (i.e., a laser with an inhomogeneous distribution of the dielectric constant lengthwise along the cavity may be considered several homogeneous joined active waveguides), each joint makes its own contribution to the total change in the threshold current and the nonlinearity of the watt-current characteristic.

6. EXPERIMENTAL EVIDENCE OF THE LONGITUDINAL INHOMOGENEITY OF A HETEROLASER AND AN ANALOG OF THE FRANCK-CONDON PRINCIPLE

Section 1 demonstrated that for a theoretical analysis of the output characteristics of a heterolaser we may use existing results obtained in quantum mechanics for particle motion in a time-dependent potential. Specifically, the Franck-Condon principle exists in quantum mechanics (see, for example, [23]) that makes it possible to identify the most probable vibrational states to be occupied by a polyatomic molecule during electron transition. Here, the electron jump from orbit to orbit occurs very quickly compared to the change in the wave function of the heavy (compared to the electron) nuclei moving in the potential generated by the electron cloud of all atoms in the molecule. As a result, the pattern appears as an instantaneous change in the potential energy of the nuclei. Such an instantaneous change in potential (i.e., the potential function $V(t)$ has a time dependence such that when $t < t_0$ it is constant over time: a steady-state potential, and with $t > t_0$ it changes irregularly over time and becomes a different potential that nonetheless remains time-stationary) in molecular spectroscopy is described by the Franck-Condon principle.

Since we have established that for the field in the laser the role of potential V is played by the dielectric constant ε (accurate to the constant), while the role of time (also accurate to the constant) is played by the coordinate on the cavity axis of a lengthwise inhomogeneous heterolaser, we have a full analogy of the pattern of a stepwise change in the properties of the active region of the heterolaser on its axis to the pattern of change in vibrational motion of nuclei in the electron transition in a polyatomic molecule. In order to use this analogy, we will use existing results in quantum mechanics to explain the various operating conditions of a heterolaser and we will emphasize that the role of the field $\tilde{E}$ is played by the wave function Ψ. In quantum mechanics, bound states exist when the probability density $|\Psi|^2$ diminishes exponentially away from the potential energy minimum. For a heterolaser, these bound states with quantum energy levels correspond to modal solutions of the wave equation. Hence, any conclusion regarding the transitions between energy levels from changes in potential over time may be reformulated analogously for the case of a redistribution of luminous energy among the modes when two (or more) different parts of the active section of a heterolaser are joined corresponding to an inhomogeneity in the dielectric constant on the cavity axis. As noted above, a continuous joint (a slow change in ε as a function of z) is possible together with a stepwise joint (a step in the ε profile in the vicinity of the joint). In quantum mechanics with a nonsteady-state potential (in certain cases) the Schrodinger equation is solved for changes in potential continuous over time, although such cases are few and they include a specific dependence of the potential on the coordinates (see [33]). For time-irregular (instantaneous) changes in potential, in quantum mechanics the behavior of a system for the case of polyatomic molecules is predicted by the Franck-Condon principle for potentials that are arbitrarily dependent on the coordinates. Hence, this formulation of the analog of the Franck-Condon principle for heterolaser operation is of interest, since the predictions obtained on the basis of this principle are independent of models describing the dielectric constant of the active region of the laser. We will now incorporate our concepts [11, 34, 35] of why it is necessary to account for the inhomogeneities of the active region lengthwise in the cavity (on the z axis) in actual heterolasers with a stripe contact. Above all we cite the recognized experimental fact that heterolasers with broad contacts ($\geqslant$ 50 μm) most often generate separate channels even at the lasing threshold. This is due to technological inhomogeneities present at the active region. The length of the heterolaser (the size on the z axis) is normally quite large (~200 μm); hence, there is no reason to ignore the inhomogeneities on the z axis. The inhomogeneous distribution of spontaneous emission on the cavity axis is the experimental confirmation of the dependence of the gain (and/or the absorption) on the coordinate z. The lateral face of the stripe heterolaser was polished in order to measure this relation (Fig. 12a). The distance from the stripe contact to the lateral face was ~20 μm. The image of the active region on the polished lateral face was focused by means of a lens onto a screen with a slit, and a photomultiplier was placed behind the slit (see the insert to Fig. 12). Scanning the slit simultaneously from the photomultiplier, we may measure the dependence of the spontaneous emission intensity on the coordinate z. In order to establish whether or not the spectrum of spontaneous emission exiting the lateral face contains scattered laser emission components that could distort the distribution of spontaneous emission on the z axis, the spectrum of emission exiting laterally (see Fig. 12b) was monitored; the image of the active region on the side of the polished lateral face was projected onto the inlet slit of the DFS-12 spectrograph for this purpose.

Other experimental evidence includes the difference in power, spectral composition and directional pattern of emission exiting the opposing ends (cavity mirrors) of a stripe heterolaser. This differential was identified in a number of studies [34, 35] where a careful experimental testing of the output characteristics from both sides of the stripe heterolaser without lateral optical confinement was carried out. The results are given in Fig. 13. These experimental results indicate that at low pumping currents, the watt-current characteristics (Fig. 13a) coincide for both sides. With an increase in the excitation level, there is a noticeable difference in the watt-current characteristic, which is accompanied by different spectra (Fig. 13c) of emission exiting from both sides, as well as different directional patterns (Fig. 13b), and the more pronounced the bend in the watt-current characteristic (the greater the deviation from linearity), the greater the deviation of the directional pattern maximum from the optical axis.

In studies [10, 11] we proposed that the influence of inhomogeneities lengthwise in the cavity of a heterolaser be accounted for by a model in which the active region is represented as a set of several complex waveguides with different dielectric constants coupled by graded junctions where the coupling between these separate sections of the active region is achieved by incorporation of the coupling constant between modes propagating in each waveguide, $C_{\bar{n}n}$. The field in the heterolaser with graded inhomogeneities on the z axis is described by equation (13). For the case of a stepwise change in $y(z)$ we may use equation (6), drafting the equation for each of the homogeneous sections of the active region in which the dielectric constant has various values (independent of z in the homogeneous section). It was demonstrated in Section 1 that the behavior of the field in the section of a stripe heterolaser homogeneous on the z axis is analogous to the behavior of a quantum mechanical particle in a single-dimension potential (distinct from the analogy to the two-dimensional case discussed above). The single-dimensional quantum mechanical equation describes vibrational motion of a diatomic molecule rather than a polyatomic molecule. Consequently, the problem of redistribution of luminous energy among the modes in a stripe heterolaser with inhomogeneities on the z axis corresponding to the jointed homogeneous active regions, is formally equivalent in this case to the quantum mechanics problem of the probability of transitions between vibrational levels of diatomic molecules in connection with the electron transition, which yields the vibronic structure of the electron lines in the absorption or emission spectra of diatomic molecules.

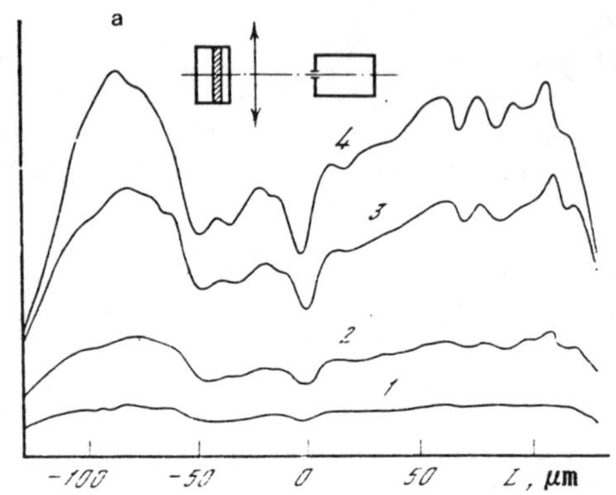

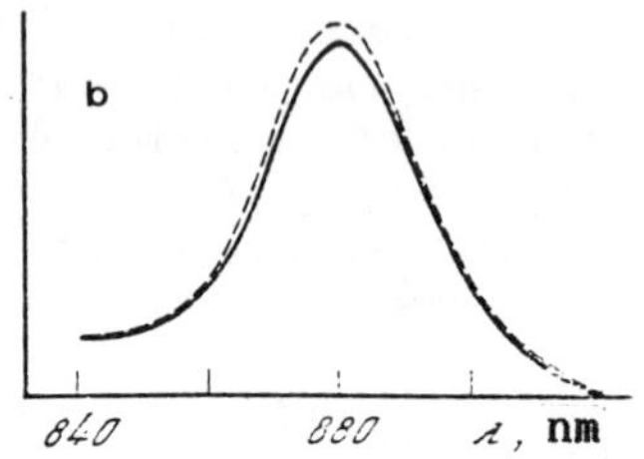

Fig. 12. The spontaneous emission intensity distribution lengthwise in the cavity (a) and its spectral composition (b) for a planar stripe heterolaser for various pumping currents.
a - J = 50 ma (1); 100 (2); 200 (3); 285 (4); b - J = 200 ma (solid line); 230 (dotted line). Threshold current: 130 ma. Above - experimental scheme.

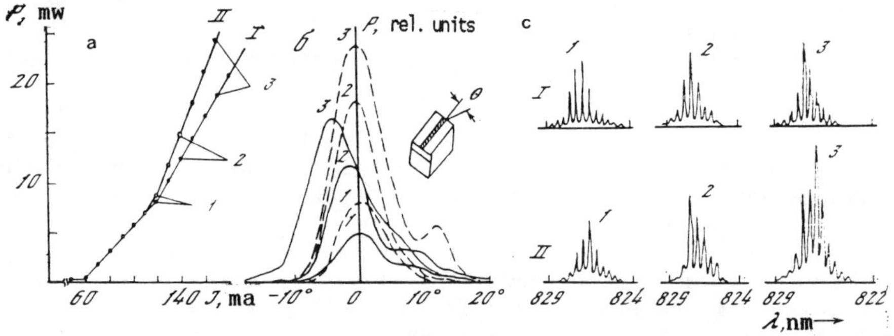

Fig. 13. The watt-current characteristic (a), directional patterns (b) and emission spectra (c) from the opposite (I, II) mirrors of the cavity of a planar stripe heterolaser.
The numbers marking the spectra and the diagram correspond to the current values in the watt-current characteristic: 120 ma (1); 140 (2); 170 (3).

In understanding this fact we will formulate the fundamental result of this study specifically for a lengthwise-inhomogeneous (on the z axis) stripe heterolaser. Let this laser consist of two end-joined (each with a homogeneous length) active regions (two lasers).

Let the electromagnetic field in the first laser propagate as a given mode with an index $\bar{n}$ (the solution of equation (23) $Y_{\bar{n}}(y)$). As the field crosses over to the second laser, modes corresponding to the characteristics of the second laser with the index n are excited (the solution of equation (23) $Y_n(y)$). In order to identify the number of the mode n in which the primary portion of energy in the second laser is concentrated, we will develop the following geometric formulation- an analog of the Franck-Condon principle. Let $-\varepsilon_I(y)$ and $-\varepsilon_{II}(y)$ be the distributions of the dielectric constant in the first and second lasers, respectively, accounting for the normalization of all constants reducing the dielectric constant $\varepsilon(y)$ to the formal potential $V(y)$ whose jump is shown in Fig. 14a. We will consider the waveguide case ignoring the imaginary parts of the dielectric constant (accounting for these requires a separate analysis).

Fig. 14. Diagram illustrating the analog of the Franck-Condon principle (a) and a diagram of the mode indices n excited in the second laser by the $\bar{n}$-mode of the first laser, for the parabolic potential (b).

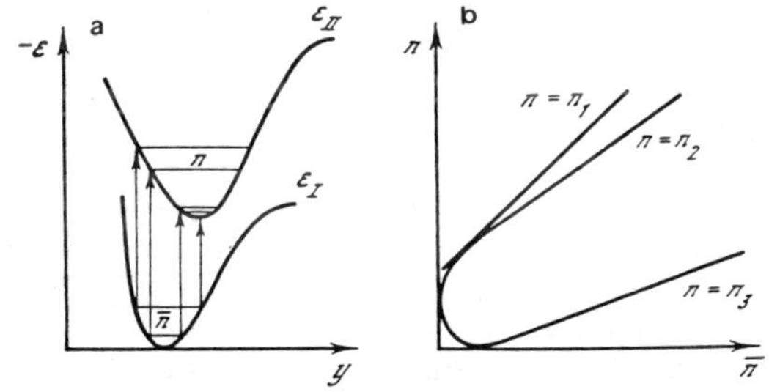

We note the corresponding "energy levels" in the second laser with mode numbers n and the "energy level" for the mode in the first laser with the index $\bar{n}$. Then in the second laser the energy is largely contributed to the mode whose n number is determined by the following rule: we will construct a perpendicular from the point of intersection of the $\bar{n}$-"level" and the curve ε_I to intersect with the ε_{II} curve. The mode of the second laser whose number n is closest to the point of intersection of this perpendicular with the ε_{II} curve of the "level" corresponding to the mode in the second laser is excited to the greatest degree. We emphasize that this rule is valid in a waveguide situation when the imaginary part of the dielectric constant does not significantly change the spectrum of mode "levels" (in quantum mechanical language, the imaginary part of the complex potential does not alter the energy level spectrum, but rather is responsible only for broadening of the levels and their finite lifetime; however, the width of the levels here are small compared to the distance between levels).

This analog of the Franck-Condon principle essentially produces a qualitative prediction for finding the maximum of the square of the modulus of the mode coupling coefficient of two lasers $C_{\bar{n}n}$,

$$C_{\bar{n}n} = \int_{-\infty}^{\infty} Y_{\bar{n}}^{*}(y)\, Y_n(y)\, dy, \tag{42}$$

i.e., an analog of the Franck-Condon factor

$$W_{\bar{n}n} = | C_{\bar{n}n} |^2 \tag{43}$$

for diatomic molecules (the overlap integral of the vibrational wave functions before and after the electron transition). This factor describes the portion of energy contained in the mode of the second laser with the index n if all energy in the first laser was concentrated in the mode with the index $\bar{n}$.

This principle formulated for a lengthwise-inhomogeneous heterolaser identifies a geometric rule for finding the maximum of $W_{\bar{n}n}$ determining the relative intensity of the n mode of the second laser with a fixed mode of the first laser with a given index $\bar{n}$. For certain types of $\varepsilon(y)$ relations we may analytically calculate the coupling coefficient such as for the parabolic relation $\varepsilon(y)$; a consideration of the Morse potential and several other potentials is possible. However, even in these cases, an explicit analysis of the maxima $W_{\bar{n}n}$ is difficult, and the principle formulated here makes it possible to easily predict the excited mode in the second laser.

We will make certain qualitative statements based on this principle. If the distribution of the dielectric constant is such that two active regions exist in the model of a heterolaser with separate sections on the z axis, and very high order modes may exist in each of the active regions, this means that any potential curve near the maximum of the distribution of the dielectric constant may be considered a parabolic curve (an analog of an approximation of minor vibrations for a particle in any potential well with a minimum, when the vibrations may be considered to be harmonic for small energies). A statement exists for the harmonic potential $\varepsilon(y)$ [36, 37] that makes it possible to state that for the case of this laser, the maxima of the square of the modulus of the mode coupling coefficient $|C_{\bar{n}n}|^2$ lies on a familiar curve: a Condon parabola with an additional line (see Fig. 14b). With a fixed index $\bar{n}$ of one mode in the first laser, the modes in the second laser most strongly excited are those with a mode index n that appears as the y-coordinate lying on the intersection of the perpendicular from the point $\bar{n}$ to the x-axis and the Condon parabola with the additional line. There are three such points for the higher-order modes. Hence it is most probable that if the field is contained in a single mode in the first laser, it will be contained in no fewer than three modes in the second laser. Consequently, there is a discrepancy in this case in the distribution of emission at opposing mirrors (ends) of the laser, since the speckle structure of the light at the end of the first laser corresponding to the $\bar{n}$-mode is accompanied by a speckle structure: the superimposition of fields from the three higher order modes. For very high order modes, the number of light and dark spots for the sum field of the three modes may be so significant that the end of the second laser will appear to be uniformly illuminated. Thus, a discrepancy in the emission distribution in the near field for the case of a strong speckle structure at at least one of the end mirrors of the cavity of the heterolaser will indicate a lengthwise inhomogeneity of the cavity of the heterolaser. This is consistent with the statement from study [21] in which kinks in the watt-current characteristic were used as the basis for concluding that a compound heterolaser structure (one consisting of two lasers) was present. The nonlinearity of the watt-current characteristic of the heterolaser may be related to the additional power losses from joining two (or more) active regions of a lengthwise-inhomogeneous heterolaser [32, 11]. Hence, the discrepancy in the near field patterns of the two opposite ends of the heterolaser (see, for example, [35]) together with the nonlinearity in the watt-current characteristic indicate that an inhomogeneity exists on the heterolaser axis. The absence of such phenomena may indicate both homogeneity of the laser along its axis and that the laser consists of various sections, one of which functions as a matching element. This possibility was identified in study [22]. Of particular interest is a structure of longitudinal inhomogeneities in which the junctions between the homogeneous sections are fabricated such that total matching of all sections in the lengthwise-inhomogeneous laser is provided. The principle of operation of a matching element will be examined in the next section.

7. AN ANALOG OF THE RAMSAUER EFFECT FOR A LENGTHWISE-INHOMOGENEOUS HETEROLASER

In the first section we discussed that the equation for electromagnetic emission propagation along the heterolaser is completely analogous to the "temporal Schrodinger equation" for a particle with two degrees of freedom; the longitudinal spatial coordinate plays the role of time. In the preceding section we used this analogy for conversion to the problem of light propagation in a heterolaser, which is known as the Franck-Condon principle in quantum mechanics. However, other effects exist in quantum mechanics whose analogs may be used in the problem of light propagation in active optical waveguides. One such effect is the Ramsauer effect which (after formulation of its analog) makes it possible to establish the possible influence on the near field structure of a lengthwise-inhomogeneous heterolaser. The purpose of this section is to discuss the possibility of the existence of an analog of the Ramsauer effect for such a laser.

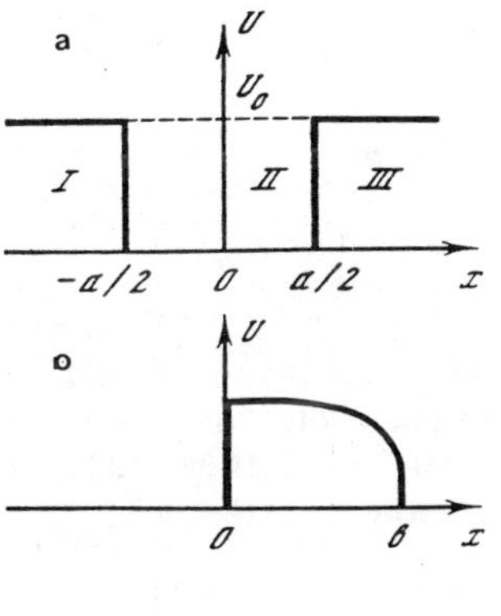

Fig. 15. Diagrams illustrating the analog of a potential well and the refractive index profile (a) and a quasiclassical approximation for the barrier refractive index profile (b).

The Ramsauer effect (see [38, p. 633]) in quantum mechanics is as follows: the potential at which a flow of particles scatters becomes "transparent" at certain particle energies; specifically, the reflection coefficient of particles at this potential as a function of the energy of the incident particles has nulls at certain energies. The idea of using this quantum mechanical effect in the problem of light propagation in a lengthwise-inhomogeneous heterolaser is as follows: for the case of an inhomogeneity (a joint inhomogeneity) resulting primarily from the joining of two (or more) lengthwise-homogeneous sections of the heterolaser, the "Schrodinger equation" corresponding to this pattern in an approximation of the parabolic profile of the refractive index in transverse coordinates may be solved exactly [39] and the answer for the mode coupling coefficient in the two joined lasers is expressed by the "transit coefficient (transparency) of a fictitious particle" through the potential determined by the dependence of the refractive index on the longitudinal coordinate. This dependence of the refractive index on the longitudinal coordinate may be such (or may be designed) to guarantee conditions in which the reflection coefficient is equal to zero, i.e., the joint between the lasers is made translucent and no emission energy is redistributed among the modes of the second laser compared to the mode distributions in the first laser.

We will first describe, consistent with [40], the situation in quantum mechanics, specifically: in which conditions does the transit coefficient of a particle of mass m through the potential vanish, using a single-dimensional rectangular well of depth U_0 and with a (see Fig. 15a). The Schrodinger equation for the case of the potential shown in Fig. 15a, takes the form

$$\left\{\frac{d^2}{dx^2}+\frac{2m}{\hbar}\left[E-U(x)\right]\right\}\Psi(x)=0. \tag{44}$$

The solution in region I (see Fig. 15a) for states with a positive energy differential $E - U_0 = \mathcal{E} > 0$ describing the superposition of the incident wave from the left and the reflected wave takes the form

$$\Psi_I=\exp i\varkappa_0 x+B\exp(-i\varkappa_0 x),\qquad \hbar\varkappa_0=\sqrt{2m\mathcal{E}}. \tag{45}$$

Only the previous wave exists in region III, i.e.,:

$$\Psi_{III}=C\exp i\varkappa_0 x. \tag{46}$$

The actions of the potential well in region II are

$$\Psi_{II}=\alpha\exp i\varkappa x+\beta\exp(-i\varkappa x),\qquad \hbar\varkappa=\left[2m(\mathcal{E}+U_0)\right]^{1/2}. \tag{47}$$

In order to determine the transient coefficient $D = |C|^2$ and the reflection coefficient $R = |B|^2$ related by the relation $R = 1 - D$, we will equate the functions and their derivatives on the boundaries of the potential well. In accordance with [40, pp. 108-119], we will write the following for the transit coefficient

$$D = [1 + {}^1/_4\,(\varkappa_0/\varkappa - \varkappa/\varkappa_0)^2 \sin^2 \varkappa a]^{-1}. \tag{48}$$

when

$$\sin \varkappa a = 0 \tag{49}$$

we find that the transit coefficient (transparency) $D = 1$, while the reflection coefficient $R = 0$, i.e., reflection is not present, and all particles pass freely through the potential. The "translucence condition" (49) may be formulated in the following manner: a whole number of half waves are accumulated within the well. The wavelength is related to the momentum of the particle and to its energy. The experimental conditions in atomic physics are normally such that the potential and the experimenter may change the energy of the incident particles; this energy then passes through its resonance value at which the potential does not scatter (the scattering cross-section decreases significantly: this is the Ramsauer effect). However, generally speaking, another formulation is possible: for a given energy of the incident particle, it is possible to select the width and depth of the well to satisfy condition (49). In the physics of lengthwise-inhomogeneous heterolasers, changing the potential means changing the nature of the inhomogeneity in the vicinity of the joint of the two lasers.

The transparency result of the potential barrier shown in the example of the rectangular well is qualitatively valid for other dependencies of the potential on the coordinate as well. This is easy to demonstrate using a quasiclassical approximation. Thus, the reflection coefficient R for a particle of energy E from the potential barrier whose boundaries (for example, at the points 0 and B) have potential values that are identical and equal to $U(0)$ (see Fig. 15b) takes the form [40, p. 101-107]:

$$R = \frac{(\varkappa_0^2 - a^2)\,\mathrm{tg}^2\,\varphi}{4\varkappa_0^2 a^2 + (\varkappa_0^2 + a^2)\,\mathrm{tg}^2\,\varphi}, \tag{50}$$

where

$$a = \frac{1}{\hbar}\sqrt{2m\,[E - U(0)]}, \qquad \varphi = \frac{1}{\hbar}\int_0^b \sqrt{2m\,[E - U(x)]}\,dx.$$

When

$$\int_0^b \sqrt{2m\,[E - U(x)]}\,dx = n\pi\hbar, \qquad n = 1, 2, \ldots, \tag{51}$$

The potential barrier is transparent. This condition may be interpreted in the following manner: a whole number of half waves $\bar{\lambda}$ are accumulated along the length of the barrier b, i.e., $b = n\bar{\lambda}/2$, where

$\bar{\lambda} = 2\pi\hbar/\bar{p}$, a $\bar{p} = \frac{1}{b}\int_0^b \sqrt{2m\,[E - U(x)]}\,dx$ — is the mean of the particle momentum. Thus, for potentials of various type we may anticipate a significant drop in the reflection coefficient R when resonance condition (51) is satisfied.

We will now proceed to a description of the model case when the dielectric constant is a quadratic expression for transverse coordinates with coefficients arbitrarily dependent on the longitudinal coordinate,

$$\varepsilon(x, y, z) = \varepsilon_0 + a(z)x^2 + b(z)y^2 + c(z)xy + d(z)x + e(z)y. \tag{52}$$

In this case, the equation for the electromagnetic field in a lengthwise-inhomogeneous heterolaser (13) is solved exactly (as the Schrodinger equation for a two-dimensional oscillator with variable frequency acted on by an excitation force (see, for example, [33]). In the general case these solutions describe modes in a heterolaser by special functions: Hermite polynomials in two variables. Then the overlap integral for the mode coupling coefficients in a lengthwise-inhomogeneous heterolaser (represented as two end-joined lengthwise-homogeneous lasers) is expressed by Hermite polynomials in four variables [33] consistent with the overlap integral describing the intensity of vibrational repetitions in electron transitions in a polyatomic molecule.

We will now deal with the important practical case where the distribution of the dielectric constant in one cross-section has a stepwise form (for the case of a heterolaser, the distribution of the dielectric constant in the direction of current flow is determined by the differential of the refractive index of the active region and the broadband layers of the heterostructure, i.e., a strong waveguide exists in this direction), while in the other transverse direction the distribution is approximated by a parabola (a symmetric waveguide; see, for example, [8]). Then there are no linear terms in formula (52), and we have a quadratic dependence on only a single transverse coordinate (the analog of a variable frequency harmonic oscillator), and the answer for the square of the modulus of the coupling coefficients, i.e., for the Franck-Condon factor (compare (42), (43)) takes the form (see [33, p. 46])

$$W_{nm} = \frac{n!}{m!}\frac{1}{|\zeta|}\,|P^{(m-n)/2}_{(m+n)/2}(\cos(\theta/2))|^2, \tag{53}$$

where $P^l_k(x)$ are Legendre polynomials (we are using their definition as given in [28]),

$$\cos\theta = 1 - 2|\eta/\zeta|^2, \quad |\eta/\zeta|^2 = R, \quad |\zeta|^2 - |\eta|^2 = 1, \quad m > n.$$

Thus, in this case, the Franck-Condon factor depends only on a single parameter R, which may be purely formally interpreted as the reflection coefficient of the potential determined by the "frequency" of the oscillator. It is clear from the preceding discussion that it is possible to select a type of dependence of the inhomogeneity along the heterolaser (the nature of the "potential" dependence $b(z)$) in order to satisfy conditions so that the Ramsauer effect will occur, specifically: $R = 0$, $W_{nm} = \delta_{nm}$ and no electromagnetic field energy redistribution will occur among the modes. In this case, the lengthwise-inhomogeneous laser will have an inhomogeneity structure that exists with the "translucence" effect at the joint between the two lasers.

We will now cite certain stipulations. Above all this "translucence" effect - the analog of the Ramsauer effect - was obtained within the scope of a model dielectric constant (52); however, we hope that qualitatively this effect will be preserved for the case of other potentials as well. Moreover, this discussion is strictly valid for the case of the real coefficient $b(z)$. Generally speaking, it is necessary to account for the imaginary additions to this coefficient in active media and media with absorption.

We will now discuss what physical influences make it possible to change the dielectric constant at the joint of the two lasers in a lengthwise-inhomogeneous heterolaser. It is possible to influence the waveguide properties of a heterolaser by varying the temperature transverse to the direction of light propagation. In our studies [41, 42] we achieved this action by means of pumping stripe heterolaser with double pulses. The double pulse pumping condition makes it possible to investigate the influence of the nonsteady-state temperature waveguide formed as current passes through the active layer on the characteristics of heterolasers. We remember that as the current heats the active region, the refractive index increases, which serves to increase the dielectric constant [43]; at the same time, as the pumping current is increased, the electron concentration in the active region increases, thereby reducing the dielectric constant (the anti-waveguide effect) [44]. In principle by combining these two methods of changing the dielectric constant, it is possible to control losses in the active waveguide, which is the active region of the heterolaser, as well as the configuration of its natural mode.

The double pulse pumping method is described in detail in studies [42, 45] and involves the use of a first prepulse with a positive (amplitude below the lasing threshold) or negative polarity to generate a dynamic temperature waveguide, while the second pulse has positive polarity (i.e., it creates forward bias) and is used as a probe pumping pulse for lasing emission. The electrical circuit for the double pumping pulses includes the following: the primary pulsed pumping current from the G5-15 generator is fed to the heterolaser through a voltage divider. At the same time it fires another G5-15 generator which produces current pulses of either positive or negative polarity. The delay of the primary pulse with respect to the prepulse was varied from 0.1-10 μs. The duration of the positive prepulse was varied from one to a few microseconds while the duration of the negative polarity prepulse was 100 ns (with an amplitude from 10-60 ma). GaAs-$Ga_{1-x}Al_xAs$ heterolasers were used as the test samples (x = 0.35 for the broadband layer, x = 0.07 for the narrowband layer); these samples were fabricated by liquid-phase epitaxy onto (100)n-GaAs:Si-substrates. The dimensions of the active region were 400 $\times$ 15 $\times$ 0.3 μm for planar geometry samples (fabricated by oxide insulation) and 250 $\times$ 10 $\times$ 0.15 μm for the fine mesa-stripe structure.

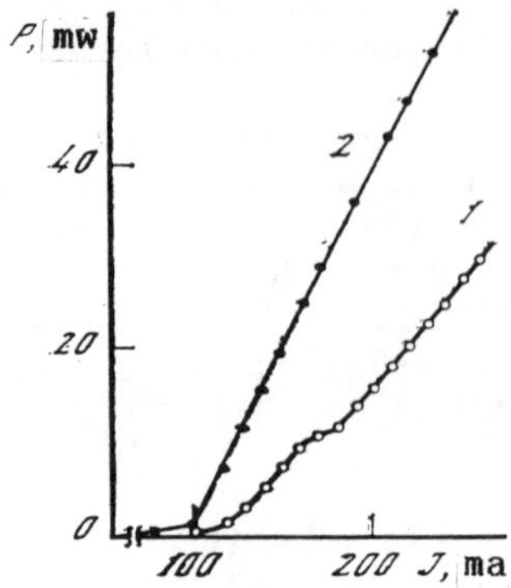

Fig. 16. The watt-current characteristic of the number 30 heterolaser without a prepulse (1) and with a prepulse (J = 30 ma, τ = 100 ns) of negative polarity (2)

The influence of the pre-pumping pulses on the output characteristics of the stripe heterolaser are demonstrated by Figs. 16-20 [41]. Fig. 16 shows a significant improvement in the watt-current characteristic of heterolaser number 30 under the action of a negative polarity prepulse with an amplitude of 30 ma and a duration of 100 ns with a delay relative to the primary pumping pulse of 100 ns. As we see from Fig. 16, the reduction in the threshold current is 20%, while the watt-current characteristic is converted from nonlinear to linear, and the output emission power is significantly increased. Figs. 17 through 20 show the change in the spectral composition of emission (Fig. 17), the dependence of emission power P on delay τ_3 between the positive primary pulse and prepulse (Fig. 18a) as well as the current amplitude of the negative prepulses (Fig. 18b); these figures also give the variation in the emission intensity distribution in the far field (Fig. 19) and oscillograms of a light pulse at temperatures of 300 and 350 K and various pumping conditions (Fig. 20). Fig. 17 gives the emission spectra when a positive prepulse is applied (curve 2), and when a negative prepulse is applied (curve 3) compared to the initial spectrum (curve 1). We see that the spectrum converts from multimode to single-mode and its maximum shifts by 2.5 nm to longer wavelengths. We note that single-mode conditions are preserved to 1.5 times lasing threshold. It follows from an analysis of the relations in Fig. 18 that with an increase in τ_Z from 0 to 0.2 μs, the power increases and then remains on the same level with τ_Z varying from 0.2 to 1.0 μs; a further increase in τ_Z to 1.8 μs causes the power to drop (in the range $0.2 < \tau_Z < 1.8$ μs the emission spectrum maintains its single-mode characteristic up through 1.5 times over the pumping threshold.) With $\tau_Z > 1.89$ μs, the output power remains fixed, although the number of spectral modes increases. When the amplitude of the negative current pulses J is increased to 30 ma, the output emission power increases and is stabilized at 30 ma < J < 50 ma; measurements of the spectral characteristics indicate that in the vicinity 10 < J < 50 ma a single mode is preserved when the pumping current exceeds the threshold by a factor of 1.5.

Moreover, noticeable changes have been observed in the intensity distribution in the far and near fields when both positive and negative prepulses are applied. Specifically, the emission distribution in the far field, which was asymmetric and whose maximum was 3.5° off axis (curve 1 in Fig. 19) is balanced relative to the axis (curves 2 and 3 in Fig. 19), when positive or negative prepulses are applied.

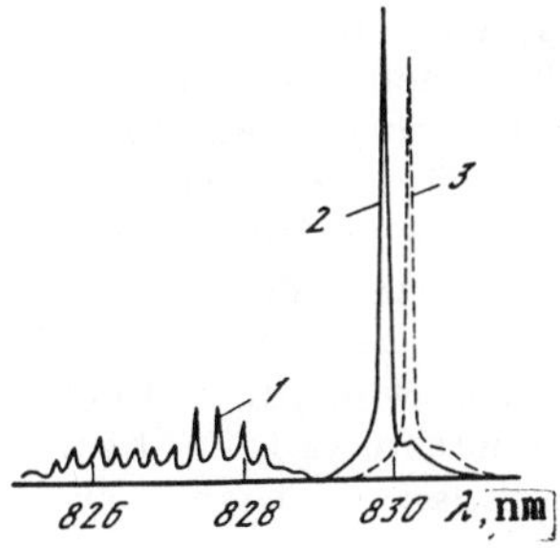

Fig. 17. The emission spectrum of heterolaser number 611 under double pulse pumping.
1 - J = 115 ma, pulse duration 100 ns; $J_{пор}$ = 120 ma;
2 - the same values with positive current prepulses of 100 ma; $\tau_{и}$ = 1 μs; $\tau_{з}$ = 200 ns;
3 - the same values with negative current prepulses of 50 ma; $\tau_{и}$ = 100 ns; $\tau_{з}$ = 500 ns.

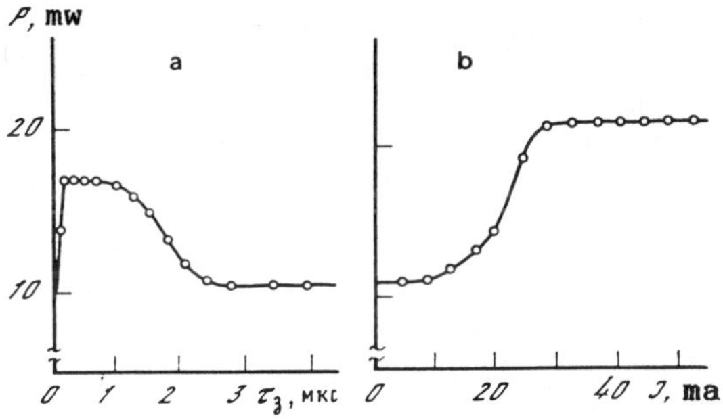

Fig. 18. The output emission power P of heterolaser number 611 plotted as a function of the delay $\tau_{з}$ of the primary pumping pulse with respect to the positive prepulse (a) and as a function of the negative current prepulse amplitude (b)

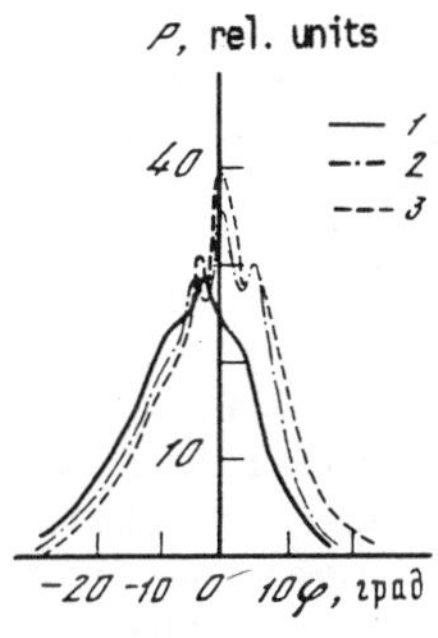

Fig. 19. The emission intensity distribution of heterolaser number 611 in the far field
1-3 correspond to the designations of Fig. 17

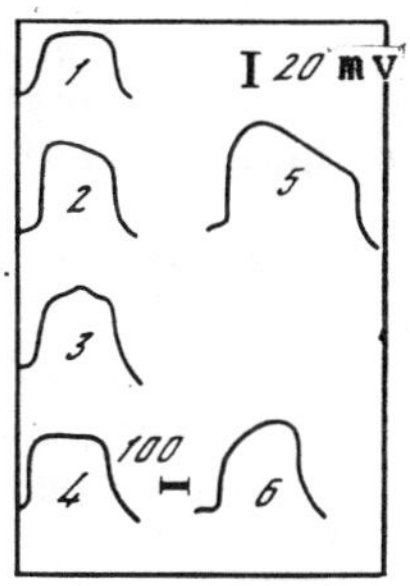

Fig. 20. Oscillograms of a light pulse with T = 300 K (1-4) and 350 K (5, 6) for various pumping conditions J
1-4: J = 120 ma; pulse duration 300 ns; 5, 6: J = 235 ma; pulse duration: 500 ns. Negative prepulses with an amplitude of 60 ma are applied, $\tau_{и}$ = 100 ns (2-6); $\tau_{з}$ = 0 (2, 5); 200 ns (3); 900 (4); 1600 (6); heterolaser number 611

An investigation of the emission at the end mirror of the heterolaser demonstrated that the far field is narrowed and the wavefront becomes less concave when double pulse pumping is used.

Fig. 20 shows the change in the waveform of the light pulse for various double pumping conditions at two temperatures: 300 and 350 K. It is clear in oscillogram 2 that the change in intensity in the form of an additional peak at the flat apex of the light pulse occurs in the first 100 to 150 ns of the lasing pulse. Increasing the delay between the prepulse and the pumping pulse $\tau_{з}$ from 100 to 600 ns causes this additional spike to shift 150 ns from the onset of lasing (oscillogram 3, Fig. 20). With T = 350 K the increase in intensity is observed over a longer time (approximately 300 ns) and is preserved with longer delays τ_3 = 300 ns (oscillogram 6, Fig. 20) than when T = 300 K; when τ_3 = 900 ns, the additional spike disappears (oscillogram 4, Fig. 20). We note that the changes in the light pulse waveform shown in Fig. 20 differ from those observed in study [46] where a stepwise lasing pulse shape was observed in a certain pumping current interval at pulse durations of greater than 2 μs. Such behavior of the light pulse waveform may be explained by the onset of an adiabatic temperature waveguide where the transverse emission distribution in the *p-n*-junction plane becomes constricted, thereby reducing internal optical losses. This serves to increase emission output power during the existence of the adiabatic waveguide.

Estimates have indicated that the 2.5 nm shift in the lasing spectrum to longer wave emission (Fig. 17) corresponds to a 10 K increase in the temperature of the active region. Such heating serves to increase the real part of the dielectric constant by a quantity equal to $\delta\varepsilon = (\partial\varepsilon/\partial T)\delta R \simeq 3 \cdot 10^{-2}$ where $\partial\varepsilon/\partial T \approx 3 \cdot 10^{-3}$ [8, 43]. This is sufficient to create a waveguide which narrows and stabilizes the electromagnetic field in the *p-n*-junction plane and makes it possible to achieve single-frequency lasing [47, 48]. Observations of the light pulse waveform with changes in temperature and delay between the prepulse and the pumping pulse (Fig. 20) makes it possible to state that an adiabatic waveguide is created during the pumping pulse; the prepulse and proper selection of laser temperature (as well as the light pulse) aid in the creation of this waveguide. The range of variation in the real part of the dielectric constant depends both on the amplitude and duration of the prepulses as well as the delay between the prepulse and the pumping pulse.

In analyzing the experimental results described above, we concluded in study [42] that the nonstationary temperature waveguide with the refractive index differential from the center of the active stripe to its periphery (equal to 10^{-2}) sufficient to stabilize and narrow the laser spatial channel was responsible for the enhancement in output characteristics of the stripe heterolaser for the case of double pulse pumping. The maximum effect from the temperature waveguide is observed 100-200 ns from the beginning of the pumping pulse, after which the temperature distribution begins to spread out. In CW conditions, a steady-state temperature waveguide is formed with a smaller refractive index differential; this is added to variations associated with the distribution of free inversion carriers, residual voltages and other factors. In adiabatic heating conditions of the active region, the nonsteady-state temperature waveguide effect predominates over other factors acting on the dielectric constant profile in stripe heterolasers for the case of zero optical confinement.

An alternative explanation is also possible for the three characteristic phenomena observed in our experiments, including the narrowing of the spot in the near field, enhanced symmetry of the directional pattern and linearization of the watt-current characteristic. Study [21] has demonstrated that the asymmetry of the directional pattern of the same order of magnitude observed in our experiments together with the kinks in the watt-current characteristic indicate that the heterolaser consists of two end-joined active regions. The output characteristics of our test heterolasers with ordinary pumping (without using prepulses) are similar to the characteristics of the lasers in study [21].

On the other hand, an analysis of the possibilities for the Ramsauer indicates that the use of a matching element in a compound heterolaser must produce the following regularities: the near field must be narrowed, the structure of light spots on opposing end mirrors must be similar, the emission distribution in the far field must be symmetric and the watt-current characteristic must be linear. Of these relations we observed the three identified above (unfortunately we did not test for the existence of identical output characteristics from opposing end mirrors of the heterolaser for the case of double pulse pumping), which supports the explanation of enhanced characteristics related to the Ramsauer effect; specifically, the double pumping pulses are responsible for a certain temperature distribution on the axis within a narrow region of the active layer at the "joint" of the two homogeneous regions, which produces this dependence of the dielectric constant on the longitudinal coordinate, where the analog of the Ramsauer effect holds. In order to test this hypothesis it is necessary to carry out an investigation of the temperature distribution on the heterolaser axis.

In light of the concepts described above regarding the possibility of using the analog of the Ramsauer effect in the physics of lengthwise-inhomogeneous resonators in heterolasers, it is interesting to carry out an experiment to control the output characteristics of heterolasers. Here, we may use the traditional advantage of semiconductor lasers where electrical current may be used as the primary control technique. For example, if we add to the stripe contact an additional small contact (compared to the length of the stripe contact) scanning on the longitudinal coordinate, then the current flowing through this contact will create a local variation in the dielectric constant in the narrow region on the longitudinal coordinate (caused by the competition of the contribution of excess carriers and the additional heating of the active region). By moving the contact along the laser axis and by altering the additional current flowing through this contact, we may change the characteristics of the additional "potential well" dividing the two regions of the heterolaser at the point where this contact is located. In the same manner it is possible to convert the uncontrolled axially inhomogeneous heterolaser into a controlled "dielectric insert" created by the "potential well." This "dielectric insert" may have characteristics where the analog of the Ramsauer effect may become manifest if the depth and width of the additional potential well are selected so that formula (49) is valid.

Above we discussed the possibility of controlling the characteristics of a lengthwise-inhomogeneous heterolaser, although the additional contact described above (the "dielectric insert") may control the operation of a lengthwise-homogeneous heterolaser as well. Indeed, with a "dielectric insert" the lengthwise-homogeneous laser becomes inhomogeneous (it is split into two homogeneous sections). By controlling the characteristics of the "insert", we may essentially control the output characteristics of the homogeneous (in the absence of current flowing through the additional contact) laser. We may select the structure of the "potential well" in order to create a different near (and far) field of the homogeneous laser at opposing end mirrors of the laser; if one spot is on one mirror, we may generate on the other mirror various light spot structures corresponding to various modes in the second laser by using the additional current to change the "insert."

This method of active control over the output characteristics of laser emission using an additional mobile contact is consistent with the method of enhancing emission characteristics of a heterolaser without lateral optical confinement proposed, implemented and investigated in study [48]; this technique employed a double stripe contact (essentially an additional mobile contact). The redistribution in the current density resulting from the addition of the second stripe contact created a refractive index profile that enhanced the properties of the active optical waveguide. The device proposed in this study employing an additional mobile contact develops the approach used in study [48] with a second fixed stripe contact.

CONCLUSION

This study developed a concept based on an analog of the quantum mechanical Schrodinger equation and an equation describing light distribution in inhomogeneous media (certain simple aspects of this analogy have been described in detail in a recent Russian translation of a monograph by M. Adams [49]). The specific dependence of the complex dielectric constant in the plane of the *p-n*-junction, analogous to the quantum mechanical Eckert potential (see [38, pp. 602-611]) is implemented in heterostructure injection lasers in the form of an Epstein layer. A new parameter of an active semiconductor waveguide (the complex reduced width $s = s' + is''$) is incorporated for the relation $\varepsilon(y)$ in the form of an Epstein layer; the real and imaginary parts of this parameter are used to tabulate, in the form of three-dimensional graphs, the parameters of the emission directional pattern of such waveguides both for the case of a symmetric and an asymmetric profile of the refractive index in the plane of the *p-n*-junction. The simplicity of incorporating the complex reduced width parameter is confirmed by the fact that this parameter unambiguously expresses the mode coupling coefficients $C_{\bar{n}n}$ in a model describing a lengthwise-inhomogeneous heterolaser in the form of two (or several) end-coupled homogeneous active waveguides. On the other hand, the analog of the Franck-Condon principle formulated in this study (see also [23]) for such a model of a compound heterolaser easily demonstrates that this coupling coefficient is expressed through the shift in the maximum of the dielectric constant (the analog of Stokes losses) and other parameters characterizing the "potential well" (the change in frequency with the parabolic relation $\varepsilon(y)$). It is these parameters that are used in the theory of biatomic molecular spectra; it is of interest to establish the relationship between the reduced width parameter and these quantities.

Another analogy that would allow using results from quantum mechanics in the physics of inhomogeneous heterolasers is the analog of the Ramsauer effect proposed in this study (see also [50]). With a specific dependence of the dielectric constant on the longitudinal coordinate at the joint between the two active regions of the heterolaser indicating that a whole number of de Broglie half waves accumulate in the "width of the potential well" of the corresponding quantum mechanical Schrodinger equation, complete matching of the two end-coupled waveguides occurs. An active control method based on this effect (an analog of the Ramsauer effect) is proposed for controlling the output characteristics of a heterolaser, one possible variant of which would include using an additional contact on the stripe contact. Further experiments are necessary to establish the effectiveness of this or other variants for zero-lag control over heterolaser characteristics.

In conclusion the authors wish to express their gratitude to P. G. Eliseev, V. I. Man'ko, and Yu. M. Popov for their valuable discussion.

BIBLIOGRAPHY

1. Vzyatyshev, V. F. "Dielektricheskie volnovody" [Dielectric waveguides] Moscow: Sov. radio, 1970, 216 p.

2. Kogel'nik, G. "Teoriya dielektricheskikh volnovodov" [Dielectric waveguide theory" In: Integral'naya optika [Integrated optics] Edited by T. Tamir. Moscow: Mir, 1978, p.27-96.

3. Unger, Kh. G. "Planarnye i volokonnye opticheskie volnovody" [Planar and fiber optic waveguides] Moscow: Mir, 1980, 656 p.

4. Keysi, Kh., Panish, M. "Lazery na geterostrukturakh" [Heterostructure lasers] Moscow: Mir, 1981, V. 1. 299 p.

5. Eliseev, P. G. "Vvedeniy v fiziku inzhektsionnykh lazerov" [An introduction to the physics of injection lasers] Moscow: Nauka, 1983, p. 54-107.

6. Anderson, W. W. "Mode-confinement and gain junction lasers" IEEE J. QUANT. ELECTRON., 1965, V. 1, p. 228-236.

7. Adams, M. J., Cross, M. "Electromagnetic theory of heterostructure injection lasers" SOLID STATE ELECTRON., 1971, V. 14, p. 865-883.

8. Cook, D. D., Nash, E. R. "Gain induced guiding and astigmatic output beam of GaAs lasers" J. APPL. PHYS., 1975, V. 46, p. 1660-1672.
9. Thompson, G. H. B., Lovelace, D. F., Turley, S. B. H. "Kinks in the light-current characteristics and near-field shifts in (GaAl)As-heterostructure stripe lasers and their explanation by the effect of self-focusing on a built-in optical waveguide" IEE PROC. SOLID STATE ELECTRON. DEVICES, 1978, V. 2, p. 12-30.

10. Eliseev, P. G., Osinskiy, M. "The application of the Epstein dielectric model to a description of modes in planar stripe heterolasers" KVANTOVAYA ELEKTRON., 1980, V. 7, p. 1407-1415.

11. Bogatov, A. P., Eliseev, P. G., Man'ko, M. A., et al. "A model of a stripe heterolaser accounting for inhomogeneities in the complex dielectric constant through the cavity length" Preprint FIAN. No. 101, Moscow, 1981, 44 p.

12. Eliseev, P. G., Man'ko, M. A., Mikaelyan, G. T., "Model of an injection laser with continuous variation in the complex dielectric constant along the *p-n*-junction" TR. FIAN, 1983, V. 141, p. 119-125.

13. Man'ko, M.A., Mikaelyan, G. T. "The properties of gain-guided modes in active semiconductor waveguides" Preprint FIAN. No. 166, Moscow, 1984, 23 p.

14. Manko, M. A., Makhsudov, B. I., Mikaelyan, G. T., et al. "Measurement and analysis of planar stripe-geometry GaAlAs/GaAs heterojunction laser wavefronts and their variation with pumping current" IEE PROC. OPTOELECTRON., 1985, V. 132, p. 64-68.

15. Dziewiecka, T., Osinski, M. "Generalized Epstein model of stripe-geometry injection lasers" APPL. OPT., 1984, V. 23, p. 94-99.

16. Streifer, W., Burnham, R. D., Scifres, D. R., "Symmetrical and asymmetrical waveguiding in very narrow conducting stripe lasers" IEEE J. QUANT. ELECTRON., 1979, V. 15, p. 136-141.

17. Paoli, T. L. "Waveguiding in stripe-geometry junction lasers" IEEE J. QUANT. ELECTRON., 1977, V. 13, p. 662-668.

18. Epstein, P. S. "Reflection of waves in an inhomogeneous absorbing medium" PROC. NAT. ACAD. SCI. USA, 1930, v. 16, p. 627-637.

19. Stolyarov, S. N. "The influence of absorption on the waveguide properties of layers with variable parameters" IZV. VUZOV. RADIOFIZIKA, 1970, V. 13, p. 749-756.

20. Osinski, M. "Dielectric slab and Epstein-layer electromagnetic models of injection laser" ARCH. ELEKTRON. UND UBERTAG., 1976, V. 30, p. 223-224.

21. Campos, M. D., Hwang, C. J., Bossi, R. I., et al. "Cavity competition in anomalous emission intensity in double-heterostructure (DH) lasers" IEEE J. QUANT. ELECTRON., 1977, V. 13, p. 687-691.

22. Man'ko, M. A., Mikaelyan, G. T. "An analog of the Franck-Condon principle for a lengthwise-inhomogeneous heterolaser" KVANTOVAYA ELEKTRON., 1986, V. 13, p. 302-309.

23. El'yashevich, M. A. "Atomnaya i molekulyarnaya spektroskopiya" [Atomic and molecular spectroscopy] Moscow: Fizmatgiz, 1962, p. 757-761.

24. Man'ko, V. I. "Vozmozhnosti primeneniya metoda kogerentnykh sostoyaniy i integralov dvizheniya dlya rassmotreniya dinamicheskikh sistem" [The possibilities for using the coherent state method and integrals of motion to examine dynamical systems] In: Kvantovaya elektrodinamika s vneshnim polem [Quantum electrodynamics with an external field] Tomsk: Izd-vo Tomsk. gos. un ta, 1977, p. 101-119.

25. Sisakyan, I. N. "Formirovanie volnovykh frontov kogerentnogo elektromanitnogo izlucheniya v prodol'no- i poperechno-neodnorodnykh sredakh" [The formation of coherent electromagnetic emission wavefronts in longitudinally- and transversely-inhomogeneous media] Dissertation for the Doctorate of Physical and Mathematical Sciences in the form of a scientific paper, Moscow: FIAN, 1982, 60 pp.

26. Leontovich, M. A., Fok, V. A. "The solution of the problem of electromagnetic wave propagation along the earth's surface using a parabolic equation method" ZHETF, 1946, V. 16, p. 557-573.

27. Nieto, M. M. "Exact wave-function normalization constants for B_0 $tanh\ z - U_0 \cosh^{-2} z$ and Pöschel-Teller potentials. PHYS. REV. A, 1978, V. 17, p. 1273-1283.

28. Beitman, G., Erdeyi, A. "Vysshie transtsendentnye funktsii" [Higher transcendental functions] Moscow: Nauka, 1965, V. 1, 294 p.

29. Asbeck, P. M., Cammack, D. A., Daniele, J. J. "Non-Gaussian fundamental mode patterns in narrow-stripe-geometry lasers" APPL. PHYS. LETT., 1978, v. 33, p. 504-506.

30. Streifer, W., Burnham, R. D., Scifres, D. R. "Current status of (GaAl)As diode laser" SPIE JOURNAL. INTEGRATED OPTICS, 1981, V. 269, p. 2-7.

31. Man'ko, M. A., Makhsudov, B. I., Fam Van Khoy "An investigation of the wavefront of planar stripe GaAlAs/GaAs-heterolasers" TR. FIAN, 1986, V. 166, (See this volume).

32. Nash, F. R., Wagner, W. R., Brown, R. L. "Threshold current variations and optical scattering losses in (Al, Ga)As double-heterostructure lasers" J. APPL. PHYS., 1976, V. 47, p. 3992-4005.

33. Mal'kin, I. A., Man'ko, V. I. "Dinamicheskie simmetrii i kogerentnye sostoyaniya kvantovykh sistem" [Dynamical symmetries and coherent states of quantum systems] Moscow: Nauka, 1979, 319 p.

34. Vu Van Lyk "Osobennosti impul'snykh izluchatel'nykh kharakteristik i elektricheskaya diagnostika rezhimov raboty planarnykh poloskovykh lazerov na osnove GaAlAs/GaAs" [The features of the pulsed radiative characteristics and the electrical diagnostics of operating conditions of GaAlAs/GaAs planar stripe lasers] Dissertation for Candidate of Physical and Mathematical Sciences Moscow: FIAN, 1982, 151 p.

35. Vu Van Lyk, Man'ko, M.A. "Vliyanie neodnorodnostey vdol' dliny rezonatora na vykhodnye kharakteristiki s protivopolozhnykh tortsov planarnogo poloskovogo reterolazera bez bokovogo opticheskogo ogranicheniya" [The influence of inhomogeneities along cavity length on the output characteristics from opposite ends of a planar stripe heterolaser without lateral optical confinement] DOKL. NATS. TSENTRA NAUCH. ISSLED. CRV, 1985, V. 1, No. 1.

36. Doktorov, Ye. V., Malkin, I. A., Man'ko, V. I. "The vibronic spectra of polyatomic molecules" TR. FIAN, 1983, V. 152, p. 202-238.

37. Doktorov, E. V., Malkin, I. A., Man'ko, V. I., et al. "Franck-Condon factor for diatomic molecule and dynamical symmetry for vibronic transitions" PreprINT. P. N. Lebedev Phys. Inst. No. 36. Moscow, 1974, 41 p.

38. Mors, F. M., Feshbakh, G. "Methods of theoretical physics" Translation from the English, 1960, V. 2, 896 p.

39. Malkin, I. A., Man'ko, V. I., Trifonov, D. A. "Linear adiabatic invariants and coherent states" J. MATH. PHYS., 1973, V. 14, p. 576-582.

40. Davydov, A. S. "Kvantovaya mekhanika" [Quantum mechanics] Moscow: Nauka, 1973, 703 p.

41. Vu Van Lyk, Kalshabekov, A. S., Man'ko, M. A., et al. "The influence of pumping prepulses on the output characteristics of a stripe heterolaser" KVANTOVAYA ELEKTRON, 1981, V. 8, p. 2697-2699.

42. Vu Van Lyk, Eliseev, P. G., Man'ko, M. A., et al. "Vliyanie nestatsionarnogo temperaturnogo volnovoda na kharakteristiki izlucheniya geterolazerov" [The influence of a nonsteady-state temperature waveguide on the emission characteristics of heterolasers]

In: III Vsesoyuz. konf. po fizicheskim protsessam v poluprovodnikovykh geterostrukturakh [The Third All Union Conference on the Physical Processes in Semiconductor Heterostructures] I. I. Mechnikova, 1982, V. 2, p. 65-65.

43. Marple, D. T. F. "Refractive index of GaAs" J. APPL. PHYS., 1964, V. 35, p. 1241-1242.

44. Nash, E. R. "Mode guidance parallel to the junction plane of double-heterostructure GaAs lasers" J. APPL. PHYS., 1973, V. 44, p. 4696-4707.

45. Vu Van Lyk, Eliseev, P. G., Man'ko, M. A. "An investigation of a planar stripe heterolaser in pulsed operation" PREPRINT FIAN No. 122, Moscow, 1982, 20 p.

46. Alaverdyan, S. A., Bazhenov, V. Yu., Bogatov, A. P., et al. "The stepwise waveform of emission pulses from injection lasers based on a double heterostructure with a stripe contract" KVANTOVAYA ELEKTRON., 1980, V. 7, p. 123-127.

47. Lang, R. "Lateral transverse mode instability and its stabilization in stripe geometry injection lasers" IEEE J. QUANT. ELECTRON., 1979, V. 15, p. 718-726.

48. Bogatov, A. P., Vu Van Lyk, Man'ko, M. A., et. al. "The waveguide properties of the active region of an injection laser with a double stripe contact" KVAT. SOOBSHCH. PO FIZIKE FIAN, 1980, No. 10, p. 26-31.

49. Adams, M. "An introduction to optical waveguide theory" Translation from the English, Moscow: Mir, 1984, 512 p.

50. Man'ko, M. A. "Controlling the spatial characteristics of emission from a heterolaser" KVAT. SOOBSHCH. PO FIZIKE FIAN, 1985, No. 12, p. 34-38.

INVESTIGATION OF THE WAVEFRONT OF PLANAR STRIPE GaAlAs/GaAs-HETEROLASERS

M.A. Man'ko, B.I. Makhsudov, Fam Van Khoy

Abstract: A modified Michelson interferometer in planar stripe GaAlAs/GaAs-heterolasers is used to investigate the wavefronts and directional patterns of emission as well as changes in these properties with variations in the pumping current. Elementary data are compared to calculated data based on a model with a linear variation in the complex dielectric constant in the form of an Epstein layer. The influence of nonlinearities in the watt-current characteristic on the wavefront is investigated together with the dependence of the slope of the wavefront on the width of the stripe contact.

INTRODUCTION

In order to employ injection lasers in recording and playback systems, precise quantitative measurements of the wavefronts of the laser light beams are required. The more complex directional pattern characteristic of planar stripe heterolasers without lateral optical confinement produces laser modes formed in the plane of the active layer by gain guiding action, and in the perpendicular plane by the difference in the refractive index of the active region and the adjacent passive layers of the heterostructure. The wavefronts of the light beams emitted by such lasers have significant astigmatism [1], and it is necessary to employ cylindrical optics to obtain light beams with a round cross-section. In turn the beam astigmatism may be used successfully to provide an automatic focusing system for playback of information from video disks.

Study [2] (see the preceding article in this volume) substantiated in detail the possibility of using dielectric waveguide theory for the case of active stripe waveguides; the active region of a stripe heterolaser may be used as such waveguides. We know (see, for example, [3]) that in lasing, amplification and optical losses are balanced, although integrally rather than locally, since the local gain is concentrated primarily in the active region while losses are concentrated in the passive surrounding regions and, of course, at the end mirrors. The consequence of this distribution of gain and losses is additional field attenuation in the cross-section outwards from the active region. The mode wavefront (accounting for the contribution of the imaginary part of the dielectric constant, $\mathrm{Im}\ \varepsilon \equiv 0$) is nonplanar: it lags in regions where absorption predominates and thus expands on the propagation side. As a result, variations in phase occur on the transverse axis, thereby resulting in amplitude oscillations. The optical gain in the center of the mode produces significant localization, and the influence of the contribution of $\mathrm{Im}\ \varepsilon$ may be decisive for the mode profile and its characteristics (the so-called gain guiding effect). Due to the gain guiding action in the plane parallel to the *p-n*-junction in planar stripe heterolasers without lateral optical confinement, the constriction of the caustic of the light beam corresponds to the imaginary source and is within the cavity at a distance Δz_0 from the end mirror, at the same time that in the plane perpendicular to the *p-n*-junction, the constriction remains at the end mirror. In the domestic scientific literature, the quantity Δz_0 is called the distance between caustic constrictions in orthogonal directions of the main laser beam cross-section (see, for example, [4]) or the astigmatic differential [5] and in English language scientific literature this term is called the astigmatism distance (see, for example, [6]).

The number of studies devoted to the investigation of injection laser wavefronts has increased recently [6-11]. Studies [6-8] have performed experimental investigations of the parameters of a light beam produced by stripe heterolasers; specifically, the radius of curvature of the wavefront *R and the quantity* Δz_0. Knowledge of these parameters is required for calculating focusing and collimating devices using injection lasers as the emission source. These studies employed an interferometer method based on a modification of a Mach-Zehnder interferometer [6], a Schmart diffraction interferometer [7] and a Wollaston interferometer [8]. Study [12] carried out microscopic measurements of the quantity Δz_0 for a wide range of single-frequency injection lasers covering the spectral range from 700-1300 nm.

This study measures the wavefront in the principal sections of a laser beam and investigates the variation in wavefront with pumping current for planar stripe heterolasers without lateral optical confinement. An interferential method analogous to that described in study [10] was used; this method is based on a modification of a Michelson interferometer (Section 1). The quantity Δz_0 is measured and its dependence on the pumping current and the width of the stripe contact is investigated (the corresponding graphs are given in Section 2). The phase distribution of the wavefront is investigated in orthogonal directions as a function of the watt-current characteristic and the pumping current. The influence of nonlinearities in the watt-current characteristic on the wavefront shape is investigated as well as the dependence of the curvature of the wavefront on the width of the stripe contact. The derived experimental data are compared to a theoretical model with a continuous variation in the complex dielectric constant in the form of an Epstein layer, which made it possible to estimate the parameters of the active waveguide layer influencing the directional pattern (Section 3).

1. METHODOLOGICAL ISSUES

The experimental configuration is given in Fig. 1. Emission from planar stripe heterolaser 1 was collimated by means of lens 2 and after beam splitting cube 3 was directed onto flat mirror 4 in one arm of the Michelson interferometer and onto convex spherical mirror 5 in the other arm of the interferometer. Emission from the axial portion of the beam, expanded by the spherical mirror, was considered the reference emission, and a ring pattern was obtained on the screen of the interferometer; the visibility of this pattern characterizes the spatial coherence of the beam and its geometry characterizes the wavefront. It is easy to show that the form of the interference rings depends on collimation of the laser beam. Following [13] we write the phase distribution φ of laser emission, accounting for astigmatism in the form

$$\varphi = {}^1/_2\,(x^2/A_x^2 + y^2/A_y^2), \tag{1}$$

where x and y are the coordinates of the observation point, while A_x and A_y are the distances from the observation point to the lens image in two orthogonal directions. The wavefront of the reference beam $\varphi_{оп}$ is determined by the formula

$$\varphi_{оп} = (x^2 + y^2)/2R, \tag{2}$$

where R is the radius of a spherical wave. The interference pattern is given by the relation

$$I = 2\,[1 + \cos k_0\,(\varphi - \varphi_{оп})], \tag{3}$$

where $k_0 = 2\pi/\lambda_0$ is the wave vector; λ_0 is the emission wavelength. The luminescence bands of the interference pattern are determined by the following expression:

$$\frac{1}{2}\left(\frac{x^2}{A_x^2} + \frac{y^2}{A_y^2}\right) - \frac{x^2 + y^2}{2R} = m\lambda_0, \tag{4}$$

where $m = 0, \pm1, \pm2, \ldots$ The interference pattern described by equation (4) is a group of ellipses. It is possible to obtain various second order curves depending on the relation between the parameters A_x, A_y and R. With $A_x = R$,

$$y = [2m\lambda_0 A_y R/(R - A_y)]^{1/2}, \tag{5}$$

and with $A_y = R$

$$x = [2m\lambda_0 A_x R/(R - A_x)]^{1/2}, \tag{6}$$

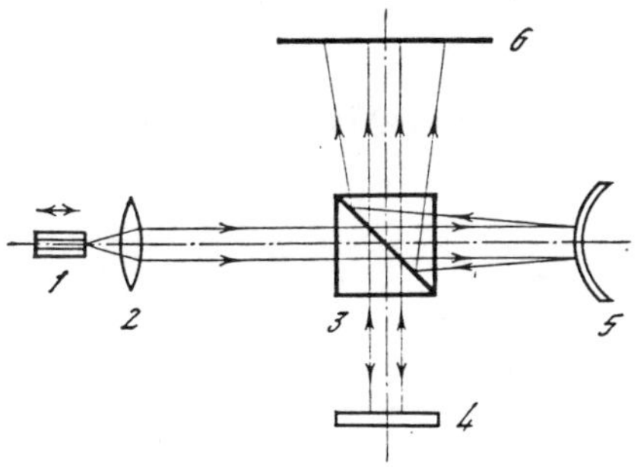

Fig. 1. Experimental configuration for measuring the wavefronts of injection lasers
1 - planar stripe heterolaser;
2 - lens;
3 - splitting cube;
4 - flat mirror;
5 - spherical mirror;
6 - image converter tube screen or photodetector

The interference pattern is a family of curves. With $A_x = A_y$ they become a family of concentric circles, and with $A_x < R$ and $A_y > R$ they become hyperbolas.

Displacement of the laser on the optical axis (shown in Fig. 1 by the arrows) makes it possible to observe a significant change in the interference pattern. This is demonstrated by Fig. 2, which gives the interference patterns for various distances from the end mirror of the laser diode to the lens. When there is a sufficient distance between the laser and the lens (L = 5.0 mm, Fig. 2), $A_x \simeq A_y$ and interference pattern 1 assumes the form of concentric circles. When the distance between the laser and the lens is reduced (L = 4.0 mm, Fig. 2.2), $A_y > A_x$ and we have a group of ellipses. With $A_x < R$, $A_y > R$, Figs. 2, 3, and 4, L = 3.0 and 2.99 mm) a family of hyperbolas is observed. A further reduction in the distance (L = 2.8 mm, Fig. 2, 5) where the condition $A_x \simeq \infty$, $A_y \simeq R$ is implemented, the interference pattern again becomes a group of ellipses. At a sufficiently short distance L = 1.9 mm (Fig. 2, 6) when $|A_x| \simeq |A_y| > R$, a group of concentric circles is observed.

Visual observations of the change in the interference pattern were used to determine the position of the laser beam constriction in one plane with respect to the constriction in the orthogonal plane or with respect to the end mirror of the laser diode. A linear relation between the diode displacement Δz and the cosine of the angle ϕ between the asymptotes of the hyperbolic bands on the interferometer screen were used to measure the position of the constriction [11]

$$\cos \phi = 1 + \Delta z / C \tag{11}$$

where C is a constant. The distance between constrictions Δz_0 is determined by the points $\cos \phi = \pm 1$, i.e., between the positions where the interference bands form a parallel system in two orthogonal directions. This method has made it possible to experimentally measure the quantity Δz_0 accurate to ±1 μm and to measure the phase distribution accurate to ±0.2 π.

These tests were carried out on planar stripe GaAlAs/GaAs-heterolasers without lateral optical confinement with the following

Table 1

Characteristics of test samples and experimentally-measured values for Δz_0

Heterolasers	*Threshold current, ma*	*Width of stripe contact, μm*	*Wave-length nm*	$J/J_{\Pi OP}$	Δz_0, μm
D.6.K	200	16	796.0	1-1.4	23*
D.9.K	200	10	795.0	1-1.5	25
D.13.K	170	12	788.8	1-2.4	14-22
D.9	225	16	799.7	1-1.9	15
D.36	140	15	885.0	1-2.0	15-24**
D.37	180	15	883.8	1-1.6	18-24**
D.38	200	16	889.6	1-1.5	15-35**
D.45	170	8	884.0	1-1.6	9-13

* *The interference pattern vanishes in the vicinity of the inflection of the watt-current characteristic.*

** *Δz_0 changes nonmonotonically with pumping current.*

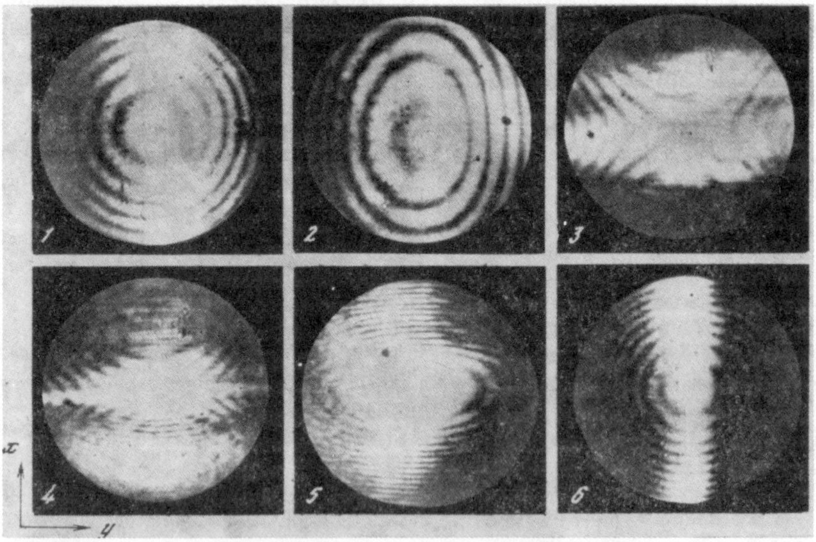

Fig. 2. The interference pattern for various distances d between the lens and the end mirror of the laser.
1-6: d = 5 mm (1); 4(2); 3(3); 2.99(4); 2.8(5); 1.9(6)

parameters varied over the following ranges: thickness of the active region - 0.15 to 0.35 μm; cavity length - from 160 to 240 μm; width of the stripe injection contact - from 8 to 16 μm. These devices operated at room temperature with a pumping level 1.6 times threshold for CW conditions and 2.4 times threshold in pulsed operation (pulse duration 400 ns, pulse repetition frequency 10^4 Hz). The wavelength range was 788-890 nm. The characteristics of the test samples are given in Table 1.

2. EXPERIMENTAL MEASUREMENT OF WAVEFRONTS OF PLANAR STRIPE GaAlAs/GaAs-HETEROLASERS AND WAVEFRONT VARIATIONS WITH PUMPING CURRENT

Studies have demonstrated that the constriction of the caustic of a light beam in the plane perpendicular to the *p-n*-junction is found at the end mirror, while constriction of the light beam in the plane of the *p-n*-junction is within the cavity of the laser diode (i.e., it is "imaginary"). This fact is demonstrated by Fig. 3 which shows the phase distribution of emission from the D.6.K laser in vertical (x axis) and horizontal (y axis) directions (we remind the reader that light is propagating on the z axis). It is clear from Fig. 3 that the wavefront is flat in the plane of the *p-n*-junction and is arched in the plane perpendicular to the *p-n*-junction. This confirms the statement in the introduction that a strong waveguide exists as a result of jumps in the refractive index at the boundary of the active region and adjacent passive layers of the heterostructure in the plane perpendicular to the *p-n*-junction, as well as a weak waveguide caused by gain guiding in the plane of the *p-n*-junction. Table 1 gives measured values of the quantity Δz_0 which is characterized by the following behavior with an increase in pumping current: in a number of samples (such as the D.9.K and D.9 samples) Δz_0 maintained its value, while in other samples (for example, D.45 and D.13.K) it increased monotonically, and the change in Δz_0 manifested a nonmonotonic nature in D.36, D.37, and D.38; there were cases where the interference pattern disappeared (sample D.6.K) at a pumping current corresponding to the discontinuity in the watt-current characteristic.

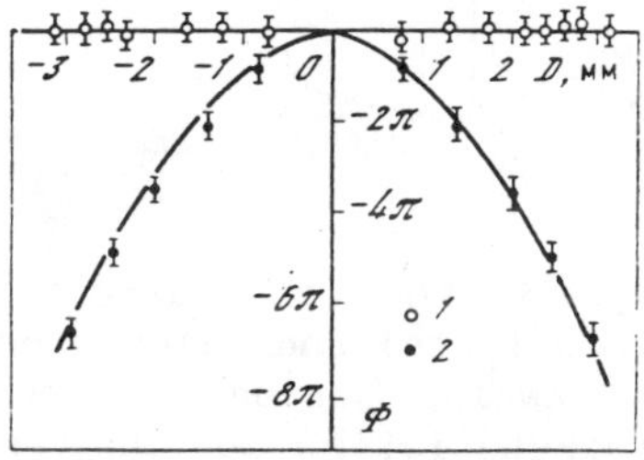

Fig. 3. Phase distribution of emission from the D.6.K heterolaser (on the image converter tube screen).
Continuous pumping current: 210 ma.
1 - In the plane perpendicular to the plane of the *p-n*-junction;
2 - In the plane parallel to the plane of the *p-n*-junction

In order to investigate the action of pumping current on Δz_0, we carried out measurements using a number of samples with different watt-current characteristics. Standard cases are shown in Fig. 4, while Fig. 4a gives the watt-current characteristic with a step (sample D.37) while Fig. 4b shows a linear relation (sample D.45). In the first case, the change in Δz (as indicated above, Δz_0 is determined by the quantity Δz in formula (7)) manifests a nonmonotonic

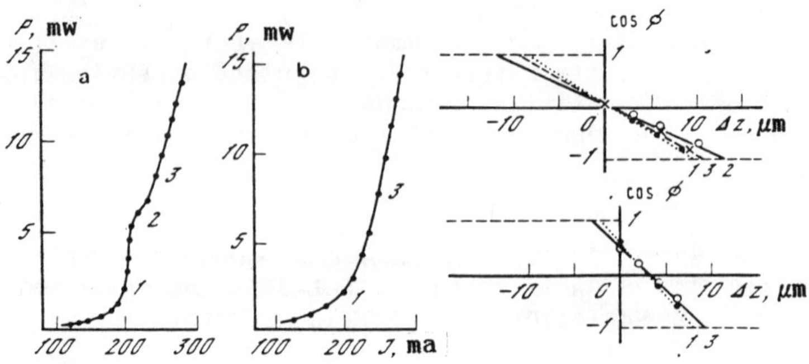

Fig. 4. Watt-current characteristics (a, b) and the cos ϕ plotted as a function of Δz (c, d)
At pumping currents corresponding to the values 1–3 on the watt-current characteristics: 200 ma (1); 220 (2); 240 (3). Samples D.37 (a, c) and D.45 (b, d).

nature with an increase in pumping current from 200 to 240 ma (see Fig. 4c) unlike the case in Fig. 4d where Δz remains virtually the same with the pumping current varied from 200 to 240 ma. The wavefront of emission from the D.37 sample for various continuous pumping currents indicated on the watt-current characteristic (Fig. 5a) is shown in Fig. 5b by the experimentally measured points. The solid

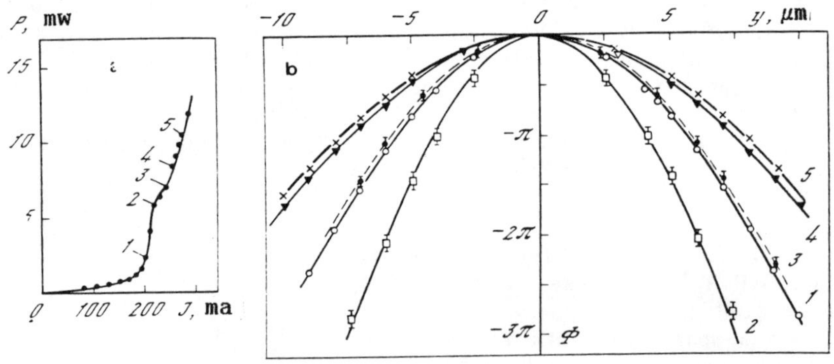

Fig. 5. The watt-current characteristic (a) and experimentally measured (dots) and calculated wavefronts (b) of the D.37 laser
At pumping currents corresponding to values 1–5 on the watt-current characteristic: 200 ma(1); 220(2); 240(3); 260(4); 280(5); CW operation. The calculation was carried out using the following parameter values:
1 – $\varepsilon_1 = \varepsilon_3 = 12.891 + i0.002$; $\varepsilon_2 = 12.786 - i0.003$; $l = 5.2$ μm; 2– $\varepsilon_1 = \varepsilon_3 = 12.891 + i0.005$; $\varepsilon_2 = 12.757 - i0.009$; $l = 3.2$ μm; 3 – $\varepsilon_1 = \varepsilon_3 = 12.891 + i0.007$; $\varepsilon_2 = 12.801 - i0.009$; $l = 5.4$ μm; 4 –= $\varepsilon_1 = \varepsilon_3 = 12.891 + i0.007$; $\varepsilon_2 = 12.770 - i0.0095$; $l = 16$ μm; 5 – $\varepsilon_1 = \varepsilon_3 = 12.891 + i0.007$; $\varepsilon_2 = 12.770 - i0.01$; $l = 16$ μm

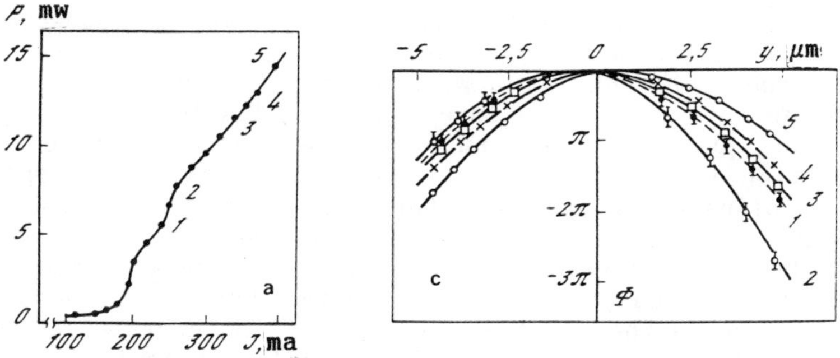

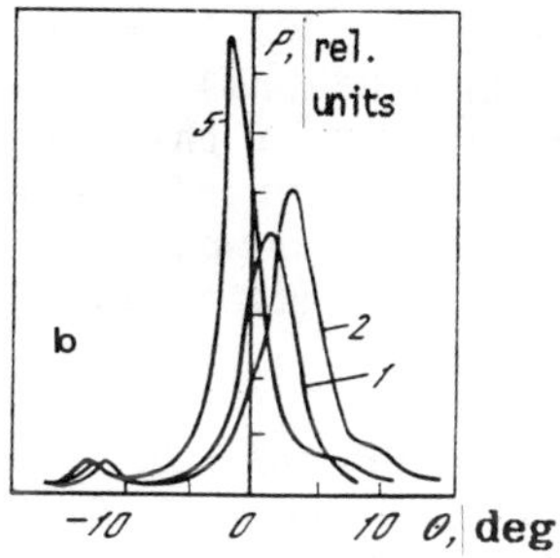

Fig. 6. The watt-current characteristic (a), the directional pattern (b) and the wavefront (c) of the D.13.K sample for various pulsed pumping currents J
J = 240 ma(1); 260(2); 320(3); 360(4); 400(5). Pulse duration - 400 ns; pulse repetition frequency - 10 kHz

Fig. 7. Experimentally-measured (dots) and calculated phase distributions of the electromagnetic field at the end mirrors of samples D.6.K (1), D.37 (2) and D.45 (3) at pumping currents 10% above threshold
The calculation was carried out with the following parameter values:
1 - $\varepsilon_1 = \varepsilon_3 = 12.891 + i0.001$, $\varepsilon_2 = 12.880 = i0.002$, $l = 4.7\ \mu m$, $\ll_0 = 800$ nm, $\Gamma_0 = 0.8$;
2 - $\varepsilon_1 = \varepsilon_2 = 12.891 + i0.002$, $\varepsilon_2 = 12.786 - i0.003$, $l = 5.2\ \mu m$, $\lambda_0 = 900$ nm, $\Gamma_0 = 0.8$;
3 - $\varepsilon_1 = \varepsilon_2 = 12.891 + i0.003$, $\varepsilon_2 = 12.867 - i0.008$, $l = 1.5\ \mu m$, $\lambda_0 = 900$ nm, $\Gamma_0 = 0.8$

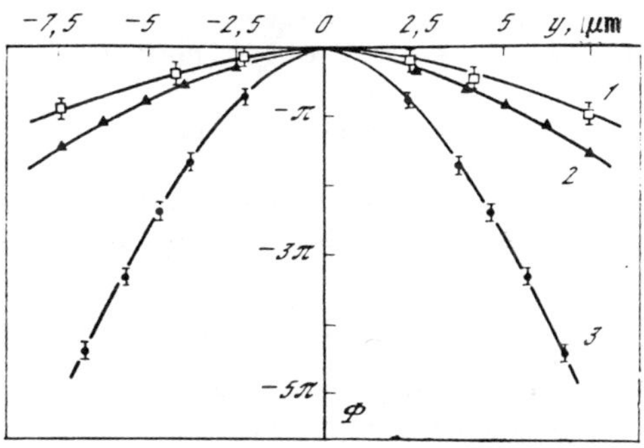

line curves are used to show the theoretically calculated phase distribution curves for emission at the end mirror of the laser diode (this is discussed in greater detail in Section 3). It is clear in Fig. 5 that as the pumping current increases, the wavefront changes. Changes in the wavefront at higher excitation levels were observed during pulsed current pumping. The characteristic behavior is demonstrated by Fig. 6 which gives the emission distribution in the far field (directional pattern, Fig. 6b)and the wavefront (Fig. 6c) for the various pumping currents indicated on the watt-current characteristic (Fig. 6a) for the D.13.K laser diode. At pumping currents 1.5 times threshold, the wavefront is asymmetrical; this asymmetry is related to the change in the emission intensity distribution in the near and far fields and depends on the form of the watt-current characteristic (the deviation from the optical axis is different for samples with linear and nonlinear watt-current characteristics). Two series of samples (one with a stripe contact 8 μm in width and the other with one 16 μm in width) were investigated in order to establish the influence of the width of the stripe contact on the wavefront. Results are given in Fig. 7 where the dots represent the experimental-measured wavefront for the D.6.K samples (width of the stripe contact - 16 μm, λ_0 = 796 μm), D.45 (width of the stripe contact - 8 μm, λ_0 = 884 nm) and D.37 (width of the stripe contact - 15 μm, λ_0 = 884 nm) at pumping currents 10% over threshold. The solid line curves in this diagram represent the calculated values of the phase distribution of the electromagnetic field at the end mirrors (results from a comparison to a theoretical model will be discussed in detail in the next section).

3. A COMPARISON OF EXPERIMENTAL DATA TO A THEORETICAL MODEL WITH CONTINUOUS VARIATION IN THE COMPLEX DIELECTRIC CONSTANT IN THE FORM OF AN EPSTEIN LAYER

Many studies (for example, [14-16]) have used a Gaussian approximation for electromagnetic emission intensity distribution outside the laser cavity. The solution of the wave equation on the x axis (across the active region) in the form

$$E(y) \sim \exp[-(y^2/2)k_0(a_r + ia_i)], \tag{8}$$

where $k_0 = 2\pi/\lambda_0$ corresponds to the case where the dielectric constant in the plane of the p-n-junction obeys a parabolic law

$$\varepsilon(y) = \varepsilon_0 + a^2y^2, \tag{9}$$

where $\varepsilon_0 = \varepsilon_{0_r} + i\varepsilon_{0_i}$ and $a = a_r + ia_i$. It was demonstrated in other studies (for example, [17-19]) that the emission intensity distribution in the near and far fields in heterolasers with a sufficiently narrow stripe contact of ≤10 μm differs significantly from a Gaussian distribution and, as a result, a model with a parabolic law is not suitable for the dielectric constant. In this regard study [17] proposed the following relation for the complex dielectric constant distribution in the plane of the p-n-junction:

$$\varepsilon(y) = \varepsilon_s + \Delta\varepsilon/\mathrm{ch}^2(y/w_0), \tag{10}$$

where ε_s and $\Delta\varepsilon$ are the complex constants, while w_0 is the effective width parameter. Relation (10) near $y = 0$ is a parabolic relation, while with large values $|y|$ it tends towards ε_s exponentially. In study [19] the $\varepsilon(y)$ relation was modeled by a complex function containing exponential and trigonometric functions

$$\eta(y) = -\eta_0 + \eta_3 \, \mathrm{th}^2\left(\frac{y}{y_3}\right) + (\eta_4 - \eta_3)\frac{\exp(y/y_3)}{2\,\mathrm{ch}(y/y_3)}, \tag{11}$$

where $\eta(0) = -\eta_0 + (\eta_4 - \eta_3)/2$, $\eta(-\infty) = =\eta_0 + \eta_3$, $\eta(\infty) = -\eta_0 + \eta_4$, y_3 is the effective width parameter.

Potential (11) may be reduced to an Epstein layer [20]:

$$\varepsilon(y) = \varepsilon_1 + 2(2\varepsilon_2 - \varepsilon_1 - \varepsilon_3)\frac{\exp(y/l)}{[1+\exp(y/l)]^2} + (\varepsilon_3 - \varepsilon_1)\frac{\exp(y/l)}{1+\exp(y/l)}, \tag{12}$$

where $\varepsilon_1 = \varepsilon(-\infty)$, $\varepsilon_2 = \varepsilon(0)$, $\varepsilon_23 = \varepsilon(\infty$, l is the width parameter of the layer with the following redesignations: $y_3 = 2l$, $\varepsilon_2 = -\eta_0 + (\eta_4 - \eta_3)/2$, $\varepsilon_1 = -\eta_0 + \eta_3$ and $\varepsilon_3 = -\eta_0 + \eta_4$. The profile of the dielectric constant on the y axis in the form of an Epstein layer (12) was used by several authors (see, for example, [21]), and was also used in our research [2, 10, 22-24]. For the case of a symmetrical distribution ($\varepsilon_1 = \varepsilon_3 = \varepsilon_0$), formula (12) takes the form

$$\varepsilon(y) = 4(\varepsilon_2 - \varepsilon_0)\frac{\exp(y/l)}{[1+\exp(y/l)]^2} = \frac{\varepsilon_2 - \varepsilon_0}{\mathrm{ch}^2(y/2l)}. \tag{13}$$

A comprehensive list of research utilizing the symmetrical Epstein layer may be found in study [25].

In accordance with [22, 24] we will write solutions of the wave equation for the case of dielectric constant relation (12) in the following form

$$Y_0(y) = \frac{\exp(a_0 y/2l)}{[\mathrm{ch}(y/2l)]^s} \quad \text{when } n = 0 \tag{14}$$

$$Y_1(y) = \frac{\exp(a_0 y/2l)\,[a_1 - s\,\mathrm{th}(y/2l)]}{[\mathrm{ch}(y/2l)]^{s-1}} \quad \text{when } n = 1. \tag{15}$$

Formulae (14) and (15) incorporate the following two parameters: the reduced width parameter s and the asymmetry parameter a_n [2, 10, 24]:

$$s = {}^1\!/_2\{-1 \pm [1 + 8(2\varepsilon_2 - \varepsilon_1 - \varepsilon_3)k_0^2\Gamma_0 l^2]^{1/2}\}, \tag{16}$$

$$a_n = k_0^2\Gamma_0 l^2(\varepsilon_3 - \varepsilon_1)(s - n), \tag{17}$$

where $\Gamma_0 = \int_{-d/2}^{d/2} |X(x)|^2 dx \Big/ \int_{-\infty}^{\infty} |X(x)|^2 dx$ — is the optical confinement parameter (on the x axis), d is the thickness of the active region of the heterolaser, $k_0 = 2\pi/\lambda_0$, λ_0 is the emission wavelength. For the case of a symmetrical distribution of $\varepsilon(y)$ (13) $a_n = 0$. We remind the reader that since the dielectric constant is a complex quantity, i.e., $\varepsilon = \varepsilon' + i\varepsilon''$, the parameters s and a_n are also complex quantities $s = s' + is''$, $a_n = a_n' + ia_n''$. We may write the phase of the electromagnetic field in the near field (at the end mirror) Φ for the fundamental mode ($n = 0$) in the following form:

$$\Phi \pm -s''\ln[\mathrm{ch}(y/2l)] + a_0'' y/2l. \tag{18}$$

The field intensity distribution in the far field (the directional pattern) is determined by the pattern of the near field (see, for example, [26, 22]).

We will attempt to select values of the parameters s, a_0 and l in formula (8)in order to satisfy the experimentally-observed wavefronts of heterolaser D.37 (Fig. 5)and heterolasers D.6.K and D.45 (Fig. 7). Good agreement between calculated curves and experimentally-measured curves is obtained with the active waveguide parameters shown in the legends to Figs. 5 and 7. We note that an anomalous change in the slope of the wavefront is observed at pumping currents corresponding to the nonlinearity on the watt-current characteristic. If we assume that this is accompanied by a degradation of the waveguide properties of the active region of the heterolaser, by proper selection of the dielectric constant values ($\varepsilon_1 = \varepsilon_3$ and ε_2, a symmetrical active waveguide) we may achieve a good correlation between experimental and calculated curves. The slope of the wavefront is reduced with an increase in pumping current, which is accompanied by an expansion of the light spot at the end mirror of the heterolaser as well as by a change in its waveguide properties.

We will now solve the inverse problem, specifically we will attempt to estimate the complex dielectric constant in the active region $\varepsilon = \varepsilon' + i\varepsilon''$ from the experimentally-observed wavefronts and amplitude distribution of the electromagnetic field at the end mirror of the heterolaser. For this we will use results from our studies [22, 23] from which the following relations are written for the fundamental mode ($n = 0$):

$$s' = -1/_2 + 2^{-3/2}\{[1 + 8k_0^2\Gamma_0 l^2\delta\varepsilon' + 16k_0^4\Gamma_0^2 l^4(\delta\varepsilon'^2 + \delta\varepsilon''^2)]^{1/2} + 1 + 4k_0^2\Gamma_0 l^2\delta\varepsilon'\}^{1/2}, \tag{19}$$

$$s'' = 2^{-3/2}\{[1 + 8k_0^2\Gamma_0 l^2\delta\varepsilon' + 16k_0^4\Gamma_0^2 l^4(\delta\varepsilon'^2 + \delta\varepsilon''^2)]^{1/2} - 1 - 4k_0^2\Gamma_0 l^2\delta\varepsilon'\}^{1/2}, \tag{20}$$

$$a_0' = \frac{k_0^2\Gamma_0 l^2}{s'^2 + s''^2}[s'(\varepsilon_3' - \varepsilon_1') + s''(\varepsilon_3'' - \varepsilon_1'')], \tag{21}$$

$$a_0'' = \frac{k_0^2\Gamma_0 l^2}{s'^2 + s''^2}[s'(\varepsilon_3'' - \varepsilon_1'') - s''(\varepsilon_3' - \varepsilon_1')], \tag{22}$$

where $\delta\varepsilon' = 2(2\varepsilon_2' - \varepsilon_1' - \varepsilon_3')$, $\delta\varepsilon'' = 2(2\varepsilon_2'' - \varepsilon_1'' - \varepsilon_3'')$.

Measurements of the wavefronts allows us to estimate s'' and a_0'', and s' and a_0' may be determined from the near field pattern. The nonmonotonic variation in the parameters of the active waveguide s and a_n as well as the values of the complex dielectric constant in the active region $\varepsilon = \varepsilon' + i\varepsilon''$ cause kinks and nonlinearities in the watt-current characteristics, and cause anomalous change in the slope of the wavefront, responsible for the nonmonotonic changes in the far field and near field patterns.

Table 2 gives the experimental values for the parameters l and s''_0 as well as the calculated values $\delta\varepsilon'$ and $\delta\varepsilon''$ for the D.37 heterolaser; the changes in the wavefront of emission from this laser with pumping current are shown in Fig. 5b.

Table 2

Experimental values of l, s''_0, and calculated values of $\sigma\varepsilon'$, $\sigma\varepsilon''$ for the D.37 heterolaser

J, ma	l, μm	s''_0	$\delta\varepsilon''$	$\delta\varepsilon'$
200	5.5	-3.5	-0.019	-0.413
220	3.2	-2.29	-0.056	-0.535
240	5.4	-3.1	-0.064	-0.335
260	16	-11.26	-0.066	-0.494
280	16	-11.26	0.066	-0.494

As we see from Fig. 5b, the wavefront of the light beam emitted by heterolaser D.37 has a symmetrical shape. For the case of a symmetrical distribution of $\varepsilon(y)$ (13) $\varepsilon_0 = \varepsilon_1 = \varepsilon_2$, $\theta = 0$ (see (22)) and the spreading of the light spot on the mirror (an increase in the parameter l) with an increase in pumping current is caused by the change in $\delta\varepsilon'$ and $\delta\varepsilon''$ in the active layer. For the case of an asymmetrical distribution of $\varepsilon(y)$ (12) we will carry out an analysis of the behavior of heterolaser D.13.K; a change in the wavefront of emission from this heterolaser with changes in the pumping current is shown in Fig. 6c.

The experimentally determined values of the parameters l, $s0$ and a''_0 as well as the calculated values of $\varepsilon'_3 - \varepsilon'_1$ and $\varepsilon''_3 - \varepsilon''_1$ are given in Table 3.

Table 3

Experimental values of l, s''_0, a''_0 and calculated values of $\varepsilon'_3 - \varepsilon'_1$, $\varepsilon''_3 - \varepsilon''_1$ for the D.13.K heterolaser

J, ma	l, μm	s_0	a_0	$(\varepsilon'_3-\varepsilon'_1)\cdot 10^3$	$(\varepsilon''_3-\varepsilon''_1)\cdot 10^3$
240	1	-0.41	-0.1	88	1
260	0.78	-0.39	-0.16	227	3
320	1.4	-0.60	-0.08	38	0.7
360	2.8	-1.79	0	-0.06	0.2
400	0.75	-0.24	0.06	-109.4	-0.7

As we see from Table 3, the asymmetry parameter a_0'' changes sign from negative to positive, passing through the zero value at pumping currents greater than 1.5 times threshold. We will attempt to calculate the complex dielectric constants ε_1, ε_2, and ε_3 using the experimentally measured wavefront of the D.13.K heterolaser (curve 1 in Fig. 6c) as well as results from our previous studies [22, 23]. In accordance with study [27], we calculated the values of $\varepsilon_3' = 12.891$ and $\varepsilon_3'' = 0.0054$ in studies [22, 23], and we determined the following values as the result of this study: $s_0'' = 0.41$, $\varepsilon_3' - \varepsilon_1' = 0.088$ and $\varepsilon_3'' - \varepsilon_1'' = 0.001$. Thus, $\varepsilon_1 = 12.803 + i0.0044$; $\varepsilon_2 = 12.743 - i0.006$; $\varepsilon_3 = 12.891 + i0.0054$.

CONCLUSION

The comparison of the experimentally-measured phase distribution of emission at the end (outcoupling) mirror of a planar stripe heterolaser without lateral optical confinement to the theoretical calculation based on a model with a continuously varying complex dielectric constant in the plane of the *p-n*-junction in the form of an Epstein layer performed in section 3 demonstrated the utility of this model. A qualitative feature of this model is the capability to vary the width of the active stripe layer with a given profile of the complex dielectric constant, as well as the capability to account for the dielectric asymmetry of the structure and the lack of correlation between the gain peaks and the refractive index peaks, together with the capability to account for the negative contribution to the real part of the complex dielectric constant caused by the excess current carriers. The use of this model has made it possible to both discuss the wavefront of emission and to investigate the change in the dielectric constant in the active layer with variations in the pumping current. Results from experimental measurements and a theoretical analysis of the wavefronts of planar stripe GaAlAs/GaAs-heterolasers led us to conclude that the change in the wavefront is closely related to the nonlinearities (kinks) in the output watt-current characteristics, while the directional pattern of emission is determined by the profile of the complex dielectric constant in the active layer, and the asymmetry in the $\varepsilon(y)$ profile increases with an increase in pumping current, while small variations in the dielectric constant ε in the active cause significant changes in the wavefront shape.

The authors are grateful to P. G. Eliseev and G. T. Mikaelyan for their useful discussion.

BIBLIOGRAPHY

1. Cook, D. D., Nash, F. R., Gain-induced guiding and astigmatic output beam of GaAs laser" J. APPL. PHYS., 1975, V. 46, p. 1660-1672.

2. Man'ko, M. A., Mikaelyan, G. T. "Modes and mode transformation in active semiconductor waveguides" TR. FIAN, 1986, V. 166 (see this volume)

3. Eliseev, P. G., "Vvedenie v fiziku inzhektsionnykh lazerov" [An introduction to the physics of injection lasers] Moscow: Nauka, 1983, 294 p.

4. Karlov, N. V. "Lektsii po kvantovoy elektronike" [Lectures on quantum electronics] Moscow: Nauka, 1983, 319 p.

5. Begunov, B. N., Zakaznov, N. P., Kiryushin, S. I., et al. "Teoriya opticheskikh sistem" [Optical system theory] Moscow: Mashinostroenie, 1981, 432 p.

6. Tatsumo, K., Arimoto, A. "Measurement and analysis of diode laser wavefronts" APPL. OPT., 1981, V. 20, p. 3520-3525.

7. Arquie, L. "Semiconductor laser wavefronts" In: Proc. Eur. Conf. Opt. Common. Cannes (France), 1982, p. 183-186.

8. Van Eck, D. C. "Wavefront measurement on semiconductor laser" IEEE J. QUANT. ELECTRON., 1983, V. 19, p. 966-968.

9. Kuntz, D. W. "Specifying laser diodes optics" LASER FOCUS, 1983, V. 20, p. 44-55.

10. Manko, M. A., Makhsudov, B. I., Mikaelyan, G. T., et al. "Measurement and analysis of planar satripe-geometry GaAlAs/GaAs heterojunction laser wavefronts and their variation with pumping current" IEE PROC. OPTOELECTRON., 1985, V. 132, p. 64-68.

11. Man'ko, M. A., Makhsudov, B. I., Fam Van Khoy "Izuchenie volnovogo fronta i actigmatizma planarnykh poloskovykh geterolazerov s λ = 7800-9000 A" [An investigation of the wavefront and astigmatism of planar stripe heterolaser at λ = 7800-9000 Å] In: Dokl. II Vsesoyuz. konf. po radiooptike [Papers of the Second All-Union Conference on Radiooptics] Tvidisi, 1985. In: Problemy razvitiya radiooptiki [Issues in the development of radioelectronics] Moscow, 1985, p. 232-233.

12. Lee, T. P., Burrus, C. A., Marcuse, D., et al. "Measurement of beam parameters of index-guided and gain-guided single-frequency InGaAsP injection lasers" ELECTRON. LETT., 1982, V. 18, p. 902-904.

13. Born, M., Vol'f, E. "Osnovy optiki" [The principles of optics] Tranaslation from the English, Moscow: Nauka, 1970, 855 p.

14. Zachos, T.H., Dyment, J. C. "Resonant modes of GaAs junction laser. III: Propagation characteristics of laser beam with rectangular symmetry" IEEE J. QUANT. ELECTRON., 1970, V. 6, p. 317-324.

15. Adams, M. "Vvedenie v teoriyu opticheskikh volnovodov" [An introduction to optical waveguide theory] Translation from the English, Moscow: Mir, 1984, 512 p.

16. Kirkby, P. A., Goodwin, A. R., Thompson, G. H. B., et al. "Observations of self-focusing in stripe-geometry semiconductor lasers and development of a comprehensive model of their operation" IEEE J. QUANT. ELECTRON., 1977, V. 13, p. 705-719.

17. Asbeck, P. M., Cammack, D. A., Daniele, J. J. "Non-Gaussian fundamental modes patterns in narrow stripe-geometry lasers" APPL. PHYS. LETT., 1978, V. 33, p.504-506.

18. Butler, J. K., sommers, H. S., Jr. "Assymmetry modes in oxide stripe heterojunction lasers" IEEE J. QUANT. ELECTRON., 1978, V. 14, p. 413-417.

19. Streifer, W., Burnham, R. D., Scifres, D. R. "Symmetrical and asymmetrical waveguiding in very narrow conducting stripe lasers" IEEE J. QUANT. ELECTRON., 1979, V. 15, p. 136-141.

20. Epstein, P. S. "Reflection of waves in an inhomogeneous absorbing medium" PROC. NAT. ACAD. SCI. USA, 1930, V. 16, p. 627-637.

21. Eliseev, P. G., Osinskiy, M. "The application of the dielectric Epstein model to a description of planar stripe heterolaser modes" KVANTOVAYA ELEKTRON., 1980, V. 7, p. 1407-1415.

22. Eliseev, P. G., Man'ko, M. A., Mikaelyan, G. T. "Model of an injection laser with a continuously varying complex dielectric constant along the *p-n*-junction" TR. FIAN, 1983, V. 141, p. 119-125.

23. Bogatov, A. P., Eliseev, P. G., Man'ko, M. A., et al. "Model of a stripe heterolaser accounting for inhomogeneities in the complex dielectric constant lengthwise in the cavity" Preprint FIAN, No. 101, Moscow, 1981, 44 p.

24. Man'ko, M. A., Mikaelyan, G. T. "Svoystva mod. napravlyaemykh usileniem, v aktivnykh poluprobodnikovykh volnovodakh" [The properties of gain-guided modes in active semiconductor waveguides] In: Integral'naya optika: Fizicheskie osnovy, prilozheniya [Integrated optics: physical principles, applications] Novosibirsk: Nauka, 1986.

25. Dziewiecka, T., Osinski, M. "Generalized Epstein model of stripe geometry injection lasers" APPL. OPT., 1984, V. 23, p. 94-99.

26. Kirkby, P. A., Thompson, G. H. G. "The effect of double-heterojunction waveguide parameters on the far field emission patterns of laser" OPTO-ELECTRON., 1972, V. 4, p. 323-324.

27. Casey, H. C., Jr., Sell, D. D., Panish, M. V. "Refractive index of $Al_xGa_{1-x}As$ between 1.2 and 1.8 ev" APPL. PHYS. LETT., 1974, V. 24, p. 63-65.

INVESTIGATION OF THE SPECTRAL AND SPATIAL CHARACTERISTICS OF NARROW MESA-STRIPE CONTACT HETEROLASERS

Yu. L. Bessonov, N.B. Kornilova, V.D. Kurnosov, M.A. Man'ko, V.N. Morozov
Chan Min'Tkhay, Fam Van Khoy, V.R. Shidlovskiy

Abstract: The spatial and spectral characteristics of GaAlAs/heterolasers with a narrow mesa stripe contact are investigated experimentally. The spectral width of emission and degree of astigmatism are measured as a function of the width of the contact. Experimental results are compared to theoretical models describing the profile of the complex dielectric constant in the p-n junction plane.

INTRODUCTION

The spectral and spatial characteristics of narrow mesa-stripe contact heterolasers are of interest in connection with the broad application of such lasers to fiber optic communication systems as well as information recording and playback equipment. The most promising lasers for application in such systems include single spatial mode lasers. The single-mode condition may be achieved by creating an inbuilt optical waveguide (the so-called lateral optical confinement) or by narrowing the width of the stripe contact (to 10 μm or less). Heterolasers of the first type have a single-frequency lasing spectrum and a linear watt-current characteristic, although they have rather complex fabrication technology and are very sensitive to external influences (temperature changes, parasitic optical feedback, etc.) [1]. Narrow mesa-strips contact heterolasers without additional lateral optical confinement are significantly simpler from the viewpoint of fabrication technology and are less sensitive to parasitic optical feedback; these devices are also characterized by a linear watt-current characteristic, although they have less stable spectral and spatial characteristics [2, 3].

An important feature of these lasers is the waveguide effect induced by the gain in the active medium (since the imaginary part of the complex dielectric constant makes a significant contribution to the formation and localization of the laser mode). Since the gain level is sensitive to emission intensity due to the saturation effect, the configuration of gain-guided modes also depends on the intensity, which is manifest as nonlinearities, hystereses and discontinuities in the watt-current characteristic of such lasers with a contact width of 15 μm and up [4, 5]. The gain-guiding effect in narrow mesa-stripe contact lasers serves to curve the phase front of an electromagnetic wave traveling in the laser [6] due to energy leakage into the passive regions. Curvature of the phase front serves to increase the spontaneous background level in the lasing mode and creates astigmatism of laser emission [6], i.e., this changes the position of the field "constriction" in the plane of the *p-n*-junction with respect to the cavity mirror. It is obvious that the characteristics of such lasers are highly dependent on the width of the stripe contact.

This study performs a detailed investigation of the characteristics of double GaAlAs heterostructure injection lasers with various widths of the stripe contact. The first section is devoted to certain methodological issues and the second and third sections present experimental results from the investigation of the laser characteristics outlined above while the fourth section discusses the derived results.

1. METHODOLOGICAL ISSUES

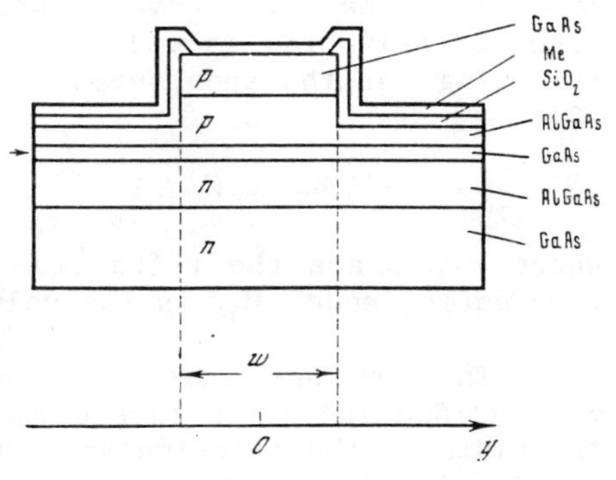

Fig. 1. Schematic representation of laser structures
The arrow indicates the active region (p = $10 \cdot 10^{17}$ cm^{-3}; w is the width of the contact)

The test heterolaser structures are given in schematic form in Fig. 1. The thickness of the active region was 0.3 μm. The width of the stripe contact varied from 2-15 μm. Current flow outside the active stripe was significantly reduced by oxide insulation. The laser diode cavity was formed by cleaving the initial wafer; cavity length was 150-200 μm. For subsequent studies only samples that operated in steady-state lasing were employed.

As noted above, the curvature to the wavefront of the field in a gain-guided laser serves to increase the spontaneous background level at the lasing mode. It has been demonstrated previously [7, 8] that such lasers may have a multifrequency emission spectrum even in steady-state lasing conditions. It is demonstrated in study [7] that the width of the emission spectrum of the laser $\delta\lambda$ may be represented as

$$\Delta\lambda = \frac{\eta_{\text{вн}} P_{\text{сп}}(\lambda_0)\,\pi\,(\Delta\Lambda)^2}{\eta_{\text{внутр}} \cdot 4\delta\lambda P_0}\,\gamma(r), \tag{1}$$

where $\gamma(r) = \frac{1}{r}\left(\frac{1-r}{\ln(1/r)}\right)^2$,

ε_{BH}, $\varepsilon_{\text{BHyTP}}$ are the internal and external quantum efficiencies, respectively; $P_{\text{сп}}(\lambda_0)$ is the spontaneous emission power at the central wavelength in the lasing spectrum of the laser, $\Delta\lambda$ is the width of the gain spectrum, $\delta\lambda$ is the mode distance of the laser cavity, P_0 is the emission power of the laser and R is the reflection coefficient of the cavity faces.

From (1) we may derive [7]

$$\Delta\lambda = \frac{\pi c \hbar\nu}{P_0} \ln\left(\frac{1}{r}\right) \gamma(r) \frac{R_{\text{сп}}}{R_{\text{ст}}} \left(\frac{\Delta\Lambda}{\lambda}\right)^2 K, \tag{2}$$

where $\hbar\nu$ is the photon energy, $R_{\text{сп}}$, R_{CT} are the rates of spontaneous and stimulated emission, respectively, λ is the emission wavelength and K is the astigmatism factor [6]

$$K = \left(\int |F(y)|^2 dy\right)^2 \cdot \left|\int F^2(y)\,dy\right|^{-2}, \tag{3}$$

where $F(y)$ is the amplitude distribution of the field in the plane of the p-n-junction (y axis, see Fig. 1). It is important to note that the spontaneous background coefficient at the lasing mode in gain-guided lasers also increases by a factor of K compared to lasers with lateral optical confinement, where $K = 1$. If we assume that a Gaussian approximation is valid for the relation $F(y)$, we may derive the following for the spontaneous background coefficient for the $\beta_{\text{сп}}$ mode [6]:

$$\beta_{\text{сп}} = \lambda^4 K / 4\pi n_a n_g V_{\text{эф}} \Delta\Lambda, \tag{4}$$

where n_a, $_g$ are the refractive indices of the active region and the propagating mode, $V_{\text{эф}}$ is the volume of the active region.

Thus we see that the spatial and spectral characteristics of gain guided injection lasers must be closely interrelated. Moreover, the width of the emission spectrum is highly dependent on the width $\Delta\lambda$ of the effective gain spectrum

$$g_{\text{эф}} = \Gamma \int g_{\text{ст}}(y) \cdot I_N(y)\,dy \Big/ \int I_N(y)\,dy, \tag{5}$$

where Γ is the optical confinement factor, $g_{\text{CT}}(y)$ is the gain distribution in the plane y, $I_N(y)$ is the field intensity distribution in the active region.

In connection with the discussion above, primary attention in this study is devoted to an investigation of the gain spectrum $g_{Э\Phi}(\lambda)$ at the lasing threshold and the spatial characteristics of laser emission. In order to investigate the gain spectra, a method based on restoring the gain spectrum from the emission spectrum at the lasing threshold is employed. We may write the following for the gain at the resonance frequency $w_{рез}$ of the natural Fabry-Perot cavity of the laser [9]

$$g(\omega_{рез}) = \alpha_{внутр} + \frac{1}{\Gamma L}\left(\frac{1}{2}\ln\frac{1}{r_1 r_2} + \ln\frac{\sqrt{I^+} - \sqrt{I^-}}{\sqrt{I^+} + \sqrt{I^-}}\right), \tag{6}$$

where $\alpha_{\Phi НУТП}$ are the internal emission losses in the in the active region, L is the cavity length, r_1, r_2 are the reflection coefficients of the cavity faces; I^+ is the emission intensity at $\omega_{рзз}$ and I_- is the emission intensity at the neighboring minimum in the spectrum. The following parameter values were used for calculation: $\alpha_{\Phi НУТП}$ = 35 cm^{-1}, Γ = 0.35 and r_1 = r_2 = 0.3. The spectral investigations were carried out using the DFS-12 diffraction spectrometer with a resolution of 0.3 Å.

The emission distribution in the far field was recorded using the standard method employing a photodiode with a slit diaphragm, while measurements in the near field were carried out using the X60 microlens and a television camera connected to a monitor for visual observation and a television scan line selector and oscillograph for direct measurements.

In order to investigate the emission astigmatism and wavefront, we use the method described previously in [11] based on the use of a modified Michelson interferometer, which makes it possible to experimentally measure the distance between the constrictions in orthogonal directions of the principal section of the laser beam (accurate to ±1 μm) as well as the distribution of the emission wavefront (accurate to $\lambda/10$). A ring pattern was obtained at the interferometer output whose geometry characterizes the wavefront shape [12]. The form of these rings depends on the collimation of the laser beam and displacement of the laser diode on the optical axis makes it possible to change their shape. It is demonstrated in study [12] that the angle Ψ between the asymptotes of the hyperbolic bands in the outcoupling plane of the interferometer and displacement of the diode with respect to the collimating lens Δz are related by the relation

$$\cos\varphi = 1 + \Delta z/C, \tag{7}$$

where C is a constant. The distance between the constrictions of the beam Δz_0 was determined by the points $\cos\Psi$ = ±1, i.e., between the positions where the interference bands form a system of parallel lines in two orthogonal directions. The radius of curvature of the wavefront was calculated by analyzing the interference pattern using the method described in studies [11, 12].

It is obvious that the spatial parameters of emission of the laser Δz_0, the radius of curvature of the wavefront R, the width of the directional pattern θ and the dimensions of the near field w_{pe3} at the cavity mirror are closely interrelated and are related to the astigmatism parameter K. Hence we will now provide a brief description of the primary models for the spatial characteristics of gain-guided lasers.

One of the most often encountered model is the model of the parabolic distribution of the complex dielectric constant $\varepsilon(y)$ under the injection contact [13, 14]

where $\varepsilon_0 = \varepsilon_0' + i\varepsilon''_0$, $a = a' + ia''$.

In this case, the field in the active region has a Gaussian profile, while the quantities Δz_0, R, θ, w_{pe3} and K are interrelated by the following relations [13, 14]:

$$R = \Delta z_0 \left[1 + \left(\frac{\pi w_0}{\lambda \Delta z_0}\right)^2\right], \qquad \mathrm{tg}\,\theta = \frac{\lambda}{\pi w_0}, \qquad K = \frac{w_{pe3}}{w_0}, \tag{8}$$

where w_0 is the beam width in the vicinity of the constriction.

Thus, it would seem that the parabolic approximation is convenient for describing the spatial characteristics of gain-guided lasers. However, in spite of the simplicity of these results, the parabolic approximation does not always yield sufficient agreement with experimentally-observed results. In our view the most universal model is the model of an injection laser with a continuous distribution of the dielectric constant through the Epstein layer, which is examined in studies [15, 16]. In this model the complex dielectric constant is represented as [15]:

$$\varepsilon(y) = (\varepsilon_3 - \varepsilon_1)\frac{\exp(y/l)}{1 + \exp(y/l)} + 2(2\varepsilon_2 - \varepsilon_1 - \varepsilon_3)\frac{\exp(y/l)}{[1 + \exp(y/l)]^2}, \tag{9}$$

where $\varepsilon_1 = \varepsilon(-\infty)$, $\varepsilon_2 = \varepsilon(0)$, $\varepsilon_3 = \varepsilon(\infty)$, l is the width parameter. In this case the expression for the intensity distribution in an analytical representation is rather complex and hence we will not perform such an analysis here (see [15]). It is important to note that this model makes it possible to describe the multimode nature of emission from stripe heterolasers. Naturally investigating the boundaries on the applicability of the parabolic model for describing the characteristics of gain guided lasers is quite important.

2. THE SPATIAL CHARACTERISTICS OF HETEROLASERS WITH A VARIABLE WIDTH OF THE MESA-STRIPE CONTACT

Results from measurements of the distance Δz_0 between the beam constriction in the y plane and the end mirror of the cavity are given in Table 1. These results indicate that the dependence of the distance Δz_0 on the width of the stripe contact has a rather complex form. Thus, a change in the width of the contact w from 10 to 5 μm initially produces a drop in the distance Δz_0, and with a further reduction in w, the distance Δz_0 sharply increases. It is demonstrated in [5] that the astigmatism of the laser emission beam is highly dependent on the gain distribution $g_{CT}(y)$ in the y plane. It is clear that a reduction in the width of the contact w to 3 μm significantly changes the profile $g_{CT}(y)$. In this case the portion of laser emission energy absorbed in the passive regions increases due to the absence of the inbuilt optical waveguide, which may cause significant curvature of the emission wavefront in a laser with w = 3 μm compared

Table 1

Experimentally measured values of the distance between the light beam constriction and the outcoupling mirror of the heterolaser cavity

Sample number	w, μm	$J_{пор}$, ka/cm^2	Δz, μm (with $J=1.2J_{пор}$)
31	15	3.3	27±1
32	15	2.8	27±1
42	10	3.4	24±1
43	10	3.3	24±1
22	7	4.5	21±1
23	7	4.5	21±1
13	5	6.1	18±1
14	5	6.1	19±1
1	3	15.6	28±1

to lasers with w = 5 or 7 μm. Moreover we note that carrier injection also produces an antiwaveguide effect in the active region, i.e., it causes a negative addition $\Delta\varepsilon_0'$ to appear in the real part of the complex dielectric constant in the active region; the absolute value of $\Delta\varepsilon_0'$ is greater the greater the injection current density [17]. Thus, an increase in Δz_0 for lasers with w = 3 μm may also result from an increase in the threshold current density when narrowing the width of the stripe contact, as observed in gain guided mode lasers (see Table 1 and [18]). The increase in the distance Δz_0 for lasers with a stripe contact width of 15 μm may be related to the appearance of a second order transverse mode in these lasers.

As noted in Section 1, for the case of a parabolic approximation for $\varepsilon(y)$ it is easy to estimate the anticipated values for emission divergence proceeding from the experimentally measured values R, Δz_0. Table 2 gives the experimental and calculated values (based on (8)) for the divergence angle θ of laser emission in the far field. It is clear that for lasers with w = 5 μm good agreement between experimental and theoretical results is observed, while for other values of w, a discrepancy exists between experimental results and those anticipated from (8); for lasers with w = 15 μm this differential is significant, and is related to the appearance of the second order transverse mode in these lasers. However, in order to estimate the astigmatism factor, it is convenient to use a parabolic approximation for $\varepsilon(y)$, and we assume that this approximation may also be used in our case for calculating K values needed to estimate the width of the emission spectrum of the test lasers. The values calculated using formula (8) for lasers with w = 3, 5, 7, and 10 μm are given in Table 3. The increase in astigmatism factor K with w = 3 μm is related to the sharp increase in Δz_0 and the degradation of the waveguide properties of the structure.

Table 2

Experimentally measured and calculated parameters of the directional pattern with J = 1.15 $J_{пор}$ (threshold)

Sample number	*w, μm*	*R, μm*	*θ, degrees*	
			Experiment	*Calculation*
31	15	43.5 ±1	3.8 ±0.5	6.3 ±0.2
42	10	35.7 ±1	5.2 ±0.5	7.2 >0.2
23	7	31.5 ±1	6.5 ±0.5	7.5 ±0.2
14	5	26.5 ±1	8.3 ±0.5	8.3 ±0.2
1	3	45 ±2	13 ±0.5	

Table 3

Calculated values of the astigmatism factor

w, μm	*K (at J = 1.15$J_{пор}$) (threshold)*	*w, μm*	*K (at J = 1.15$J_{пор}$) (threshold)*
3	6.2	7	4.8
5	4.8	10	4.5

Because of the significant differential in the spatial distribution of emission in the far field for a laser with an injection contact width of 15 μm, we used a model of a continuous distribution of the complex dielectric constant through the Epstein layer [16] to estimate the shape of the far field, using the following parameters $\varepsilon_1 = \varepsilon_3 = 12.891 + i0.001$, $\varepsilon_2 = 12.88 - i0.004$, $l = 5$ μm, $\lambda_0 = 0.83$ μm and $\Gamma = 0.3$.

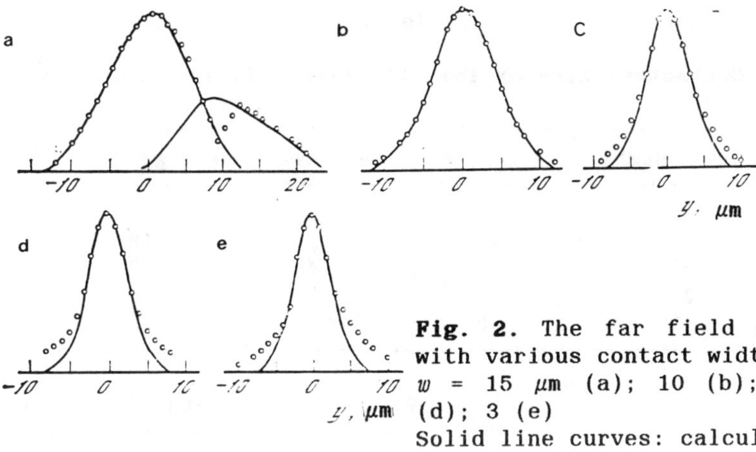

Fig. 2. The far field for lasers with various contact widths w
w = 15 μm (a); 10 (b); 7 (b); 5 (d); 3 (e)
Solid line curves: calculation

Fig. 2 gives the experimentally measured intensity distributions and intensity distributions derived from the parabolic approximation of $\varepsilon(y)$ for laser emission in the near field for lasers with w = 3-10 μm and from the Epstein model with w = 15 μm, as well as experimentally measured values. However, it is important to note that the Epstein model does not account for the interaction of spatial modes nor the influence of spatial inhomogeneities in the laser structure [4, 5] which may produce imprecise results (in our case it will eliminate the symmetrical additional peak in the experimentally-measured directional pattern).

3. SPECTRAL CHARACTERISTICS OF VARIABLE STRIPE CONTACT WIDTH INJECTION LASERS

We investigated the spectral characteristics of variable contact width lasers in two stages: initially we investigated the superluminescence spectra for restoring the spectral profile of the effective gain $g_{э\Phi}$ and then measured the spectra of the various samples in lasing conditions. Results from an investigation of the effective gain spectrum at the lasing threshold are given in Table 4. Here $\Delta\hbar\omega$ is the width of the gain spectrum at half amplitude, $\delta g/d$ is the relative gain deficit in the mode next to the fundamental longitudinal mode [19]. The trend towards an increase in the width of the gain spectrum is related to the fact that as the width of the stripe contact increases, the influence of spatial inhomogeneities on the form of the effective gain $g_{э\Phi}$ increases, and with a decrease in the width of the contact to 3 μm, the threshold injection current density sharply increases [18].

Table 4

Characteristics of the effective gain spectrum

Sample number	w, μm	$\Delta\hbar\omega$, *mev*	$(\delta g/g)\cdot 10^3$
13	5	11	8.9
43	10	12	7.8
31	15	14	6.8
22	7	12	7.5
1	3	13	7.0

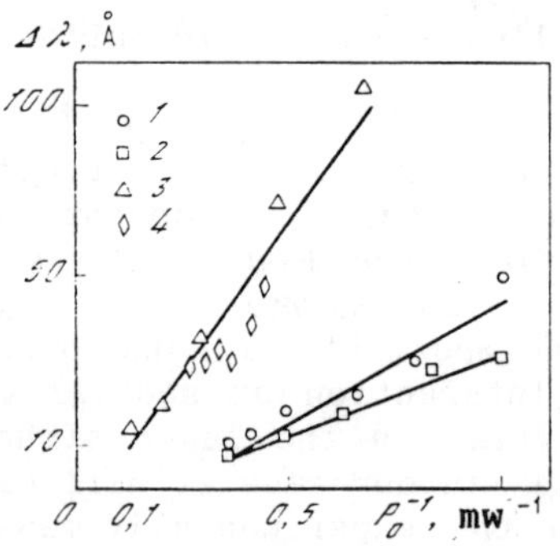

Fig. 3. The width of the emission spectrum plotted as a function of lasing power for samples with variable contact widths w = 7 μm (1); 5(2); 3(3); 10(4)

The dependencies of the emission spectral width on the lasing power are shown in Fig. 3. The curves given here correspond to the relations derived from (2) and the $R_{\text{сп}}/R_{\text{ст}}$ was calculated from experimental results as [19]

$$\frac{R_{\text{сп}}}{R_{\text{ст}}}=\left[1-\exp\left(\frac{\Delta F-\hbar\omega}{kT}\right)\right]^{-1}.$$

We see that the experimental points fall within the predicted relations (the values of the factor K for the lasers are estimated in section 3) and the emission spectrum is narrowed in near inverse proportion to the power; with identical emission power levels, the spectral width for lasers with w = 3 μm is significantly greater than for lasers with w = 5 or 7 μm; this is related to the increase in the factor K and the value $\Delta\Lambda$. An increase in the factor K reflects an increase in the spontaneous background level at the lasing mode for lasers with w = 3 μm. However, for lasers with an injection contact width of 10 and 15 μm, we did not observe $\Delta\lambda \sim P_0^{-1}$ (one possible $\Delta\lambda(P)$ relation with w = 10 μm is given in Fig. 3). Moreover, the spectral width of emission from lasers with w = 10-15 μm varied significantly (by a factor of 1.5) from sample to sample, which indicates significant influence of the additional spatial modes and inhomogeneities on the form of the emission spectrum in lasers with such widths of the stripe contact. The discrepancy between experimental and theoretical results for $\Delta\lambda(P)$ in lasers with a contact width w = 3, 5 and 7 μm is related, first, to inaccuracies in calculating the factor K from the parabolic approximation (see Section 2) and, secondly, to the change in the astigmatism of laser emission with changes in the injection current [5] which will not be considered in detail in this study.

We see that an unambiguous relation exists between the spatial and spectral characteristics of lasers with a stripe contact 3-7 μm in width; the emission spectrum of such lasers is narrowed in inverse proportion to the emission power and is a multifrequency spectrum with developed lasing when the pumping current exceeds the threshold level by up to 30%. For lasers with a contact width of 10-15 μm, spatial inhomogeneities and high order transverse modes begin to have a significant influence on the width of the emission spectrum.

4. DISCUSSION AND CONCLUSIONS

This study has carried out a detailed experimental investigation of the characteristics of narrow contact injection mesa-stripe heterolasers. It is demonstrated that the spectral characteristics of emission in such lasers are closely related to the spatial field distribution in the cavity. In this regard, the possibility of employing a parabolic approximation for the distribution of the complex dielectric constant in the plane of the *p-n*-junction was tested; this predicts a Gaussian intensity distribution at the mirror of the laser cavity. It is discovered that this model is not suitable for describing the experimentally-observed pattern of the near field emission of the laser, and a good correlation between experimental and theoretical results was obtained only with a contact width of 5 μm. Hence in the general case in order to describe the spatial characteristics it is necessary to use a model of a continuous distribution of the complex dielectric constant through the Epstein layer. From this viewpoint it is necessary to carry out a careful comparison of the spatial and spectral characteristics of mesa-stripe lasers and to calculate the astigmatism factor K as the primary characteristic used to characterize the relationship between the width of the emission spectrum and the spatial field distribution in the laser cavity.

BIBLIOGRAPHY

1. Miles, R. O., Dandridge, A.B., Giallorenzi, T. G., et al. "Low frequency noise characteristics of channel substrate planar laser diodes. APPL. PHYS. LETT., 1981, V. 138, p. 848-851.

2. Streifer, W., Scifres, D., Burnham, R. "Analysis of diode laser properties" IEEE J. QUANT. ELECTRON., 1982, V. 18, p. 1918-1929.

3. Streifer, W., Scifres, D., Burnham, R. "Longitudinal mode spectra of diode laser" APPL. PHYS. LETT., 1982, V. 40, p. 305-307.

4. Nash, F. R. "Mode guidance parallel to the junction plane of double-heterostructure lasers" J. APPL. PHYS., 1973, V. 44, p. 4696-4707.

5. Manko, M. A., Makhsudov, B. I., Mikaelyan, G. T., et al. "Measurement and analysis of planar stripe-geometry GaAlAs/GaAs heterojunction laser wavefronts and their variation with pumping current" IEE PROC. OPTOELECTRON., 1985, V. 132, p. 64-68.

6. Petermann, K. "Calculated spontaneous emission factor for double-heterostructure in injection lasers" IEEE J. QUANT. ELECTRON., 1979, V. 15, p. 566-570.

7. Arnold, G., Petermann, K., Schlosser, E. "Spectral characteristics of gain-guided semiconductor lasers" IEEE J. QUANT. ELECTRON., 1983, V. 19, p. 974-981.

8. Bessonov, Yu. L., Kurlenkov, S. S., Morozov, V. N., et al. "Investigation of multimode lasing mechanisms in double heterostructure AlGaAs narrow stripe contact injection lasers" KVANTOVAYA ELECTRON., 1985, V. 12, p. 347-352.

9. Hakki, B. W., Paoli, T. L. "Gain spectra in GaAs double-heterostructure lasers" J. APPL. PHYS., 1974, V. 46, p. 1299-1306.

10. Henry, C. H., Logan, R. A., Merritt, F. R. "Measurement of gain and absorption spectra in AlGaAs BH-lasers" J. APPL. PHYS., 1980, V. 51, p. 3042-3047.

11. Man'ko, M. A., Makhsudov, B. I., Fam Ban Khoy "An investigation of the wavefront of emission from planar stripe GaAlAs/GaAs-hetero-lasers" TR. FIAN, 1986, V. 166. (see this volume).

12. Tatsumo, K., Arimoto, K. "Measurement and analysis of diode laser wavefront" APPL. OPT., 1981, V. 20, p. 3520-3525.

13. Cook, D. D., Nash, F. R. "Gain-induced guiding and astigmatic output beam of GaAs lasers" J. APPL. PHYS., 1975, V. 46, p. 1660-1672.

14. Kirkby, P. A., Goodwin, A. R., Thompson, G. H. B., et al. "Observations of self-focusing in stripe-geometry semiconductor lasers and the development of a comprehensive model of their operation" IEEE J. QUANT. ELECTRON., 1977, V. 13, p. 705-719.

15. Eliseev, P. G., Osinskiy, M. "The application of the dielectric Epstein model to describing the modes of planar stripe hetero-lasers" KVANTOVAYA ELEKTRON., 1980, V. 7, p. 1407-1415.

16. Eliseev, P. G., Man'ko, M. A., Mikaelyan, G. T. "A model of an injection laser with a continuous distribution of the complex dielectric constant" TR. FIAN, 1983, V. 141, p. 119-125.

17. Eliseev, P. G. "Vvedenie v fiziku inzhektsionnykh lazerov" [An introduction to the physics of injection lasers] Moscow: Nauka, 1983, 294 p.

18. Keysi, Kh., Panish, M. "Lazery na geterostrukturakh" [Heterostructure lasers] Moscow: Mir, 1981, V. 2, 256 p.

19. Bogdankevich, O. V., Darznek, S. A., Eliseev, P. G. "Poluprovodnikovye lazery" [Semiconductor lasers] Moscow: Nauka, 1976, 330 p.

VOLTAGE SATURATION ACROSS THE INJECTION CONTACT IN A LASER DIODE AND THE PHENOMENON OF NEGATIVE PHOTO-EMF

Vu Van Lyk, P.G.Eliseev, M.A. Man'ko, G.T. Mikaelyan, O.G. Okhotnikov, G.T. Pak

Abstract: A phenomenological theory of an electrical signal at the lasing threshold is given and experiments on the electrical diagnostics of operating conditions in various types of injection lasers are described. Practical applications of the test phenomena in displacement, information optical reading and optical communications devices are discussed.

1. INTRODUCTION
THE VOLTAGE SATURATION EFFECT ACROSS THE INJECTION CONTACT AND ATTENDANT PHENOMENA

The electrical state of an injection contact in a laser diode is sensitive to the lasing conditions, since the injection current is a diffusion current and the stimulated emission fundamentally increases the gradient of the concentration of diffusion carriers (by reducing their lifetimes). This sensitivity is manifest to a greater or lesser degree in the saturation of voltage V applied to the injection contact (the p-n-junction or the p-n-heterojunction) after passing through the lasing threshold. This effect was discovered and interpreted in studies [1-4] using sample low threshold GaAs CW lasers and later was reproduced in experiments employing lasers based on narrowband A^4B^6 compounds [5]. It is worth noting that attempts to identify the relationship between the electrical characteristics and the laser characteristics occurred earlier [6, 7]; however, this effect was not discovered evidently due to the masking effect of the resistance of the diode. Study [6] describes a nonreproducible anomaly in the watt-current characteristic at the lasing threshold; more detailed investigations are performed in study [7] and the threshold effect is reduced to a change in the series resistance of the diode cross-section.

The voltage V saturation phenomenon at the lasing threshold is as follows. The value of V determines the difference of the Fermi quasilevels ΔF at the boundary of the active region (in certain cases eV may be identified with ΔF). This differential in turn determines the optical gain level in the laser mode, which, in steady-state lasing conditions, remains unchanged. This corresponds to gain saturation at the threshold level and, consequently, results in voltage saturation. An increase in current with a fixed V is equivalent, as noted in [3] to the disappearance of the incremental resistance of the injection contact. Thus, the total injection contact of the laser diode at threshold more or less drops irregularly by a value corresponding to the incremental resistance of the injection contact. This effect is demonstrated in Fig. 1 by the incremental resistance graph $R_d = dU/dI$ of the diode plotted as the function of the inverse of the pumping current. At the threshold I_t the experimental points convert from a sloping subthreshold branch to a horizontal branch corresponding to a fixed incremental resistance equal to the ohmic resistance of the diode width R_s. The region of constant V on the V-I characteristic (Fig. 2) shows the saturated value V_t of the voltage in lasing conditions used to estimate ΔF (V_t is the threshold value of V).

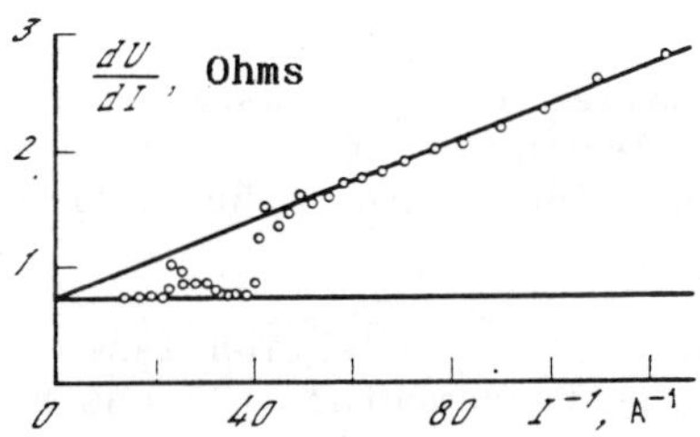

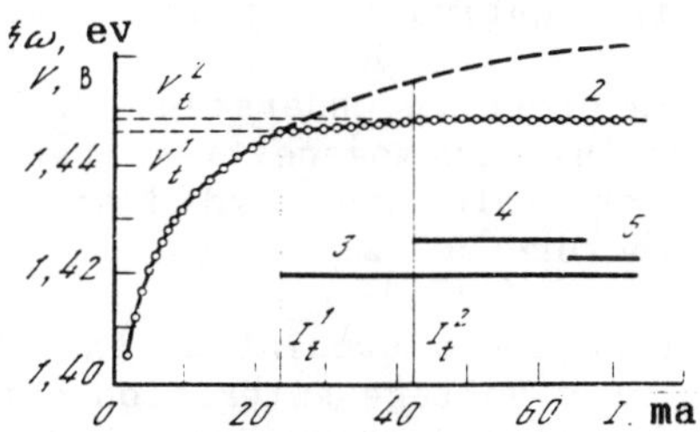

Fig. 1. The incremental resistance of an injection laser plotted against a value inversely proportional to the current [3].
Sample No. 557, CW conditions, 77 K; lasing threshold, 41.7 Å^{-1}; appearance of the second mode 23.5 Å^{-1}. The sloping region is extrapolated with $I^{-1} \to 0$ to a value of 0.74 ohms.

Fig. 2. The V-I characteristic (1) and the voltage applied to the p-n-junction (2).
3, 4, 5 - positions of the first, second, and third lasing modes, respectively. Sample No. 557, CW operation, 77 K; I^1_t = 24 ma, V^1_t = 1.446 v; I^2_t = 42.5 ma, V^2_t = 1.449 v.

The development of low-threshold CW lasers operating at room temperature has renewed interest in measuring the incremental resistance simply because this effect may be used to perform a purely electrical measurement of the threshold current (without using optical or spectroscopic instruments) based on a single form of the differential V-I characteristic. The corresponding differential methods and sample applications are given in studies [8-14]. Moreover, an automatic power control device [15] has been proposed that is based on this effect. Studies [13, 14] have also demonstrated that the form of the differential V-I characteristic may be used to determine and estimate deviations of the watt-current characteristic from linearity (kinks, etc.).

The broad utilization of these techniques has been hindered by occasional anomalies in the behavior of electrical characteristics at and beyond threshold. These anomalies are classified in the following manner:

1) incomplete voltage saturation or incomplete loss of R_d as the particular case of the absence of threshold indicators on the V-I and differential V-I characteristic;

2) a positive increment in R_d beyond threshold, i.e., an effect of opposite sign;

3) irregular changes in the voltage at the threshold, including changes of anomalous sign;

4) negative incremental resistance.

In order to understand the anomalies it is necessary to account for spatial inhomogeneities, carrier leakage from the active region, nonsteady-state and hard lasing conditions, etc. We will discuss these factors below.

From the viewpoint of the physics of an injection laser, observation of a voltage saturation effect is fundamental. It was noted in study [2] that the locking of V reflects the locking of ΔF, i.e., the points on the energy scale corresponding to the absence of both amplification and absorption of emission (with $\hbar = \Delta F$). This point lags the laser photon energy $\hbar\omega_{\Pi}$ by a quantity on the order of kT (~6.6 Mev at 77 K). If locking of the gain at the laser frequency locks such a remote point as ΔF, this means that the gain (amplification) band is homogeneously saturated, i.e., it is a homogeneously broadened band. This conclusion was obtained in study [3] together with the conclusion that it is this property that allows a semiconductor laser to produce intensive single-mode emission with the attendant mode selection [4].

Voltage saturation also confirms the concept of a diffusion pumping current. In this case the Fermi quasilevel goes over to a Fermi quasilevel in the vicinity of the majority carriers. The response of the Fermi level to the conditions in the active region is analogous to the interrelationship of levels in connected vessels. If there were to be a differential in the Fermi quasilevel between the active region and the ground region (carrier motion would be analogous in this case to a waterfall), the upper level would be insensitive to the state of the lower level. Consequently, testing the voltage saturation effect in new conditions such as over a broad temperature range and with new laser structures (various types of heterostructures) may serve as a tool for identifying the type of carrier transfer mechanism to the active region.

The differential technique [9, 12-14] has found the most extensive practical application; this technique has made it possible to investigate a large number of laser diodes including AlGaAs/GaAs and InGaAsP/InP low-threshold diodes. Accumulated experience has demonstrated that the voltage saturation and reduction in the incremental resistance occur more as a trend rather than as a hard rule and appear as the background for the "anomalies" identified above. An analysis of this behavior will be given below. The relationship between the electrical characteristics and the dynamics of emission from injection lasers as well as nonlinear phenomena in such lasers is of significant interest.

The method based on the voltage saturation effect in the form we discussed above is the particular case of electrical diagnostic methods for a laser diode deriving from its physical mechanism. We have already discussed the possibility of identifying nonlinearities in the watt-current characteristics based on the behavior of the incremental resistance of the diode beyond threshold [12, 13]. In principle electrical diagnostics may also be used to estimate the differential efficiency of the laser [16, 17] but only to the extent that it is related to the internal conversion or pumping efficiency but not to nonresonant reabsorption. Another aspect is revealed by diodes with a negative incremental resistance occurring due to hard laser firing conditions. Such diodes generate electrical instability in the pumping network and may create electrical oscillations accompanied by emission pulsations. Transient conditions and sustained pulsations in the laser are also accompanied by nonsteady-state electrical phenomena in the pumping network and may be investigated using electrical methods. Bistable phenomena in laser diodes have also been identified [14, 18]. In experiments with a power hysteresis in a AlGaAs/GaAs-laser with an external cavity, lasing breakdown was observed when a voltage jump appeared across the diode; a linear dependence between the voltage jump ΔV and the emission power jump ΔP was discovered with a slope of 0.3 v/w [18].

Study [14] describes the influence of diode immersion on the characteristics, including the differential V-I characteristic. These measurements may be used to determine the change in the threshold current for a given change in the cavity Q (due to a reduction in the reflection coefficient of the end mirrors). This technique may be used to determine such an important parameter of the active medium as the differential gain, i.e., the ratio of the increment of optical gain to the increment of pumping current.

On a broader scope, the electrical diagnostics of laser diode operating conditions based on the voltage saturation effect includes a consideration of the operating conditions of complex laser diodes: lasers with coupled cavities, with external cavities, with multiple stable states, laser sensors with compound resonators as well as the investigation of the photoresponse in laser amplifiers (and lasers), i.e., the electrical sensitivity under photon injection action. In these conditions the electrical signal arises due to a change in the rate of stimulated radiative transitions, which produces a change in ΔF. The primary reason is the reduction in threshold in laser conditions (i.e., in self-consistent conditions) when the Q of the cavity changes, including the case of "external" feedback (external with respect to the laser diode). The increase in Q reduces ΔF and V. This may be caused by an increase in the external reflection and thus may be used in a reflected signal sensor or for reading information from an optical medium. It should be remembered that external feedback in the coherent case may be both positive and negative depending on the phase of the reflected signal. In the latter case, the increase in feedback will increase V.

The external injection of photons to the cavity (i.e., unlike the pumping feedback from an external source) also serves to reduce the threshold if the injected emission is resonant with the laser mode. A locking condition may occur where the external emission controls the lasing properties. This is quite clear with some detuning from resonance, when the intensity of external photon injection must exceed a certain threshold to achieve locking conditions. Since photon injection reduces the pumping threshold, the conditions below the nominal pumping threshold (the initial threshold) correspond to so-called resonant (regenerative) amplification; these conditions do not compete with natural lasing as is the case when pumping above the nominal threshold. In conditions far from the lasing threshold yet with optical amplification present in the medium, we achieve traveling wave amplification. In order to increase the gain, we reduce the feedback to the greatest extent possible (by coating the ends, by using Brewster configuration, etc.). In these cases the electrical signal depends on the emission intensity and reflects a drop in V due to the more rapid demand for excess carriers. Naturally when the nominal threshold is achieved, the signal amplitude drops, since voltage saturation occurs by natural emission.

The photoresponse in the amplifier is opposite in sign to the photo-EMF. This reflects the physical state where the stimulated emission is a process inverse to optical absorption. However, during absorption (in the photodiode, photoresistance) the optical signal disappears and in an amplifier it is preserved and amplified. The laser diode may serve as a photoelectric signal detector.

2. AN ANALYSIS OF A HOMOGENEOUS MODEL

We will examine a homogeneous injection contact and recombination in a base (active) region of limited thickness d. We will assume that the carriers are completely reflected off of the second boundary of the base (total electron "confinement"). We will represent the V-I characteristic of the injection contact in the form [14]

$$j = (eDN/L_D)\ \text{th}\ (d/L_D)\ \text{experiment}\ (eV/\varepsilon), \tag{2.1}$$

where D and L_D is the coefficient and length of diffusion of the injected carriers; N is the concentration of these carriers at the emitter with sufficient energy to overcome the potential barrier with a height equal to the contact difference of potentials in the injection contact. We will ignore the surface recombination on both boundaries of the base region (study [14] gives expressions accounting for this recombination). The quantity j is the current density; V is the voltage applied to the injection contact, ε is the energy constant characterizing these injection conditions. Formula (2.1) for fixed values of D, L_D, N and ε correspond to linear recombination ignoring effects caused by the influence of dynamic degeneration.

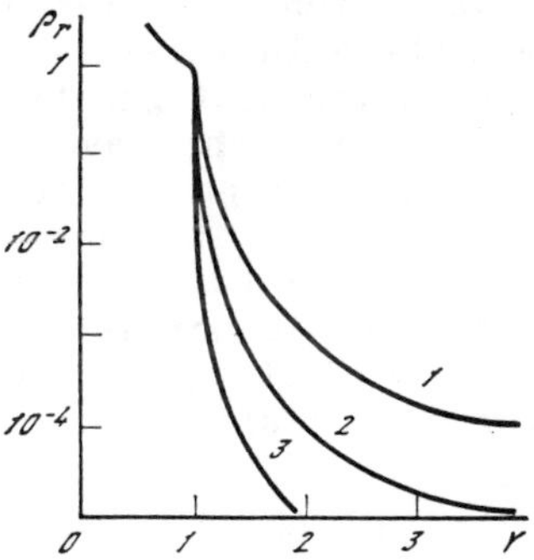

Fig. 3. The "natural" residual incremental resistance ρ_r of an injection laser contact normalized for the threshold value plotted as a function of relative pumping $Y = I/I_t$ for various values of the spontaneous emission factor β
1 - $\beta = 10^{-3}$; 2 - 10^{-4}; 3 - 10^{-5}

When passing through the lasing threshold, the recombination balance changes, since the superthreshold pumping is expended on stimulated emission. This is equivalent to shortening the lifetime and the diffusion length. From formula (2.1) we may obtain

$$V - V_t = \frac{\varepsilon}{e} \ln\left[\frac{j}{j_t}\,\frac{L_D\,\text{th}(d/L_{Dt})}{L_{Dt}\,\text{th}\,(d/L_D)}\right], \tag{2.2}$$

where V_t is the threshold voltage ($j = j_t$, L_{Dt} is the diffusion length at threshold. For these lasing conditions study [14] examined a system of rate equations accounting for the spontaneous emission factor β representing the relative portion of spontaneous emission expended on the laser mode. In this model ΔF asymptotically approaches the threshold value. Analogously the carrier concentration N_e also asymptotically approaches N_e^{∞}:

$$N_e(Y) = N_e^{\infty}\,\frac{1 + Y - [(Y-1)^2 + 4\beta Y]^{1/2}}{2\,(1-\beta)}, \tag{2.3}$$

where $Y \equiv I/I_t \equiv j/j_t$ is the normalized pumping current.

For the effective diffusion length, study [14] has derived

$$L_{эф}(Y) = L_0\left\{\frac{1 + Y - [(Y-1)^2 + 4\beta Y]^{1/2}}{2Y\,(1-\beta)}\right\}^{1/2}, \tag{2.4}$$

where L_0 is the value of L_D with $Y \ll 1$.

If we substitute $L_{эф}$ in place of L_D and L_{Dt} into formula (2.2), we derive the voltage increment over threshold, $\Delta V = V - V_t$. The corresponding expressions are significantly simplified if we limit our examination to a situation typical of laser heterostructures where $d \ll L_{\text{э}\Phi}$, L_D. *Then using the approximation th* $x \approx x(x \to 0)$, we obtain

$$\Delta V = \frac{\varepsilon}{e} \ln \frac{1 + Y - [(Y-1)^2 + 4\beta Y]^{1/2}}{2(1-\sqrt{\beta})}. \tag{2.5}$$

For the incremental resistance we obtain with $I > I_t$[1]

$$\frac{dV}{dI} = \frac{\varepsilon}{eI_t} \frac{1 - (Y - 1 + 2\beta)[(Y-1)^2 + 4\beta Y]^{-1/2}}{1 + Y - [(Y-1)^2 + 4\beta Y]^{1/2}}. \tag{2.6}$$

The quantity (2.6) is the "natural" residual incremental resistance of the injection contact with maximum total saturation ΔF; this resistance value is other than zero only due to the influence of spontaneous emission. This influence somewhat softens the watt-current characteristic of the laser and smooths out the variation in incremental resistance as shown in Fig. 3. We note that the actual values of β amount to 10^{-4}-10^{-5} for a single mode depending on its configuration (the slopes of the wavefront, astigmatism, etc.). For the case of deliberate multimode emission, it is possible to use a higher effective β. Thus, in a homogeneous model accounting for the influence of spontaneous emission, a somewhat irregular drop in R_d to very low values (but not to zero) may be anticipated. In order to demonstrate the correctness of the limiting process with $\beta \to 0$ yielding $\Delta V \to 0$, according to formula (2.5), and $R_d \to 0$ according to formula (2.6), we return to the V-I characteristic in the form

$$j = (eDNd/L_D^2) \exp(eV/\varepsilon) \tag{2.7}$$

and use $L_{\text{э}\Phi} = L_0/\sqrt{Y}$, which follows from (2.3) with $\beta = 0$.

It is easy to obtain

$$j = (eDNd/L_0^2) Y \exp(eV/\varepsilon), \tag{2.8}$$

or

$$1 = (eDNd/L_0^2 j_t) \exp(eV/\varepsilon),$$

where the pumping current is dropped from the equation. We obtain

$$V = (\varepsilon/e) \ln (j_t L_0^2/eDNd), \tag{2.9}$$

i.e., V is the constant (independent of current).

The laser emission power P in the same model may be represented as

$$P = \frac{P_2}{1 + \sqrt{1 + 8\beta}} \{Y - 1 + [(Y-1)^2 + 4\beta Y]^{1/2}\}, \tag{2.10}$$

---

[1] A typographical error occurs in the formula corresponding to this formula in study [14]: the minus sign in front of the 1/2 in the numerator has been excluded.

where P_2 is the normalizing factor equal to the emission power (in the laser mode) with $Y = 2$, i.e., at twice threshold. We may derive the following formula:

$$eV = (\varepsilon/e)\ln[(Y - Z)/(1 - \sqrt{\beta})], \tag{2.11}$$

where   is the dimensionless emission power. In the limit of significant pumping, $Y \to \infty$, we have with $\beta \ll 1$

$$\Delta V^{\infty} \approx (\varepsilon/e)(\sqrt{\beta} - \beta). \tag{2.12}$$

Formulae (2.11) (2.12) do not reflect new information compared to the preceding formulae. These expressions are interesting in that they demonstrate, based on the influence of spontaneous emission, the relation between the increment ΔV and incomplete pumping. Indeed, with $\beta = 0$ we have $Z = Y - 1$ and the quantity under the logarithm sign tends towards one, while ΔV vanishes.

An interesting approach to analyzing the relationship of ΔV and the internal quantum efficiency of the laser is proposed in study [19]. This method uses the measured value of the superthreshold voltage increment ΔV to determine the increment of the nonlaser current component using an extrapolation of the subthreshold V-I characteristic. Then the superthreshold pumping current may be divided into the laser component $I_{л}$ and the component that is an extension of the subthreshold relation

$$I - I_t = I_{л} - I_t[1 - \exp(e\Delta V/\varepsilon)], \tag{2.13}$$

and for internal quantum efficiency we have

$$\eta_i = \frac{I_{л}}{I - I_t} = 1 - \frac{I_t}{I - I_t}\left(\exp\frac{e\Delta V}{\varepsilon} - 1\right). \tag{2.14}$$

From here we have

$$\Delta V = (\varepsilon/e)\ln[1 + (Y - 1)(1 - \eta_i)], \tag{2.15}$$

where $Y > 1$ again designates a relative increment of the threshold current. The quantity η_i may be compared to the external differential efficiency of the laser η_d, which is expressed as

$$\eta_d = \eta_i \eta_P f(\alpha, L, R), \tag{2.16}$$

where η_P is the pumping efficiency, $f(\alpha, L, R)$ is the emission outcoupling factor from the cavity related to its length L *and mirror reflection* R and not equal to one due to internal optical losses with an index α. If η_P and $f(\alpha)$ are constant, the measured value η_d characterizes the relative variations in η_i, for the case of lasers with nonlinearities in the watt-current characteristic [14]. Such a comparison is given in study [14] based on experiments with planar stripe AlGaAs/GaAs heterolasers with active stripe width of 10-20 μm. Significant fluctuations in the slope of the watt-current characteristic are often encountered in such samples, even including a change to a minus sign in certain sections. An analysis demonstrated good agreement a correlation in the behavior of η_d and the incremental resistance (or ΔV) in qualitative accord with the anticipated nature of the interrelationship (specifically, incomplete pumping power demand is accompanied by an increase in ΔV). We note, however, that the steps and inflections in the watt-current characteristic are also related to variations in f (α, L, R) and are also possibly due to changes in α or a change in the configuration of the laser mode.

This phenomenological approach does not discuss the causes of variations in η_id and ΔV beyond threshold. Since the mechanism of these variations is related to spatial variations and inhomogeneities (self-focusing, spatial mode switching), a "homogeneous" model is limited in application. However, we may examine a single case that is strictly allowed within the scope of a homogeneous model in developing the phenomenological approach. Here we are discussing the variation in the threshold pumping level in superthreshold conditions. This may be caused by a change in temperature, mechanical stresses or other more or less "homogeneous" factors. For example, let the value I_t be

$$I_t = \begin{cases} I_{t0} \text{ when } I < I_0 \\ I_{t1} \text{ when } I > I_0 \end{cases} \quad (I_{t0} < I_{t1}).$$

Then in passing through the value I_0 the laser emission power drops in a stepwise fashion so that the new watt-current characteristic is extrapolated to a new threshold value, while the intensity of spontaneous emission increases in a stepwise manner. The voltage increment is

$$\Delta V = V_{t1} - V_{t0} = (\varepsilon/e) \ln (I_{t1}/I_{t0}) \tag{2.17}$$

which is fully consistent with the homogeneous model considered above. As another particular case we will consider the linear increase in the threshold current with pumping current above I_{t0}, say, in the form

$$I_t (Z) = I_{t0} + a (I - I_{t0}), \tag{2.18}$$

where $a < 1$. If the initial value η_1 at threshold is equal to one, beyond threshold it is equal to $1 - a$. Regarding the voltage increment, this will follow the threshold increment in accordance with the formula

$$\Delta V = \frac{\varepsilon}{e} \ln \frac{I_{t0} + a (I - I_{t0})}{I_{t0}} = \frac{\varepsilon}{e} \ln [1 + a (Y - 1)], \tag{2.19}$$

which coincides with the substitution of $\eta_1 = 1 - a$ into formula (2.14).

These types of formulae, particularly formula (2.17), are suitable for the case where the external feedback reduces or increases threshold current as well as when the external reflection coefficient changes in the optical feedback network (due to motion of the data carrier containing the 1 and 0 digits, for example). If the two states of the external feedback coefficient corresponding to 1 and 0 on the data medium have threshold currents of L_{t0} and I_{t1}, we may use formula (2.17) to determine the amplitude of the electrical signal across the injection contact. Studies [20, 21] examined the possibility of reading data by means of a laser diode whose electrical signal is divided into three components:

1) the ΔV effect itself across the injection contact;

2) photoconduction in the adjacent layers of the semiconductor induced by the reflected laser light;

3) the signal related to the modulation of the leakage current.

The photoconduction effect is normally small compared to ΔV. Regarding the influence of leakage, this issue is worthy of a more detailed analysis [22] which will be given below.

3. THRESHOLD EFFECT ACCOUNTING FOR MODULATION OF THE CUBIC RESISTANCE OF THE SEMICONDUCTOR

Since the injected carriers reduce the resistance of the recombination region, R_s (the series resistance of the diode minus the nonlinear resistance of the injection contact) is not a constant and as a result an additional contribution is made to the nonlinear incremental (differential) resistance of the diode R_d. We will see how the electrical threshold effect is modified by modulation of the cubic resistance of the active region and the emitters (for the case of excess carrier leakage from the active region to the emitters). We will assume, in order to simplify the calculation, that the change in conduction σ under injection action is small, i.e., $\Delta\sigma/\sigma < 1$. We will represent the resistance R_a of the active region as

$$R_a = \frac{d}{(\sigma + \Delta\sigma) S} \approx R_{a0}\left(1 - \frac{\Delta\sigma}{\sigma}\right), \tag{3.1}$$

where d is the thickness, S is the area and R_{a0} is the initial resistance of the active region. The conduction increment $\Delta\sigma$ may be related to the pumping current I assuming linear recombination with a carrier lifetime τ

$$\Delta\sigma = e\mu\,(1 + b)\,\Delta N = \mu\,(1 + b)\,I\tau/Sd, \tag{3.2}$$

where μ is the whole mobility, b is the ratio of electron and whole mobility, ΔN is the excess concentration. We may now represent R_s in the form

$$R_s = R_{s0} - IR_{a0}\frac{(1 + b)\,\tau}{e\,dN_0S}, \tag{3.3}$$

where N_0 is the equilibrium hole concentration; we will assume that the fixed part R_{s0} also includes R_{a0}. The voltage across the diode will be equal to

$$U = IR_{s0} - I^2 R_{a0} \frac{(1+b)\tau}{e\, dN_0 S} + \frac{\varepsilon}{e} \ln \frac{I}{I_0}, \tag{3.4}$$

where the last term gives the voltage V applied across the injection contact. For the incremental (differential) resistance of the diode R_d, we have

$$R_d = \frac{dU}{dI} = R_{s0} - 2IR_{a0} \frac{(1+b)\tau}{e\, dN_0 S} + \frac{\varepsilon}{e} \frac{1}{I}, \tag{3.5}$$

thus, with low currents at threshold $R_d(I_t)$ is

$$R_d(I_t) = R_{s0} - 2I_t R_{a0} \frac{(1+b)\tau}{e\, dN_0 S} + \frac{\varepsilon}{e} \frac{1}{I_t}. \tag{3.6}$$

After passing through threshold with complete voltage saturation, the diode resistance remains the same and the differential resistance beyond threshold will be equal to

$$R_d(I > I_t) = R_{s0} - I_t R_{a0} \frac{(1+b)\tau}{e\, dN_0 S}. \tag{3.7}$$

Subtracting (3.7) from (3.6), we obtain the jump in R_d at threshold

$$\Delta R_d = \frac{\varepsilon}{e} \frac{1}{I_t} - I_t R_{a0} \frac{(1+b)\tau}{e\, dN_0 S} = \frac{\varepsilon}{e} \frac{1}{I_t} - R_{a0}(1+b) \frac{\Delta N_t}{N_0}, \tag{3.8}$$

where ΔN_t is the concentration of excess carriers at the lasing threshold. Since in the ideal case the jump in incremental (differential) resistance is equal to the first term in the right half of formula (3.8), the actual jump is reduced by a magnitude equal to the left half, dependent on R_{a0} and the relative increment of the carrier concentration at threshold, $\Delta N_t/N_0$. The jump in R_d vanishes when

$$I_t = [N_0 S \varepsilon d / R_{a0} (1+b) \tau]^{1/2}, \tag{3.9}$$

while at higher threshold currents it changes sign, i.e., R_d becomes larger after passing through threshold. This corresponds to the anomaly we discussed above. We may express the anomaly condition by the inequality

$$j_t \geqslant [\sigma_0 N_0 \varepsilon / (1+b) \tau]^{1/2}, \tag{3.10}$$

where j_t is the threshold current density. In principle j_t may refer to both the threshold current density as well as other points on the watt-current characteristic where the slope increases, i.e., secondary "thresholds" related to possible nonlinearities.

We note that with anomalies in R_d, i.e., when condition (3.10) is satisfied, the subthreshold portion of the V-I characteristic is distorted. Specifically, the subthreshold slope of the relation $R_d(1/I_t)$ is not equal to ε/e, as in the ideal case, and if we use this relation, the quantity is formally defined

$$\varepsilon^* = \varepsilon\,[1 + (e/\varepsilon)\, I_t R_{a0} \Delta\sigma/\sigma]. \tag{3.11}$$

Substituting a jump in R_d into this quantity yields

$$\Delta R_d = \frac{\varepsilon}{e}\frac{1}{I_t}\left(2 - \frac{\varepsilon^*}{\varepsilon}\right), \tag{3.12}$$

i.e. R_d increases at threshold if $\varepsilon^*/\varepsilon > 2$, and with $\varepsilon^* = 2\varepsilon$ no jump in R_d is observed. For the superthreshold voltage increment V_r across the injection contact, we derive

$$V_r = (\varepsilon/e)\,(\varepsilon^*/\varepsilon - 2), \tag{3.13}$$

or if we formally designate $\varepsilon^* = nkT$ we will have

$$V_r = (kT/e)\,(n - 2\varepsilon/kT). \tag{3.14}$$

This formula may be compared to the empirical formula found in study [14]

$$V_r = (kT/e)\,(n - 1), \tag{3.15}$$

and any discrepancy from this value is attributable to the fact that the second quantity in the parentheses in formula (3.14) is not a modulo 1 quantity, but more often a modulo 2-4 quantity, since ε in the V-I characteristics is not always less than 1 and is most often near 2. It is difficult to perform a detailed examination of this discrepancy, since study [14] does not give separate values of ε^* and ε. However, a qualitative similarity between formulae (3.14) and (3.15) is clear.

The contribution of resistance modulation of the active region with leakage to the emitter layers is supplemented by the contribution of the modulation of these layers. Study [22] examines the influence of carrier leakage from the active region on the incremental (differential) resistance of the laser diode. Using heterobarriers to confine the injected carriers within the active region (so-called electron confinement) is an important advantage of laser heterostructures that makes it possible to obtain a low threshold current density compared to homolasers. Carrier leakage through the heterobarrier is undesirable, although it appears to some degree or another near the upper boundary of operating temperatures. Since leakage through the barrier increases with temperature, it makes a contribution to the temperature relation of the threshold current and reduced the parameter T_0 characterizing the rate of the exponential temperature dependence of the threshold.

Since the penetration of excess carriers into the emitter modulates its resistance, it may be detected by electrical measurements. Evidently the largest incremental (differential) resistance may be produced by modulation of the p-emitter, since its initial resistance is greater than that of the n-emitter and of the active layer.

The model used in [22] does not include the assumption of low resistance modulation, and the resistance $R_Э$ of the emitter is calculated by integration with respect to the width accounting for the spatial variation in the distribution of excess carriers:

$$R_Э = \frac{1}{S} \int_0^h \frac{dx}{\sigma(x)}, \tag{3.16}$$

where h is the width of the emitter layer, $\sigma(x)$ is its electrical conductivity. In calculating $\sigma(x)$ it is assumed that the excess concentration obeys an exponential law

$$\Delta N(x) = \frac{I_{yT} l_D}{eDS} \exp\left(-\frac{x}{l_D}\right), \tag{3.17}$$

where I_{yT} is the leakage current, l_D is the length of carrier diffusion in the emitter, D is the diffusion coefficient. The carrier leakage rate through the barrier may be represented in approximate form as a function of the concentration of excess carriers in the active N_a,

$$I_{yT} \sim N_a^q, \tag{3.18}$$

where $I_{yT} \approx 5.5$ [23]. In turn N_a is related by a power relation to the pumping current I (in the general case), $I \approx N_a^p$, where $p \approx 1\text{-}3$ depending on the predominant recombination mechanism. Strictly speaking q and p are not constants, although in order to simplify the calculation we will assume that I_{yT} is a power function of the pumping current, and we will assume in the result that the excess concentration at the boundary of the emitter layer is equal to

$$\Delta N(0) = N^*(I/I^*)^m, \tag{3.19}$$

where N^* and I^* are the adjustment constants, and m is found between 2 and 3 (while the leakage current accounts for only a small portion of the total current). Then we have

$$R_Э = \frac{d}{\sigma_0 S}\left[1 + \frac{l}{d} \ln \frac{\sigma_0 + \sigma_1 \Delta N(0) \exp(-d/l)}{\sigma_0 + \sigma_1 \Delta N(0)}\right], \tag{3.20}$$

where $\sigma_1 = e\mu_h(1 + b)$, μ_h is the whole mobility. The voltage U across the diode is represented as

$$U = IR_s + IR_Э(I) + (\varepsilon/e) \ln(I/I_0), \tag{3.21}$$

where I_0 is the preexponential factor in the formula for the V-I characteristic. We obtain

$$\frac{dU}{dI} = R_s + R_Э(I) + I\frac{dR_Э(I)}{dI} + \frac{\varepsilon}{e}\frac{1}{I}. \tag{3.22}$$

In lasing we will have

$$R_d = R_s + R_Э(I_t) \tag{3.23}$$

and the jump will be

$$\Delta R_d = I_t \frac{dR_э(I)}{dI}\bigg|_{I_t} + \frac{\varepsilon}{e}\frac{1}{I}. \tag{3.24}$$

As for the case of modulation of the base resistance, the jump may be both positive and negative. In the general case

$$\Delta R_d = I_t \frac{dR_s}{dI}\bigg|_{I_t} + \frac{\varepsilon}{e}\frac{1}{I_t}, \tag{3.25}$$

where the first term is other than zero if R_s is not a constant, and in the general case is negative, since injection reduces resistance. Thus, accounting for the modulation of the resistance of the base or the emitter layers produces an addition to the jump in the incremental (differential) resistance with an opposite sign from the addition related the voltage jump across the injection contact. This may compensate for the regular jump or produce an effect of opposite sign. In certain applications, which will be discussed below, compensation is undesirable. In principle, since the resistance modulation effect described above is caused by the threshold saturation of the concentration, it is possible to deliberately utilize high-resistance layers (active or emitter layers) for diagnostics of the threshold and in order to achieve a higher (positive) increment ΔR_d and V_r, since the conditions for achieving the increment ΔR_d have not been formulated here. We note that in calculating the leakage current in a more comprehensive model it is necessary to bear in mind the drift component of the excess carrier current in the emitter (this component continues to increase in the superthreshold conditions due to the increase in the d.c. voltage field in the emitter).

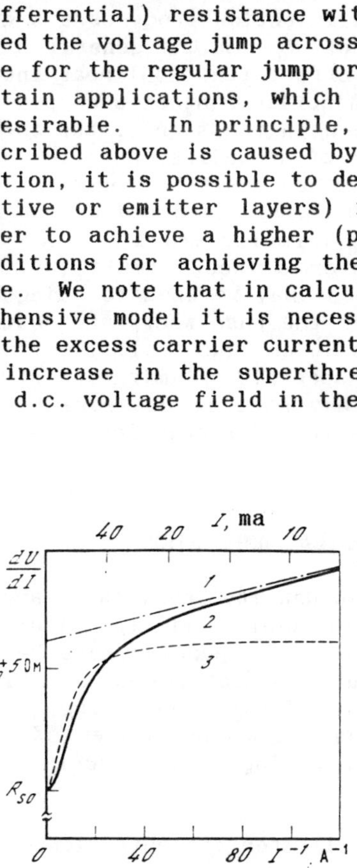

Fig. 4. Calculated dependencies of the incremental (differential) resistance dU/dI of the laser diode on a quantity inverse to the pumping current.
1 - ignoring resistance modulation of the semiconductor (nonlaser conditions); 2, 3 - accounting for modulation, nonlaser (2) and laser (3) conditions. The following parameters were used in the calculation: diode area - $3 \cdot 10^{-5}$ cm^2; thickness of p-emitter - $3 \cdot 10^{-4}$ cm; diffusion length in emitter - 10^{-4} cm; electron and hole mobility in the emitter 10^3 and 10^2 cm^2/w·sec, respectively, initial hole concentration 10^{17} cm^{-3}, parameters: $I^* = 10$ ma, $N^* = 3 \cdot 10^{14}$ cm^{-3}, $m = 2$

A numerical calculation using formulae (3.20), (3.22) is given in Fig. 4. The solid curve shows the calculated value of R_d using (3.22), i.e., for subthreshold conditions. When $I \to 0$ ($1/I \to \infty$) the curve flattens out, and extrapolation of the rectilinear section to

$I \to \infty$ gives the initial value of the series resistance R_S on the y axis (the linear resistance in the absence of resistance modulation). Thus, the slope of ε/e may be determined at a low current (if the exponent is preserved in this current range). In lasing conditions the incremental (differential) resistance is equal to a quantity (3.23) corresponding to the dotted line. A more or less sharp transition from the solid line to the dotted line occurs at threshold. It is clear that the sign of the change in R_d will depend on whether the threshold is to the right or left of the point of intersection. At high currents (to the left of intersection) the sign anomaly will occur.

In summary: accounting for the modulation of the cubic resistance of the semiconductor resulting from injection produces a significant addition to the threshold effect of R_d; in the general case, its sign is opposite that of the regular effect (the nonlinear incremental (differential) resistance of the injection contact disappears). As a result, both compensation and a change in the sign of the threshold are possible. This explains a number of anomalies characteristic of heterolasers with lightly doped active and emitter regions. In such diodes, the V-I characteristic is partially distorted by the influence of resistance modulation and the formally measured slope of this curve serves to increase the parameter ε (see formula (3.11)). In principle, the anomalous threshold effect may be used for diagnostics of leakage from the active layer (for the case where the contribution of modulation of its resistance is inconsequential, i.e., with a narrow width d and with heavy doping). Specific optimization of the electrical structure of the laser diode may make it possible to increase the observed electrical threshold effect for practical applications.

4. THE PHOTORESPONSE OF AN OPTICAL AMPLIFIER

Unlike a laser-oscillator, the photon density level in an amplifier is established by external photon injection and amplification, with the exception of gain saturation, where the pumping rate begins to play a role, as in a laser. Nonsaturation conditions should be examined on the basis of the balance of carrier concentration, i.e., on the basis of an electron equation from a standard system of rate equations. A nonsaturated amplifier is an analog to a laser in which the threshold level changes.

We will write the equation for the electron concentration N_e in the form

$$\frac{dN_e}{dt} = \frac{j}{ed} - \frac{N_e}{\tau} - KN_eN_\omega, \tag{4.1}$$

where K is the proportionality factor for stimulated recombination, N_ω is the photon density in the active medium. Ignoring spatial relations, in steady-state conditions we obtain

$$N_e(N_\omega) = \frac{N_{e0}}{1 + KN_\omega\tau}, \quad (4.2)$$

where $N_{e0} = j\tau/ed$ is the concentration of excess carriers with $N_\omega = 0$. This expression corresponds to the introduction of the effective lifetime

$$\tau_{эф} = \tau/(1 + KN_\omega\tau), \quad (4.3)$$

where τ_0 is the lifetime of carriers with $N_\omega = 0$. Then the effective diffusion length will be equal to

$$L_{эф} = L_0/\sqrt{1 + KN_\omega\tau} \quad (4.4)$$

and substitution into the V-I characteristic (2.1) with $d \ll L_D$ yields

$$V = \frac{\varepsilon}{e}\ln\left(\frac{j}{j_0}\frac{1}{1 + KN_\omega\tau}\right) = V_0 - \frac{\varepsilon}{e}\ln(1 + KN_\omega\tau), \quad (4.5)$$

where V_0 is the voltage across the injection contact with $N_\omega = 0$. Thus, the second term in formula (4.5) gives the change in voltage with a photon density N_ω. *If* $KN_\omega\tau \ll 1$,

$$\Delta V = -\varepsilon KN_\omega\tau/e. \quad (4.6)$$

If we incorporate the emission intensity $\Phi = \hbar\omega cN_\omega/n$, where n is the refractive index, we obtain

$$dV = -\varepsilon n\tau K\Phi/e\hbar\omega c \quad (4.7)$$

and

$$dV/d\Phi = -\varepsilon n\tau K/e\hbar\omega c. \quad (4.8)$$

If the power at the amplifier output is equal to P, and the beam cross-section is equal to s,

$$dV/dP = -\varepsilon n\tau K/e\hbar\omega cs. \quad (4.9)$$

This estimate gives the following values for small voltage variations $dV/dP \approx -0.5$ v/w with the following parameters used in formula (4.9): $\hbar\omega/\varepsilon = 6$, $n = 5$, $\tau = 10^{-9}$ sec, $K = 3 \cdot 10^{-7}$ cm^3/sec, $s = 10^{-7}$ cm^2 (for a planar AlGaAs-laser at 300 K). Since a standard power for a diode amplifier is 5-10 mw, the maximum electrical signal is 2.5-5 mw. With a gain of 10 dB the sensitivity of photoresponse for the input signal is approximate 10 times greater, i.e., near -5 v/w. It is possible to achieve a correspondingly larger sensitivity for an input signal in a resonance amplifier where a gain of ~30 dB is achieved with a certain limited maximum amplitude of the electrical signal.

We will apply certain concepts relating to the role of resistance modulation to the amplifier conditions. This means that in samples with an anomalous variation ΔR_d at the lasing threshold, an analogous anomaly of the photoresponse sign may be anticipated in amplification conditions as well (i.e., the "photoelectric" anomaly: the sign of the voltage coincides with the sign of the photo-EMF). However, with a reduction in the pumping current, in principle, the anomalous photoresponse should be replaced by normal photoresponse: a decrease in V with an increase in emission power. A more strict photoresponse model for a laser amplifier will include an accurate consideration of the spatial variation in emission intensity, which is more significant than in lasers. This is related to the increase in intensity from gain and produces an intensity several orders of magnitude higher at the output (in traveling wave conditions). When approaching threshold, i.e., in the conversion to resonant amplification, the inhomogeneity factor becomes similar to that in lasing conditions. We may note that in the inhomogeneous case with a regular effect ($dV/dP < 0$) the primary contribution to the electrical signal will derive from the region with maximum intensity, since it will shunt other regions of the injection contact.

5. A QUASISTATIC SIGNAL IN A LASER WITH AN EXTERNAL CAVITY WITH MIRROR MOTION

We have already identified the possibility for optoelectronic playback of optically-recorded information on a data medium if laser emission is reflected off of this medium onto the active region of the laser. The corresponding signal may be estimated by formula (2.15) if the threshold current values with critical states of the medium (reflection coefficient) are determined. An injection laser with an external mirror may also be used as a detector of longitudinal (on the cavity axis) displacement of the external mirror, since this displacement modulates the optical feedback sign. It is assumed that the feedback is coherent, i.e., the length of the external circuit falls within the coherence length of the emission. Here, the reflected signal interferes with the field inside the cavity, and either amplifies or suppresses its oscillations. In the noncoherent case, such as in multimode lasing, the reflected signal amplifies certain modes and suppresses other modes, so that the effect averages out, as a rule, in favor of amplification.

If positive feedback is achieved at a distance L_3 from the external mirror, displacement of the mirror by $\lambda/4$ changes the optical path to the external feedback circuit by half waves, which will correspond to antiphase modes in the resonator, i.e., negative feedback. The lasing conditions are influenced by the level of feedback. We will reduce the external mirror effect to an addition δR to the reflection coefficient of the end face of the diode, which is positive with a synchronous phase reflected signal and is negative with an antiphase reflected signal. We will represent the threshold current as

$$I_t = j_0 S + \frac{S}{\beta}\left\{\alpha - \frac{1}{2L}\ln\left[R\left(R \pm \delta R\right)\right]\right\}, \tag{5.1}$$

where j_0 is the inversion threshold current density, S is the area of the diode, β is the differential gain, α is the internal optical loss factor, L is the length of the diode resonator, and R is the reflection coefficient of the end of the diode; the variable sign in front of δR designates two cases when the length of the external feedback circuit changes by $\lambda/4$. Variations in the threshold current amount to

$$\Delta I_t = \frac{S}{2\beta L} \ln \frac{R - \delta R}{R + \delta R}, \tag{5.2}$$

while the regular electrical signal will be equal to

$$\Delta V = \frac{\varepsilon}{e} \ln \left(1 + \frac{S}{2\beta L I_{t0}} \ln \frac{R - \delta R}{R + \delta R}\right), \tag{5.3}$$

if the operating current is higher than the upper threshold I_{t0}. With $\delta R/R \ll 1$ formula (5.3) is reduced to

$$\Delta V = -\frac{\varepsilon}{e} \frac{S \delta R}{\beta L R I_{t0}}. \tag{5.4}$$

For example, with ε/e = 26 mv, I_{t0}/S = 1 ka/cm^2, $\delta R/R$ = 0.1, β = $2 \cdot 10^{-2}$ cm/a, we have ΔV = 5.7 mev and, since ΔL_3 = $\lambda/4$, i.e., ~0.2 μm, the sensitivity of the sensor to small displacements is

$$\Delta V/\Delta L_3 = 28.5 \text{ mv}/\mu\text{m}. \tag{5.5}$$

This corresponds to a capability for submicron measurements with appropriate stabilization of lasing conditions. For example, if the voltage noise makes it possible to record ~10 μm, displacements of ~0.35 nm may be measured, i.e., on the level of a single atom layer.

Such a sensor (with a collimator or as a simple compound resonator [24]) may provide information on inertial or thermal displacement, or striction displacement (for example, for measuring a magnetic field), and to record vibrations, to operate as a microphone or hydrophone, etc. The primary technical difficulty lies in stabilizing pumping conditions and laser conditions along the length of the external cavity.

The first experiments to investigate the sensitivity of an injection laser to the position of an external flat mirror in the immediate vicinity of the end of the laser diode were performed in studies [24-27]. These studies showed that spectral changes were observed at a distance of only a few millimeters from the diode to the external mirror; the majority of experiments on spectral selection were performed at a distance $L_3 \approx 50$ μm. At short distances, perturbations in lasing conditions become significant (up through loss of lasing). Study [25] measured variations in the threshold current for the case of continuous tuning of the compound resonator by micrometer displacements of the external mirror (on the order of up to 10% of the threshold current without an external mirror).

The electrical detection of mirror vibrations (in the longitudinal, i.e., on the cavity axis) may also be based on measuring the magnitude and sign of external feedback, since oscillations in the voltage across the diode are related to these phenomena. The spectral composition of the electrical response changes as a function of the amplitude of vibrations in the mirror; specifically, even order vibration frequencies arise (harmonics of the fundamental frequency f). This is caused by the feedback passing through the maxima and minima on one oscillation cycle. If the amplitude of the vibrations is less than $\lambda/4$, the frequency of voltage oscillations is equal to the vibration frequency.

In addition to this signal, another signal may appear simultaneously in the diode; this second signal is caused by the beating of the natural field in the cavity with the reflected wave that has experienced a Doppler shift upon reflection off of the moving mirror. this wave consists of frequency-modulated optical oscillations with a frequency deviation determined by the maximum mirror velocity and with a frequency equal to the vibration frequency. This frequency falls within the electrical signal spectrum, since it determines the periodicity in the generation of electron density perturbations in the active region due to beating.

In experiments with a vibrating external mirror as described in study [28], a signal at the vibration frequency was observed in the electrical network; this was interpreted as an autodyne effect, i.e., the result of beating with the wave that has experienced a Doppler frequency shift. A more general interpretation would account for both effects described above. These effects may easily be differentiated since the feedback modulation effect is quasistatic and is reproduced in steady-state measurements, at the same time that the autodyne effect disappears with a drop in the vibration frequency.

6. EXPERIMENTAL INVESTIGATION OF THE INCREMENTAL (DIFFERENTIAL) RESISTANCE OF LASERS INCLUDING CLEAVED COUPLED CAVITY LASERS

As noted previously, study [14] has investigated the incremental (differential) resistances of laser diodes with various deviations from linearity on the watt-current characteristics. It was demonstrated that in samples with a normal reduction in R_d it is possible to qualitatively or semiquantitatively characterize the type of nonlinearity by the form of the $R_d(I)$ function at a current above threshold current. Moreover, samples of the anomalous behavior of $R_d(I)$ were identified where the threshold jump is positive (i.e., R_d increases beyond threshold). This study uses the same technique for measuring the differential characteristics to investigate anomalies of this type and experiments are performed using more advanced laser samples with transverse stabilization of the spatial laser channels together with "C-3 lasers."

The comparatively simple methods of transverse stabilization (or lateral optical confinement) in stripe injection lasers includes designs incorporating guiding expansions in the active waveguide along the stripe axis in the structure, since the expansion causes a favorable (waveguide) configuration of the effective refractive index (i.e., it introduces continuous lateral variations corresponding to a two-dimensional waveguide). This makes it possible to enhance the linearity of the watt-current characteristic of the laser by increasing the power at which perturbations of stable lasing occur (displacement of the channel, the appearance of new modes, self-focusing, etc.). Our new example will be a terraced heterostructure (type TR) fabricated analogously to a regular double planar heterostructure with the single difference that steps (a few micrometers in height) are first etched on the substrate; thus, this is a sample structure grown on a relief substrate. The guide expansion is formed near this step by displacement of the step (occurring simultaneously with flattening) in the subsequent layers of the heterostructure during the burying process. The expansion (from 0.2 μm in the planar sections to 0.4-0.5 μm near the steps) runs parallel to the step in the (110) substrate. The stripe contact is applied directly over the expansion in the same direction. A differentiating characteristic of such a heterostructure is that the mode related to the guide expansion is characterized by a certain tilt of the polarization plane of the laser emission (the *TE*-wave) with respect to the base plane of the substrate in accordance with the inclined geometry of the expansion. This deviation may serve as an indicator of the guiding action of the expansion, since the correlation between the power at the first injection of the watt-current characteristic and the angle of deviation of the polarization plane has been identified.

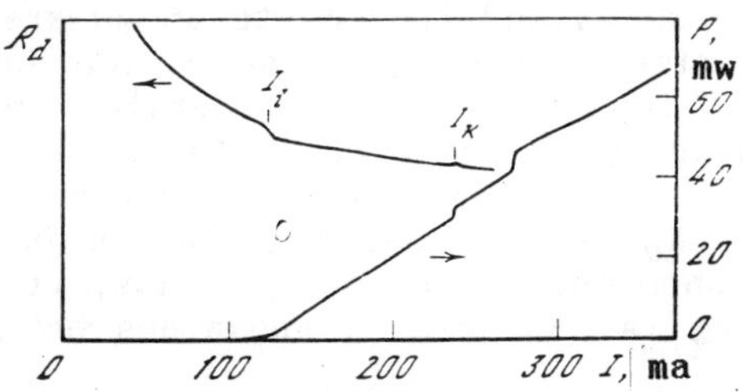

Fig. 5. The watt-current characteristics (right scale) and the differential resistance of a terraced GaAlAs laser diode (left scale)
Contact width: 10 μm, cavity length: 325 μm, room temperature.

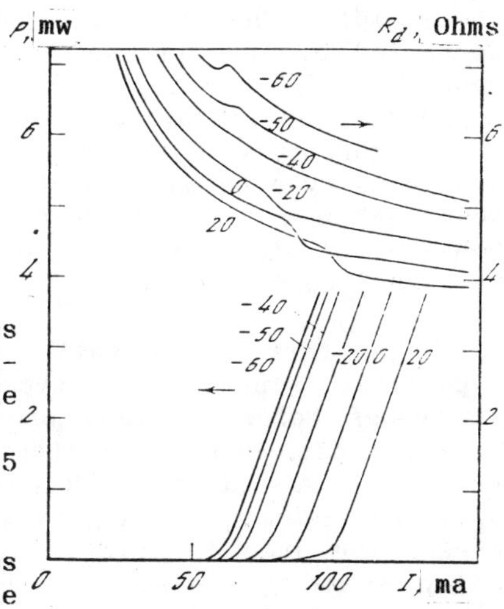

Fig. 6. The watt-current characteristics (below) and the differential resistance of a stripe contact GaAlAs laser diode at various temperature (°C) demonstrating the change of sign of the threshold differential resistance effect at -40°C.

Fig. 5 gives the watt-current characteristic $P(I)$ of the terraced diode; this characteristic is a piecewise-linear characteristic

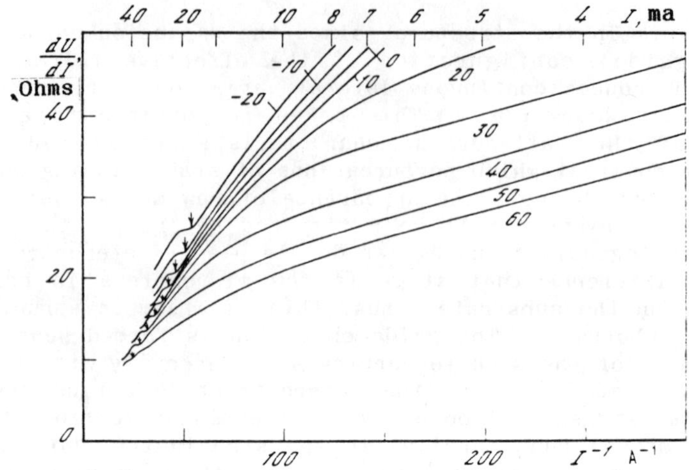

Fig. 7. Standard nonlinear dU/dI versus $1/I$ relations in a sample with a "positive" differential resistance anomaly
Stripe contact laser: 15 μm in width; threshold at room temperature: 25 ma. The figures next to the curves give the temperature in °C.

with staircase nonlinearities that differ from the regular kinks found in the characteristics of the planar diodes. The first linear section beyond threshold is rather broad, up to approximately 30 mw (at twice threshold). The steps from 30 to 32 mw and from 40 to 45 mw are accompanied by an increase in the size of the radiative region in the near field of the laser (the addition of new bright spots). In the upper section (45-64 mw) the linearity is noticeably disrupted. The lasing threshold I_t on the $R_d(I)$ differential resistance curve is accompanied by a normal reduction in R_d, at the same time that a jump in power (when $I = I_K$) produces an anomalous peak in R_d. Thus, regular behavior of R_d at moderate pumping is replaced by anomalous behavior at significant pumping.

The form of the watt-current characteristic shown in Fig. 5 is typical of lasers of this type with enhanced emission power (the maximum power before thermal breakdown is 50-100 mw). In samples with relative displacement of the active (subcontact) stripe and a guide expansion, a reduction in power was observed at the first step (in I_K); the slowing of its growth up through saturation more clearly proceeded the jump in power. In other words, the form of the nonlinearity became more similar to the form of a regular kink.

Fig. 6 also gives the characteristics of another sample with comparatively linear $P(I)$ relations in a temperature range of -60 to +20°C. In this case the evolution of a $R_d(I)$ relation from normal at +20°C (i.e., with reduction at threshold) to anomalous at -60°C is illustrated. Near -40°C the threshold feature of $R_d(I)$ is virtually compensated. We note that the appearance of the anomaly is accompanied by an increase in R_d, which, may be associated with an increase in the resistance of the bulk region of the semiconductor in which it is sensitive to injection (i.e., in the active layer or in the adjacent emitter layer). It is not likely that in this case the anomaly is not likely that in this case the anomaly is related to carrier leakage to the emitter, since such leakage must be eliminated with cooling. Evidently the cause of the anomaly is the freezout of conduction of the undoped active layer. We note that among the many examples of anomalous behavior of R_d we may find both cases of an increase in the anomalous peak with an increase in temperature as well as cases of an opposite temperature dependence (see Fig. 6). We will now focus on the following regularities in the behavior of anomalies in $R_d(I)$:

- a positive jump in R_d at threshold is most often observed in samples that differ from analogous samples (both in geometry and dimensions) in their higher absolute values of R_d (i.e., with a larger contribution from the regions of modeled resistance);

- A nonlinear behavior of the $R_d(1/I)$ relation is characteristic of such samples, as demonstrated in Fig. 7 in the subthreshold measurement region; This makes it impossible to identify the parameter ε relating to the natural V-I characteristic of the injection contact. A higher value of ε^* will follow from the slope of $R_d(1/I)$ such that the ε^*/kT relation may be as high as 5-10. These observations correspond, at least qualitatively, to anticipated behavior based on a model of a diode with modulated cubic resistance (see Section 3). Indeed, the high value of R_d and its nonlinearity on the $R_d(1/I)$ graph indicate a significant contribution of modulated cubic resistance to the nonlinear resistance of the diode.

We may also expect that there is a relation between the anomaly of $R_d(I)$ and the laser threshold current, since the carrier leakage to the emitter may both be the cause of the anomaly and the cause of exceeding the threshold at the same time. However such a relation breaks down if the anomaly in $R_d(I)$ is caused by modulation of the resistance of the active layer. In the standard case, the modulated resistance area of the p-emitter is approximately ten times greater than the thickness of the active layer and, consequently, with leakage to the p-emitter, the emitter will produce the dominant contribution to the modulated cubic resistance. In this case, leakage to the p-emitter is likely to play a role in samples with a significant anomaly in the differential resistance at threshold. Evidently this viewpoint is confirmed by data regarding the anticorrelation between the temperature parameter T_0 (characterizing the "slowness" of the exponential temperature rise of the threshold current) and the jump ΔR_d in the differential resistance at threshold (at ~25°C). Fig. 8 shows experimental points for several series of AlGaAs laser diodes with different p-emitter compositions (with AlAs concentrations from 0.2-0.35) as well as different doping levels (10^{17} - $3 \cdot 10^{18}$ cm^{-3}). It is clear that with positive values of ΔR_d (i.e., in the presence of the threshold anomaly) comparatively low values of T_0 (below 120 K) are realized, at the same time that high T_0 are typical of samples with normal (negative) ΔR_d. Thus, the method for electrical diagnostics of injection lasers may be used successfully for analyzing leakage.

The next stage in the investigation was to observe the incremental (differential) resistance in cleaved coupled cavity lasers (see the insert in Fig. 9). The particular case shown in Fig. 9 applies to a pair of laser diodes (designated *A* and *B*) with a planar stripe structure and a contact width of 10 μm separated by a gap 7 μm in width. The cavity lengths were 62 (*A*) and 90 μm (*B*). Emission was observed from diode *A*. Separate pumping of the diodes produced watt-current characteristics as shown by curves 1 and 2; separate pumping produced more complex characteristics, one of which is shown by curve 3. This corresponds to the fixed current through diode *B* (70 ma, near its individual threshold) as well as the a.c. through diode *A*. Curve 1 is typical of lasers of this type and has a step at 60-70 ma. Unlike this case, curve 3 has several small steps reflecting the relative detuning of both coupled cavities, i.e., the variation in the Q of the coupled cavity. The power peaks correspond to tuning to a com

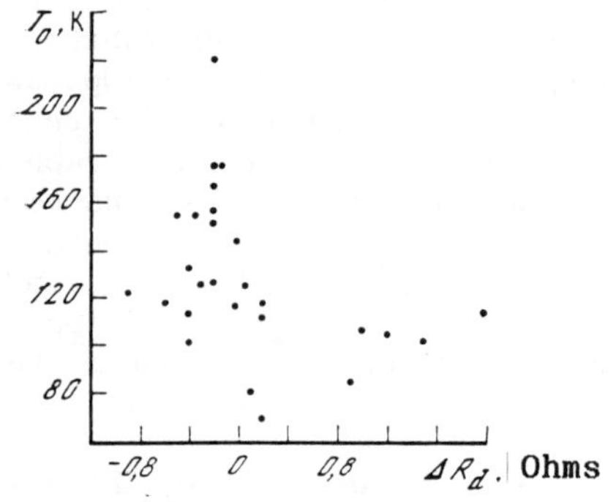

Fig. 8. The anticorrelation of the threshold jump in the differential resistance ΔR_d and the parameter T_0 of the temperature dependence of threshold current (0-60°C) in stripe GaAlAs lasers

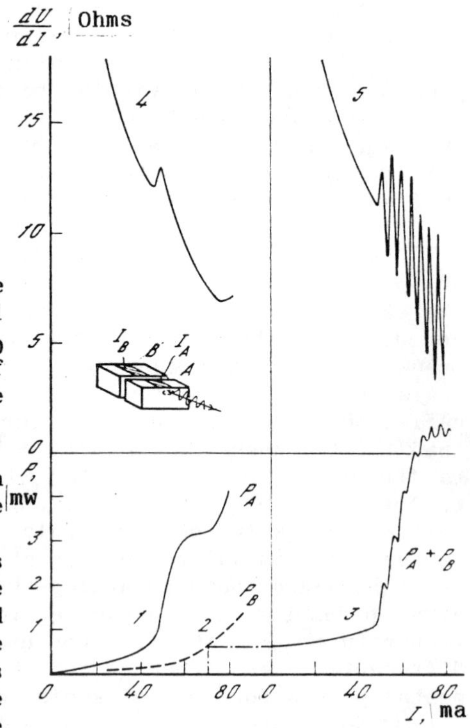

Fig. 9. The characteristics of a coupled compound cavity laser (double diode, see insert)
1, 2 - watt-current characteristics of diode *A* and *B* for separate pumping, respectively (power measured on the side of diode *A*); 3 - the same characteristics for simultaneous pumping fixed current through diode *B*, measured current *I* through diode *A*); 4, 5 - differential resistance of diode *A* at I_B = 0 (4) and 70 ma (5)

mon mode, at the same time that the minima indicate cavity detuning. This evolution is explained by the shift of the spectral modes of the cavity of diode *A* with a change in pumping current due to the influence of the carriers and temperature on the refractive index.

Curve 4 shows the differential resistance dependence of diode *A* at $I_B = 0$. A peak is visible at the threshold corresponding to the case of anomalous behavior. Curve 5 gives the differential resistance corresponding to pumping conditions on characteristic 3. At least eight differential resistance peaks are observed corresponding to the steep linear sections of the watt-current characteristic. Minima in R_d are observed in the negative slope regions; these pass below the extrapolated level of R_d for the injection contact outside of lasing conditions. This evidently indicates that in these sections the carrier concentration increases more rapidly than for the case of an increase in current without lasing; this in turn indicates a simultaneous increase in the carrier lifetimes (due to a reduction in emission intensity). The span of oscillations in R_d is rather large: up to 7 ohms with an absolute mean of 7-10 ohms.

Since the peaks in R_d correspond to a tuning condition, it is easy to use electrical diagnostics for automatic tuning of the cavities of a compound laser in order to maintain the tuned state with variations in temperature and pumping. The error signal may be taken from the differential signal of diode *A*, while tuning may be controlled in the current network of diode *B*.

The experiments described in this section continued research on the electrical diagnostics of operating conditions in injection lasers. The features of various types of lasers were demonstrated and additional experiments were carried out on the positive anomaly of the differential resistance at threshold and beyond threshold. A relationship was demonstrated to exist in C^3-lasers between the electrical characteristic and the watt-current characteristic by using the capability to significantly change both characteristics by varying operating conditions using two degrees of pumping freedom.

7. EXPERIMENTAL INVESTIGATION OF THE POSSIBILITY OF ELECTRICAL DIAGNOSTICS OF LASER DIODE OPERATING CONDITIONS

7.1. Methodological issues

As demonstrated in Section 4, laser diodes may be used both as emission oscillators and as amplifiers or emission detectors. A block diagram of a unit used in the electrical diagnostics of operating conditions of a laser diode is shown in Fig. 10. Laser diodes 2, 4 are planar GaAlAs/GaAs heterolasers with a mesa-structure and active region dimensions of $250 \times 15 \times 0.2$ μm. These regions were attached on the p^+-side to a copper holder and operate in pulsed and CW conditions at room temperature. In pulsed conditions, the G5-15 (1) generator was used to pump laser diode 2; the pulse duration was varied from 3-

5 μs with a pulse repetition frequency of 10 kHz. Emission from laser diode 2 was focused on incoupling mirror of laser diode 2 with a certain magnification to provide coupling of 10^{-2}-10^{-3} by means of microlens 3 with an aperture of 0.4. Laser diode 4 was pumped by a stabilized d.c. rectifier whose values could be varied from 0 to levels exceeding the lasing threshold. Oscillograph S1-72 (6) was used to observe the voltages across laser diode 4; the signal was fed to oscillograph S1-72 through electronic amplifier V6-4 (5). Electrical diagnostics of the operating conditions of laser diode 4 were performed in this case (identification of the inversion threshold, traveling wave conditions, regenerative amplification, electrical signal lasing threshold; the level of the latter changes sign with an increase in pumping current from positive in photodiode conditions to negative in amplification conditions), i.e., laser diode 4 functioned as a laser amplifier-monitor. Laser diodes 2 and 4 were selected from a series of samples based on their absence of anomalies in the pumping current dependencies of the differential resistance (during the transition through lasing threshold). Two laser-oscillators (LG-1) and (LG-2) function as laser diode 2; these had an emission wavelength and threshold currents of 836 and 823 nm and 250 and 140 ma, respectively. Laser diode 4 had a threshold current 230 ma and a natural lasing wavelength of 829 nm. Since this study is devoted to the electrical diagnostics of operating conditions of laser amplifier-monitors, no special measures were taken to optimize its design for maximum optical gain or for maximum signal-to-noise ratio. In order to achieve regenerative amplification, reflection off of the natural surfaces of the laser crystal was used. This simplified the experimental conditions, since all operating conditions of the laser amplifier-monitor described above could be recorded with d.c. pumping.

1 - signal pulse source;
2 - transmission laser;
3 - lens;
4 - receiver laser;
5 - electronic amplifier;
6 - oscillograph;
7 - fiber lightguide

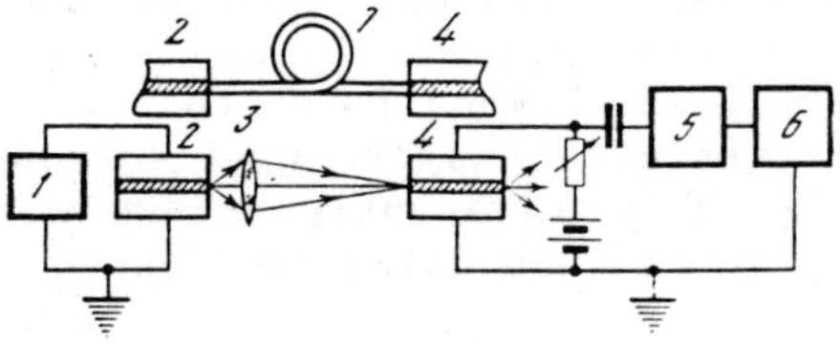

Fig. 10. Experimental configuration for the electrical diagnostics of operating conditions of laser diodes and for signal transmission on a fiber optic lightguide.

After we succeeded in obtaining an adequate electrical signal in the laser amplifier-monitor network (the interpretation of this signal in various pumping conditions and its use for diagnostics of amplification conditions will be described in detail in the next section), the problem of investigating the possibility of using the saturation effect of the voltage applied to the laser diode for a two-way optical communications system through a fiber optic lightguide with planar heterolasers used as the terminal devices was discussed. In this regard, a device whose block diagram is given in Fig. 10 was modified in some details. We replaced microlens 3 with multimode gradient fiber

1 - laser diode;
2 - microlens;
3 - semitransparent wafer;
4 - disk with data medium;
5 - photodiode;
6 - d.c. power supply;
7 - wideband amplifier;
8 - oscillograph

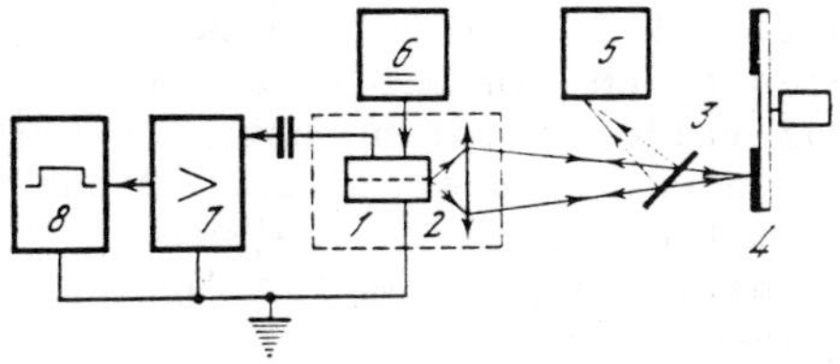

Fig. 11. Experimental configuration for investigating electrooptic signal.

optic lightguide 7 1.5 m in length with a core diameter of 50 μm and a numerical aperture of 0.2. The ends of the lightguide were connected by means of spark discharge to improve matching to the lasers. The diameter of the end microlenses was 150 μm with an optimum distance between the microlens and the mirror face of the laser diode of ~80 μm. The emission coupling efficiency to the lightguide was ~0.5. The emission spot for emission exiting the lightguide was ~30 μm on the end mirror of the laser diode. Since lasers 2 and 4 were selected to have close emission wavelengths (within the amplification band of ~kT), they were capable of establishing an optical link by introducing feedback (in our case a lightguide); as a result, two-way signal transmission was achieved in the coupled system. Fig. 10 gives the optical scheme for one-way transmission; laser diodes 2 and 4 changed places for transmission in the opposite direction. Signal pulses 3-5 μs in duration with an amplitude somewhat greater than threshold and amounting to 60 ma (at room temperature) were injected to the transmission laser. The optical signal striking the narrowband layer of the receiver laser had a power level 0.5% of the emission power of the transmitting laser. This low level resulted from the poor emission coupling efficiency to the receiving laser from the multimode fiber optic lightguide. The a.c. voltage component ΔU at the receiver laser was measured for various d.c. bias levels for signal recording.

The optoelectronic reading conditions, i.e., obtaining an electrical signal at the laser diode by information modulation of the Q of the external portion of the cavity (see Section 1 of this article) was investigated using the experimental assembly whose block diagram is shown in Fig. 11. Laser diode 1, microlens 2 and modulator 4 form a laser with a Q-switched external cavity. Planar GaAlAs/GaAs heterolasers with a fine mesa-structure were used as the laser diodes; these had an active region thickness and width of 0.2 and 10-15 μm, respectively, and operated in CW conditions at room temperature at 830-860 nm.

The studies were carried out for a number of samples from various series; the depth of the active region for the number 2, 6 and 28 samples was 1 μm below the etching depth of the channel in the mesa-structure (for sample number 4 the active region was 0.3 μm deeper); hence, in sample number 4, the depth of penetration of the field in the transverse direction is comparable to the thickness of the p-emitter material in the mesa-structure channel, which creates additional losses in the periphery regions of the stripe contact and causes a jump in the effective refractive index and the effective lateral optical confinement. A LOMO 10X0.4 microlens (2) was used to receive emission from laser diode 1 and to focus this emission on the information medium. The emission reflected off of the surface of the information medium was to be reflected back to the active region of the laser diode.

Information in the form of dark and light bands was applied to film attached to the modulator (a disk 10 cm in radius made from duralumin and rotating at a speed of 300 rpm), i.e., the information took the form of absorption and reflection bands with a film reflection coefficient R_H ~ 3% and a width of 70 and 40 μm, respectively. Such a design made it possible to obtain a signal (i.e., a change in the voltage across the laser diode resulting from the steady-state switching of the Q of the cavity, which was accompanied by modulation of the threshold current density) in the form of a series of pulses 6 μs in duration with a pulse repetition frequency of 200 kHz. The optoelectronic was amplified by the U2-8 electronic amplifier (8) and was displayed on the S1-72 oscillograph (9). In order to monitor the emission power, semitransparent plate 3 was introduced in the external cavity; this plate directed the reflected emission onto photodetector 5; the FD-7K silicon photodiode was used as this photodetector. In a number of experiments, when a significant amplitude of the optoelectronic signal was required, a totally-reflecting mirror was used as the external cavity. In this case, a darkened disk with 16 holes was used as modulator 4.

The dimensions of the reading spot on the film which determine the system resolution depend on the dimensions of the active region of the laser, the parameters of the lens and the distance d_1 from the film to the laser. In our experiments the dimensions of the reading spot on the film were 10 × 200 μm with d_1 = 30 cm and 4 × 80 μm with d_1 = 13 cm. In order to investigate the influence of temperature on the optoelectronic signal, the laser diode was attached to a copper holder which then could be heated from room temperature to 80°C (a U3 thermostat was used manufactured by MLW, the German Democratic Republic).

The optical measurements require special mention. The problem is that the width of the active region of the stripe heterolaser is 10-15 μm with a width of 0.1-0.3 μm. At the same time, focusing of emission from one laser diode onto another requires that the emission strike the active region of the second laser (for the case of the electrical diagnostics) and that this emission be reflected off of the data medium back to the active region of the initial laser (for the case of optoelectronic reading). The small thickness of the active region of the laser imposes difficult requirements on the optical tuning accuracy of the system, all of whose elements (laser diode, lenses, lightguide, modulator) were attached to special supports and were mounted in optical mounts. The latter had a micrometer adjustment scale and were mounted on optical benches fastened to a solid support plate (steel); The plate lay on a thick porolon cushion in the optical bench.

Tuning was performed in the following manner. First the mirror faces of the diode (or diodes) were placed perpendicularly to the optical axis using the LG-55 helium-neon laser. Then a modulator or another laser diode and lens were also placed on this axis. Further tuning was performed in conditions where a pulsed current was passed through the laser diode and the diode operated in spontaneous emission or superluminescence conditions. Approximate tuning could be achieved by visual monitoring and by combining the incident signal with the reflected signal or by projecting the image of the active region of one laser onto another laser by small rotations of the laser diode, the modulator or the lens. For the case of optoelectronic information reading, precise tuning was achieved by employing feedback, i.e., at a fixed pumping level, the devices were tuned to maximum emission power or minimum threshold current. The optimum tuning criterion was maximum optoelectronic signal level with a fixed pumping level.

An optical signal with two coupled laser diodes for electrical diagnostics of the operating conditions of one diode was tuned in an analogous manner. In this case, a small current is passed through both diodes to generate spontaneous lasing and tuning was performed by projecting the image of one laser onto the luminescence spot on the mirror of the other laser. In this case the precise tuning criterion was maximum negative signal level (rather than the photocurrent level, since the photocurrent may reach its maximum value outside of the stripe contact in the active region). The dimensions of the luminescent spot after the focusing lens was measured by means of a helium-neon laser; these measurements were carried out with an expanded mirror size of the laser diode, since if we know the magnification factor, it is easy to determine the size of the emission spot.

For the case of two laser diodes coupled by a lightguide with microlenses at both ends, tuning did not present particular difficulties. In the first stage it was necessary to match the first laser diode to the lightguide containing a microlens at the end; the other end of the lightguide contained a photodiode, and optimum matching was provided by achieving a maximum photocurrent from this photodiode. For matching of the other end of the lightguide to the second laser diode, a small current was passed through this diode to observe the luminescence region. Then the second laser diode and the lightguide were tuned so that emission from the first laser coincided with the luminescence region of the second laser. Precision tuning was based on the negative polarity signal from the second laser. The dimensions of the luminescent spot in this case was measured by scanning the end of the lightguide in a direction perpendicular to the *p-n*-junction.

7.2. Laser diodes in optoelectronic reading

We know that for reading information from digital disk systems, it is possible to use a light-emitting diode in superluminescence conditions; this device is integrated with the photodetector [29]. We may also use a Q-switched injection laser that employs an external cavity with a separate photodetector [30]. We employed the unique property of an injection laser to function simultaneously as a source and receiver of natural emission reflected off of the information medium. This optoelectronic reading method [31], as noted previously in Section 1 of this article, is based on the saturation effect of voltage applied to the injection contact in the laser diode during the transition through the lasing threshold [2-4]. A block diagram of the experimental assembly is shown in Fig. 11. Fig. 12a gives an oscillogram of an optoelectronic signal observed in reading information in the form of absorption (with emulsion) and reflection (without emulsion) bands on film (Fig. 12b). The signal amplitude was at a maximum at a current of 112 ma corresponding to the natural lasing threshold, amounting to 0.5 mw. With reflection off of the duralumin surface of the modulator disk ($R_H \approx 50\%$), the amplitude of the optoelectronic signal was ~3 mv.

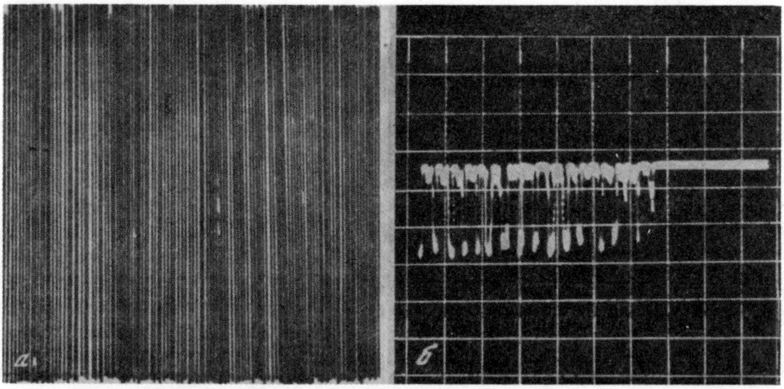

Fig. 12. Oscillogram of the optoelectronic signal (a), recorded from reading information from medium (b) in the form of absorption bands (with emulsion) and reflection bands (without emulsion) on film 13 cm from heterolaser
Pumping current 112 ma. X-axis graduated in 50 μs, y axis in 0.2 mv.

An investigation of the change in the amplitude of the optoelectronic signal with pumping current demonstrated that the behavior of the signal is closely related to the watt-current characteristic, the power increment and variations in the differential resistance of the diode $R_d = dU/dI$. Fig. 13a gives the watt-current characteristics for natural lasing (1), with external feedback using a flat mirror with a reflection coefficient of 75% 30 cm from the end of the laser diode (2) and the pumping current dependencies of the amplitude of the optoelectronic signal (3) as well as the pumping current dependencies of the differential resistance of the laser diode (4) for sample number 4. It is clear from the figure that ΔU has a negative polarity, while the differential resistance R_d drops off sharply near the lasing threshold. The optoelectronic signal appears at the lasing threshold of a laser with an external cavity I_t^1 and increases to a certain value ΔU_0 at the natural lasing threshold I_t; a further increase in the pumping current causes the amplitude of the optoelectronic signal to stabilize, remaining equal to ΔU_0.

An analysis of the watt-current characteristics of lasers in natural lasing and lasing with an external cavity demonstrates that the amplitude of the optoelectronic signal is stabilized when these watt-current characteristics are parallel. The curve showing the pumping current dependence of the amplitude of the optoelectronic signal traces the curve of the difference ΔP of emission power of a laser with an external cavity and in natural lasing conditions. Fig. 14 shows the variations ΔP and ΔU with current for sample number 6 whose watt-current characteristics demonstrate a behavior analogous to that of sample number 4 (see Fig. 13a) for various temperatures. The maximum values of the optoelectronic signal were not observed at the natural lasing threshold, but rather at a current where the maximum ΔP is observed. An analogous correlation of ΔP and ΔU existed in all test samples in which the form of the watt-current characteristic changed when feedback was applied.

Experiments have demonstrated that an optoelectronic signal of positive polarity was observed in a number of samples; for such laser diodes, the differential resistance near the lasing threshold increased significantly. An example of such behavior is demonstrated by Fig. 13b showing the characteristics of sample number 2. It is clear from this figure that the optoelectronic signal has positive polarity and is not stabilized when the pumping current exceeds the threshold level; the differential resistance of the diode increases sharply in the transition through the lasing threshold. It would be appropriate to mention that in this case as well, the sign of the optoelectronic signal retraces the sign of differential resistance. Thus, the optoelectronic signal quantity ΔU may be both positive (Fig. 13b) and negative (Fig. 13a); here, in the transition through threshold, the differential resistance of the laser diode R_d demonstrates sharp changes, increasing in the first case and decreasing in the second case.

Samples whose watt-current characteristics had clearly expressed nonlinearities were also investigated; changes in the sign of the jump in differential resistance at pumping currents exceeding the natural lasing threshold were also observed in these samples. Typical behavior for this series of samples is demonstrated by laser diode number 28 whose characteristics are shown in Fig. 15. It is clear from this figure that a complex change in the polarity of the optoelectronic signal and the sign of the jump in differential resistance occur; the changes in both quantities follow the same path.

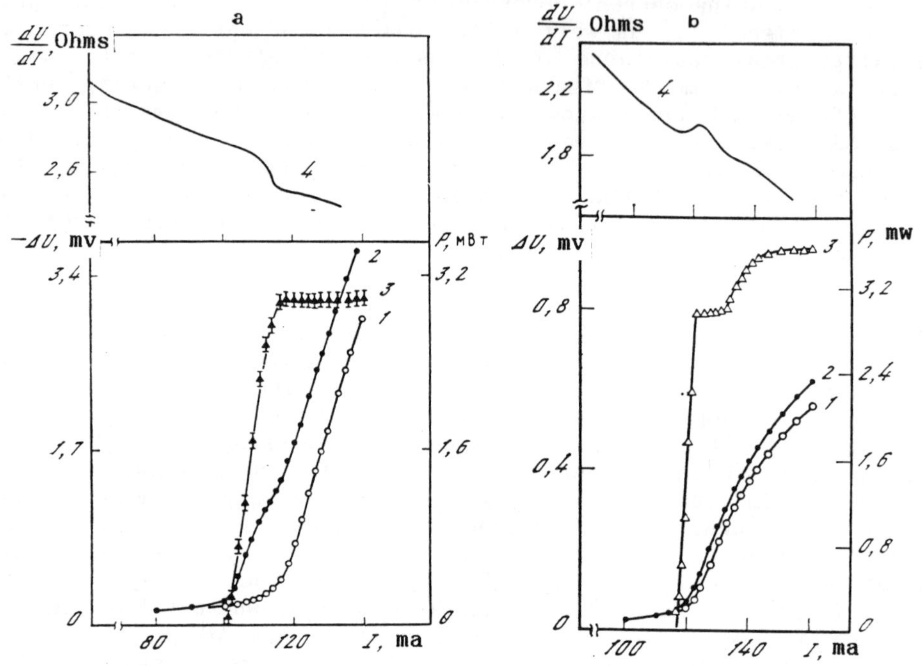

Fig. 13. Watt-current characteristics (1, 2) and the pumping current dependencies of the optoelectronic signal amplitude (3) and differential resistance of the laser diode (4) for heterolaser number 4 (a) and number 2 (b)
1 - natural lasing; 2 - with external feedback using a flat mirror with a reflection coefficient of 75% at a distance of 30 cm. Modulator chopping frequency: 1 kHz.

These experiments demonstrated the possibility of obtaining a rather strong signal (approximately 3 mv) in the electrical circuit of the laser diode for optical information reading with a surface reflection coefficient on the information medium of ~50%. In this stage of research, the noise was determined by the amplifier, and the signal-to-noise-ratio was $5 \cdot 10^2 - 5 \cdot 10^3$. The optical scheme which utilizes mechanical motion of the information medium may be the fundamental source of noise and interference. Specifically, lasing conditions are sensitive to the external length of the optical feedback circuit, i.e., to the distance from the end of the laser diode to the reflecting surface. This sensitivity may be used to observe irregu-

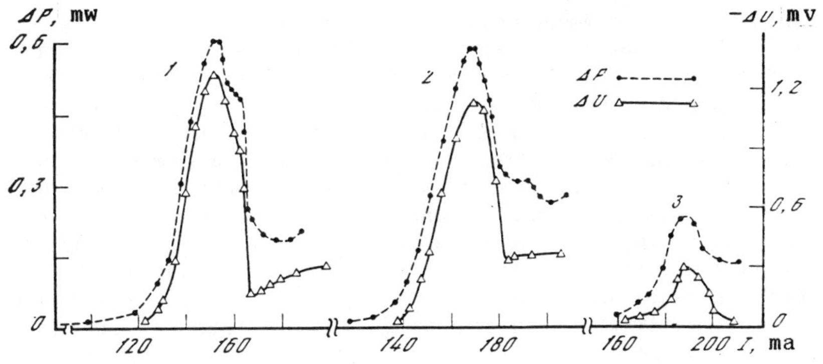

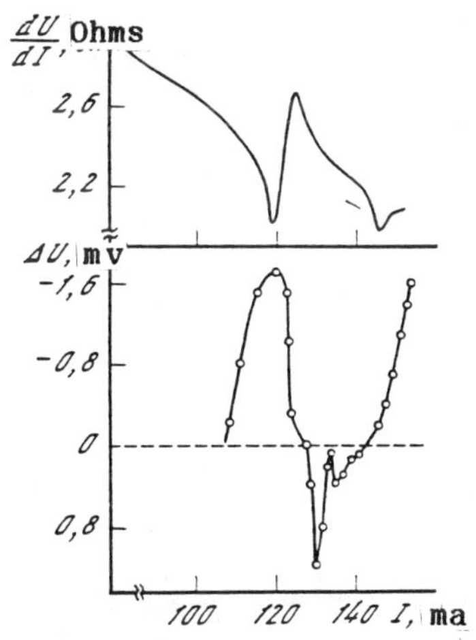

Fig. 14. Emission power and optoelectronic signal relations for heterolaser number 6 at various temperatures
1 - T = 16°C; I_t = 135 ma;
2 - T = 32°C; I_t = 148 ma;
3 - T = 45°C; I_t = 168 ma

Fig. 15. Variations in the amplitude of the optoelectronic signal and in the differential resistance of heterolaser number 28.
Threshold current: 120 ma

larities in this surface or to read information written in the form of geometric variations in the surface. In these designs, however, vibrations of a disk containing information may be a source of noise. In order to eliminate this effect it is preferable to use lasers with a coherence length less than the external coupling length; specifically, multimode planar strip heterolasers.

In concluding this subsection we note that the relationship between the optoelectronic signal and the differential electrical characteristics highlights a criterion that may be used to select laser diodes for optoelectronic information reading applications.

7.3. Laser diodes in amplification

Amplification conditions in a laser diode, depending on the input signal level, are characterized by high or low gain saturation, which in turn is accompanied by a reduction in the difference of the Fermi quasilevels in the active region compared to its nonsaturated value. The investigation of optoelectronic reading conditions in the preceding section, i.e., obtaining an electrical signal at a laser diode for information modulation of the Q of the external portion of the cavity has demonstrated that the sign and amplitude of the electrical signal correlates with the differential resistance of the laser diode. Before proceeding to a description of the amplification conditions, we note that there has been increasing interest recently in using injection lasers as emission amplifiers in a regenerative scheme [32-34], particularly for retranslation of signals in fiber-optic information transmission systems and for preamplification for signal reception [34]. An attractive feature of a laser diode-amplifier is the capability to use the electrical signal to evaluate the optical signal without using photoelectric monitors [34, 36]. Since two laser diodes are normally used in narrowband laser transmitters (a laser-oscillator and a laser-amplifier possibly in an integrated configuration), the latter may also assume the function of a monitor by producing a test signal to establish the operation of the laser-oscillator. In practical circuits it is not always convenient nor advisable to optically isolate the amplifier from the oscillator (for example, in integrated optical designs). As a result both laser diodes (or elements of an integrated optical scheme) may essentially form an integrated laser with a compound resonator including two active elements.

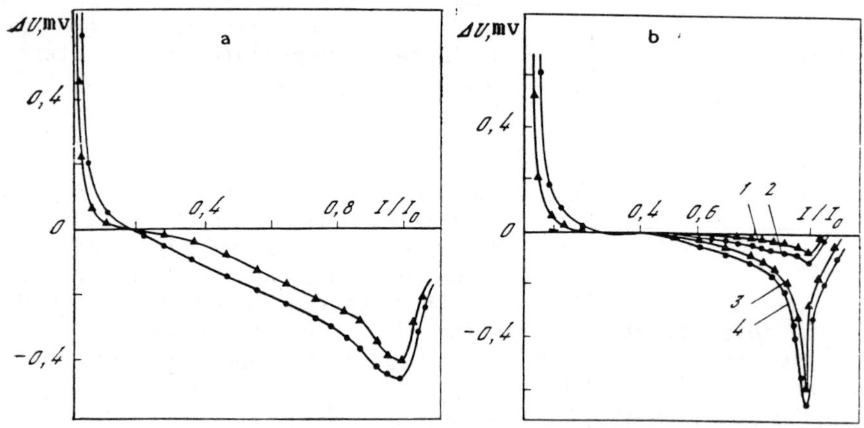

Fig. 16. The pumping current dependence of the electrical signal ΔU in a laser amplifier-monitor with optical signal injection from the LG-1 (a) and LG-2 (b).
a - LG-1 at a power level of $6 \cdot 10^{-5}$ w (1) and $2 \cdot 10^{-4}$ w (2); optical coupling coefficient: $5 \cdot 10^{-3}$; b - LG-2 at a power level of $5 \cdot 10^{-5}$ w (1); $6 \cdot 10^{-5}$ w (2); 10^{-4} w (3); $1.2 \cdot 10^{-4}$ w (4).

In this subsection we will describe results from an investigation of the variation in the electrical voltage across a laser diode in amplification (below the natural lasing threshold) as a function of pumping current and the input optical signal at lasing wavelengths 7 nm above (LG-1) and 6 nm below (LG-2) the natural lasing wavelength

(829 nm). A block diagram of the experimental set-up is given in Fig. 10. We investigated a system of relatively weakly-coupled (optical coupling on the order of 10^{-2}) laser diodes, one of which is a modulated laser-oscillator and the other is a laser-amplifier monitor. The possibility for obtaining an adequate electrical signal in the network of the laser-amplifier monitor and its interpretation form the basis for diagnostics of the operation of the laser diode. Fig. 16a shows the pulsed voltage variation ΔU in the laser-amplifier monitor with an emission pulse from the LG-1 as a function of the current flowing through the laser-amplifier monitor (for various power levels of the LG-1). At low currents the signal is positive and corresponds to the appearance of photo-EMF in the diode of the laser-amplifier monitor; then at a current of somewhat less than 20% of the threshold current the signal changes sign, which indicates that the laser-amplifier monitor is converting from the absorption of external pumping to amplification of this emission. A certain superlinear rise in the negative signal which is noticeable when approaching the threshold current is caused by the contribution of resonant amplification, particularly the accumulation of amplified emission in the cavity of the laser-amplifier monitor. In the transition through threshold the signal drops due to the competition of the saturation effect caused by the natural lasing the laser-amplifier monitor.

Fig. 16b shows analogous relations for pumping by the LG-2. Curves 3 and 4 demonstrate a sharp increase in the negative signal ΔU caused by regeneration. Unlike the curves shown in Fig. 16a the changeover point of the sign of the signal occurs at a current equal to 30-35% of the threshold current. Thus, the experimental results shown in Fig. 16 indicate the possibility of identifying the inversion threshold (the current at which the electrical signal changes sign), traveling wave and regenerative amplification conditions as well as the lasing threshold all based on the electrical signal in the laser diode. This signal changes sign with an increase in pumping current (positive in photodiode conditions and negative in amplification conditions). The maximum amplitude of the negative signal is 0.621 mv.

Near threshold the regenerative conditions introduce a certain irregularity to the signal related to the fact that in order to achieve maximum gain it is necessary to tune the wavelength of the laser oscillator to one of the longitudinal modes of the laser-amplifier monitor. However, when the pumping current of the laser-amplifier monitor changes, these modes shift, thereby disrupting resonance. Hence distortions may arise in the current dependence of ΔU related to the uncontrolled tuning and detuning of the laser-amplifier monitor and the laser oscillator. Here this problem was simplified by the fact that with long pulses the modes of the laser oscillator traverse a certain band so that resonance may occur during a portion of the pulse. It is clear, however, that there is the fundamental capability to identify regenerative conditions. These data were not compared to data on the gain of transient emission, since measuring this emission is made difficult due to the error in determining the input signal, (i.e., the coupling error). The actual gain of the optical flux is noticeable only in regenerative conditions; with low gain it is masked by emission passing the active region of the laser-amplifier monitor. The quantity ΔU in regenerative amplification was approximately 1 mv with the input power of 0.12 mw.

It may be most suitable to use this method of diagnostics for a laser diode in monitoring the operation of laser amplifiers in integrated optical designs, in laser repeaters (retranslators) (in information transmission systems) as well as a method for receiving a weak signal if it is sufficient in its amplified state for the saturation effect. Thus, it is possible to establish two-way optical communications on a single channel using the same laser diodes as signal receivers and transmitters (in the transmission intervals). The next section is devoted to a description of this capability. Here we will simply note that a laser diode operating without d.c. bias or with negative bias may receive an optical signal at a closer or shorter wavelength operating as a photodiode, while a laser diode under d.c. bias may operate most efficiently as an amplifier-monitor. In the latter case there are no optical losses and the optical signal may be detected using either method or may be directed to another section in the optical circuit. This makes it possible to implement several distributed access designs for accessing the communications link.

7.4. An investigation of signal transmission in a system of two laser diodes optically coupled by a fiber optic lightguide

The comprehensive investigation of amplification conditions (the dependence of the optoelectronic signal on the pumping current of the amplifier, the wavelength and the power of the input signal, etc.) described in the preceding two sections has demonstrated that identical laser diodes (i.e., those with similar emission wavelengths within an amplification band of width $\sim kT$) interact with optical coupling established between them. As a result it has become possible to transmit a signal in a system consisting of two planar heterolasers interconnected by a multimode fiber optic lightguide [37]. The experimental configuration was given in Fig. 10. The d.c. dependence of the electrical signal ΔU in the receiver laser diode is shown in Fig. 17 (solid line curve). The dotted line curve in Fig. 17 corresponds to the opposite transmission direction when the laser diodes change roles. It is clear from Fig. 17 that two sensitivity regions exist on the bias current axis: a photodiode region under negative bias extending to a forward current of 20-25 ma, and an amplification region with forward current through the lasing threshold and somewhat higher. In the photodiode region the signal results from the absorption and separation of electrons and holes by the field of the *p-n*-junction. In the amplification region the signal results from light amplification and the reduction in inversion population from the acceleration of stimulated transitions. Hence the electrical signal has a positive sign. In the first region ΔU has a large amplitude; the diode has a high internal resistance, while in the second region the amplitude of ΔU is reduced together with the differential resistance of the diode

Fig. 17. The d.c. dependence of the electrical signal ΔU in the receiver diode (solid line curve) The dotted line curved corresponds to the opposite direction of transmission when the laser diodes change roles. The insert shows the electrical signal in the low current region.

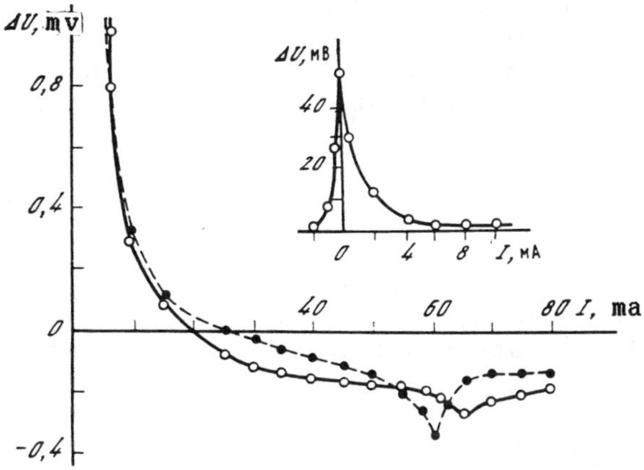

This creates certain difficulties in the electronic amplification of the signal. The peak of the negative signal near the lasing threshold is caused by regenerative amplification in the diode. We note that in the amplification region, the optical signal is not lost but is rather amplified and may be used for further transmission.

If the emission spot from the lightguide is scanned lengthwise across the laser diode, the optimum positions in the photodiode and amplification regions differ somewhat. In order to obtain a negative signal, careful tuning to the active region under the stripe contact is required since the maximum of the photodiode signal is normally observed along the periphery of the diode where no amplification signal generally is found. The negative signal is also observed in the vicinity above the lasing threshold; in other words, it is also possible to transmit a signal with electrical recording when both diodes are operating in a lasing condition. Thus, there are several possible interaction designs for laser diodes in supporting signal transmission; moreover, such transmission may occur simultaneously in both directions using a single fiber optic lightguide by changing the operating conditions of the laser diodes. The transmission laser diode may also operate as a receiver in the intervals between the signal pumping pulses. Photodiode operation is implemented with pulsed pumping with full waveform pulses (i.e., without an a.c. pumping component). A broader transmission band may be achieved by eliminating the delays in the transmission laser diode by adding an a.c. pumping component on the threshold current level. In this case the current in the interval does not return to zero and amplification reception of signals may be implemented. Finally, two-way transmission may be achieved when both laser diodes are in a coupled lasing state; these conditions depend on the currents in both laser diodes.

The experiments have demonstrated the possibility of two-way communications on a fiber optic lightguide utilizing fundamentally identical terminal devices (identical laser diodes). The transmission direction may be changed by switching the conditions of laser diodes and may be implemented in the intervals between signal transmission.

One important requirement for implementing such conditions is maintaining identical operating conditions of both laser diodes, including stabilization of identical temperatures, etc. We also note that in order to increase the coupling factor it is advisable to use single-mode light guides.

7.5. Concluding comments

The results in Section from the experimental investigation of the electrical diagnostics of operating conditions of laser diodes in amplification and lasing have demonstrated that the optoelectronic signal in a laser diode changes sign from positive to negative (with an increase in the pumping current), reaches a maximum value at the lasing threshold, amounts to ~1 mv and vanishes at a current that depends on the wavelength of the input optical signal. The possibility for electrical diagnostics of the operating conditions of a laser diode is demonstrated. Such operating conditions include photodetection, traveling wave amplification, regenerative wave amplification, locking and lasing. With optical coupling established between the two laser diodes, mutual electrical diagnostics of conditions is also possible, i.e., it is possible to obtain an electrical signal in one laser diode by modulating the current in another. The possibility for two-way transmission with identical terminal devices based on laser diodes is demonstrated.

The investigation of operating conditions of a laser diode with variable external feedback has demonstrated that in this state the threshold pumping current and voltage (or the optoelectronic signal, applied to the laser diode are modulated. The amplitude of the optoelectronic signal depends on the reflection coefficient of the external mirror and amounts to approximately 5 mv for a reflection coefficient R_H = 75%. A correlation is found between the optoelectronic signal and the differential resistance of the *p-n*-junction of the laser diode. Optoelectronic reading of information on film in the form of dark and light bands is employed as a sample practical application of the variable external feedback method; with a reflection coefficient of the film surface of R_H = 30%, a signal with an amplitude of 0.5 mv and a pulse repetition frequency of 250 kHz is achieved. The advantage of this reading method over other methods is that the laser diode operates simultaneously as an emission source and an emission receiver.

8. CONCLUSION

This study has developed a design for generating an electrical signal at the lasing threshold of an injection laser (as well as at the inclined sections of the nonlinear watt-current characteristic). Accounting for modulation of the resistance of the active layer and the emitter layer (with carrier leakage through the heterobarrier) has made it possible to explain the anomaly of the change in sign of the differential resistance at threshold in a number of samples. With both a normal and an anomalous sign of the threshold effect, the cause is saturation of the carrier concentration (as well as optical gain) at threshold. The difference is which contribution predominates: the contact contribution (normal) or the volumetric contribution (anomalous). Unfortunately these contributions tend to cancel one another, which complicates the practical application of the electrical threshold effect, since both contributions may eliminate one another.

Nonetheless, the further development of electrical diagnostic methods for lasing and amplification conditions in diodes is of interest for further practical applications in displacement sensors, for information reading and in optical communications. By virtue of the uniformity of the properties of stimulated transitions with absorption or emission of photons, we achieve a change in sign of the photo-EMF with inversion of the population of the working states in laser diodes, i.e., we observed "negative" photo-EMF for the first time. In experiments with lasers having two coupled cavities, electrical diagnostics may be useful for electronic tuning of cavities.

The application of laser diodes in optical communications engineering has two new capabilities demonstrated in this study based on the phenomenon of "negative" photo-EMF. The first method is as follows: If a diode amplifier is installed in the optical circuit for use as a repeater, it will be possible to achieve an adequate electrical signal without optical signal losses (width amplification), i.e., it will be possible to recover information or monitor link operation (to verify the optical signal level, to verify amplification based on the sign of the electrical signal, to calculate information volume, etc.). Another possibility involves using identical laser modules in both optical link terminal devices and implementing two-way transmission on a single light guide. We demonstrated such a capability in [35-37]. Recently reports have appeared regarding the use of this principle for demonstration two-way fiber-optic communications systems with transmission speeds of ~100 Mbits/sec [38]. Thus, these new capabilities are already of practical interest.

BIBLIOGRAPHY

1. Eliseev, P. G., Manko, M. A. "Semiconductor laser radiation properties: Rept III Polish Conf. on Radiospectroscopy and Quantum Electronics" Poznan, 1968, POST. FIZ., 1969, V. 20, p. 331-337.

2. Eliseev, P. G., Ismailov, I., Krasil'nikov, A. I., et al. "Porogovye yavleniya v inzhektsionnykh lazerakh" [Threshold phenomena in injection lasers] In: Tr. IX Mezhdunar. konf. po fizike poluprovodnikov [Transactions of the Ninth International Conference on Semiconductor Physics] Leningrad: Nauka, 1969, V. 1, p. 549-552.

3. Eliseev, P. G., Krasil'nikov, A. I., Man'ko, M. A., Strakhov, V. P. "Issledovanie inzhektsionnykh lazerov na postoyannom toke" [Investigation of d.c. injection lasers] In: Fizika *p-n*-perekhodov i poluprovodnikovykh priborov [The physics of *p-n*-junctions and semiconductor devices] Leningrad: Nauka, 1969, p. 131-134.

4. Eliseev, P. G. "Investigation of injection lasers" TR. FIAN, 1970, V. 52, p. 3-117.

5. Rode, D. L., Dawson, L. R. "Differential 1/*V* of heterostructures correlates with laser threshold. APPL. PHYS. LETT., 1972, V. 21, p. 90-93.

6. Nannichi, Y. "On the forward V-I characteristics of the GAALAS/GAAS laser diode" JAPAN. J. APPL. PHYS., 1964, V. 3, p. 233-235.

7. Gladkiy, B. I., Nasledov, D. N., Tsarenkov, B. V. "An investigation of the V-I characteristic of a GaAs laser in transition from amplification to lasing" FTT, 1966, V. 8, p. 2382-3288.

8. Paoli, T. L., Barnes, P. A. "Saturation of the junction voltage in stripe-geometry (AlGa)As double heterostructure junction lasers" APPL. PHYS. LETT., 1976, V. 28, p. 714-715.

9. Joyce, W. B., Dixon, R. W. "Fundamental and harmonic response voltage of a sinusoidally current-modulated ideal semiconductor laser" J. APPL. PHYS., 1976, v. 47, p. 3510-3513.

10. Joyce, W. B., Dixon, R. W., "Electrical characterization of heterojunction lasers" J. APPL. PHYS., 1978, V. 49, p. 3719-3728.

11. Ishikawa, H., Hanamitsu, K., Takasugawa, M. "Lasing-induced change in the differential resistance of stripe-geometry GaAlAs DH lasers" JAPAN. J. APPL. PHYS., 1979, V. 18, p. 333-338.

12. Eliseev, P. G., Okhotgnikov, O. G., Pak, G. T. "The properties of planar stripe heterolasers. V. 2. Analysis of electrical characteristics" KVANTOVAYA ELEKTRON., 1980, V. 7, p. 1679-1676.

13. Eliseev, PO. G. "Stripe-geometry lasers and their properties" CZECH. J. PHYS. B, 1980, V. 30, p. 300-317.

14. Eliseev, P. G., Okhotnikov, O. G., Pak, G. T., Vu Van Lyk "An investigation of AlGaAs-GaAs planar stripe heterolasers" TR. FIAN, 1983, V. 141, p. 89-117.

15. Albanese, A. "An automatic bias control (ABC) circuit for injection lasers" BELL SYST. TECHN. J., 1978, V. 57, p. 1533-1544.

16. Sommers, H. S., North, D. O. "Experimental and theoretical study of the spatial variation of junction voltage and current distribu-

tion in narrow stripe injection lasers" J. APPL. PHYS., 1977, V. 48, p. 4460-4467.

17. North, D. O., Sommers, H. S. "Saturation of the junction voltage in stripe-geometry AlGaAs double heterostructure junction lasers: a comment" APPL. PHYS. LETT., 1977, V. 30, p. 116-117.

18. Bogatov, A. P., Eliseev, P. G., Okhotnikov, O. G., et al. "An investigation of a CW injection laser with an external dispersive cavity" TR. FIAN, 1983, V. 141, p. 62-88.

19. Sommers, H. S. "Algorithm for measuring the internal quantum efficiency of individual injection lasers" APPL. PHYS. LETT., 1978, V. 32, p. 547-549.

20. Mitsuhashi, Y., Morikawa, T., Sakurai, K., et al. "Self-coupled optical pick up" OPT. COMMUNS, 1976, V. 17, p. 95-97.

21. Mitsuhashi, Y., Shimada, J., Mitsutshika, S. "Voltage change across the self-coupled semiconductor laser" IEEE J. QUANT. ELECTRON., 1981, V. 17, p. 1216-1225.

22. Eliseev, P. G., Okhotnikov, O. G., Pak, G. T. "The influence of carrier leakage from the active layer on the differential resistance of a laser diode" KVANT. SOOBSHCH. PO FIZIKE FIAN, 1984, No. 3, p. 21-25.

23. Su, C. B., Schlafer, J., Manning, J., Olshansky, R. "Measurement of the radiative recombination coefficient and carrier leakage in 1.3 μm InGaAsP lasers with lightly doped active layers" ELECTRON. LETT., 1982, V. 18, p. 1108-1110.

24. Eliseev, P. G., Ismailov, I., Man'ko, M. A., Strakhov, V. P. "An injection semiconductor laser with a compound resonator" PIS'MA V ZHETF, 1969, V. 9, p. 594-595.

25. Bakhert, Kh., Eliseev, P. G., Raab, Z. "An investigation of the characteristics of injection lasers with a tunable compound resonator" ZHURN. PRIKL. SPEKTROSKOPII, 1972, V. 16, p. 810-813.

26. Eliseev, P. G., Khaydarov, A. V. "The electrooptical and piezoelectric tuning of a compound laser resonator" OPTIKA I SPEKTROSKOPIYA, 1973, V. 34, p. 343-346.

27. Bogatov, A. P., Eliseev, P. G., Man'ko, M. A., et al. "Odnomodovyy inzhektsionnyy PKGAs perestrayvaemoy dlinoy volny generatsii" [A single-mode injection laser with tunable emission" In: Lazery s perestraivaemoy chastotoy [Tunable lasers] Kiev: In-t fiziki AN USSR, 1973, p. 173-179.

28. Dementienko, V. V., Godik, E. E., Gulyaev, Yu. V., et al. "Cohe-

rent recording of emission by an injection laser" PIS'MA V ZHTF, 1979, V. 5, p. 1349-1351.

29. Amann, M. C., Suechmider, A., Boeck, J. "Optical memory readout by superluminescent diode with integrated photodetector" ELECTRON. LETT., 1980, V. 16, p. 58-60.

30. Seko, A., Mitsuhashi, Y., Morikawa, T., et al. "Self-quenching in semiconductor lasers and its application in optical memory readout" APPL. PHYS. LETT., 1975, V. 27, p. 140-141.

31. Vu Van Lyk, Eliseev, P. G., Man'ko, M. A., et al. "Optoelectronic reading by means of injection laser" KVANTOVAYA ELEKTRON., 1982, V. 9, p. 1825-1830.

32. Yamamoto, Y. "Noise and error rate performance of semiconductor laser amplifiers in PCM-1M optical transmission systems" IEEE J. QUANT. ENECTRON., 1980, V. 16, p. 1073-1081.

33. Kobayashi, S., Kimura, T. "Gain and saturation power of resonant AlGaAs laser amplifier" ELECTRON. LETT., 1980, V. 16, p. 230-232.

34. Yamamoto, Y., Tsuchiya, H. "Optical receiver sensitivity improvement by a semiconductor laser preamplifier" ELECTRON. LETT., 1980, V. 16, p. 233-235.

35. Vu Van Lyk, Eliseev, P. G., Man'ko, M. A., Mikaelyan, G. T. "The electrical diagnostics of laser diode-based amplifier monitor" KVANTOVAYA ELEKTRON., 1982, V. 9, p. 1851-1854.

36. Vu Van Luc, Eliseev, PO. G., Manko, M. A., Mikaelyan, G. T. "Electrical disgnostic of the amplifier operation and a feasibility of signal registration on the basis of the voltage saturation effect in junction laser diodes" IEEE J. QUANT. ELECTRON., 1983, V. 19, p. 1080-1084.

37. Vu Van Lyk, Eliseev, P. G., Man'ko, M. A., Mikaelyan, G. T. "Investigation of signal transmission in a system of two injection lasers optically coupled by fiber optic lightguide" KVAT. SOOBSHCH. FIZIKE FIAN, 1982, No. 10, p. 31-35.

38. Alping, A., Eng, S. T. "Detection at Gbit/s rates with a TJS GaAlAs laser" OPT. COMMUNS, 1983, V. 44, p. 381-383.

SUBJECT INDEX